Mahmoud Khalifeh · Arild Saasen

Introduction to Permanent Plug and Abandonment of Wells

Mahmoud Khalifeh
Kjølv Egenlands hus
Stavanger, Norway

Arild Saasen
Kjølv Egenlands hus
Stavanger, Norway

https://doi.org/10.1007/978-3-030-39970-2

Acknowledgements

The book is based on research and literature review conducted on permanent plug and abandonment of hydrocarbon wells. Special thanks to the Norwegian Oil and Gas Plug and Abandonment Forum (PAF) for supporting the publication and helping the authors to publish the book as open access. We are also grateful for several colleagues helping and encouraging us to continue with the work and finally to publish it.

We hereby would like to thank them by listing their names:
Arne G. Larsen
Atle J. Sørhus
Dave Gardner
Egil Thorstensen
Farzad Shoghl
Helge Hodne
Ivar Blaauw
Lars Hovda
Martin K. Straume
Michael T. Skjold
Nils Opedal
Odd G. Taule
Øystein Arild
Rune Godøy
Steinar Strøm
Tove Rørhuus
Alex Osorio
Arild Rasmussen
Colin Beharie
John Dale
John Donachie
Laurent Delabroy
Luca Carazza

Matus Gajdos
Max B. Baumert
Steffen K. Lindsø
Steffen Kristiansen
Sylvain Bedouet
Tu N. Tran
Vidar Rygg

Contents

About the Authors

Mahmoud Khalifeh is Associate Professor at University of Stavanger (UiS). He holds a Ph.D. in plug and abandonment of wells with a focus on materials optimized for permanent P&A. He has developed the permanent P&A (Plug and Abandonment) subject (MSc level) at UiS. He has published several articles in scientific journals and conferences. He represents UiS as observer at Norwegian Plug and Abandonment Forum (PAF) since 2012.

Arild Saasen has a position as Adjunct Professor at the University of Stavanger. Parallelly, he works for other companies including his own company ALKAS Atlantic AS. He has previously worked as drilling and well fluid specialist in Det norske oljeselskap (now named AkerBP) and Statoil (now named Equinor). Earlier, he worked for Rogaland Research (now named Norce). He holds an MSc in fluid mechanics from the University of Oslo, Norway, and a Ph.D. from the Technical University of Denmark. Throughout the last three decades, he has published numerous articles about drilling and drilling fluids. He has been awarded the Nordic Rheology Society's Carl Clason Rheology Award in 2012 and the SPE North Sea Regional Drilling Engineering Award in 2018.

Abbreviations

ΔP_{f2}	Frictional Pressure Drop Between Displacing and Displaced Fluids
τ_{av}	Average Shear Bond Strength
A	Pre-exponential Constant
AFE	Authority for Expenditure
AHT	Anchor Handler Tug
AHV	Anchor Handling Vessel
ANP	The Brazilian National Agency of Petroleum, Natural Gas and Biofuels
A_p	Surface Area of Plug
AP	Annular Preventer
API	American Petroleum Institute
ARO	Asset Retirement Obligation
ASV	Annular Safety Valve
AWJC	Abrasive Water-Jet Cutting
BHA	Bottom Hole Assembly
BHTC	Bottomhole Circulating Temperature
BOP	Blowout Preventer
BSEE	The American Bureau of Safety and Environmental Enforcement
BSR	Blind Shear Ram
C_2S	Dicalcium silicate
C_3A	Tricalcium aluminate
C_3S	Tricalcium silicate
C_4AF	Tetracalcium aluminoferrite
CBL	Cement Bond Log
CBM	Conventional Buoy Mooring
CCL	Casing Collar Locator
CDF	Cumulative Distribution Function
CLT	Central Limit Theorem
C-S-H	Hydrated Silicate and Calcium
CSR	Casing Shear Ram
DAS	Distributed Acoustic Sensing

DC	Direct Current
DEA	Danish Energy Agency
DHSV	Downhole Safety Valve
D_i	Inner Diameter of the Geometry Plug Placed Inside
DNRM	The Australian Department of Natural Resources and Mines
DP	Dynamic Positioning
DPR	The Nigerian Department of Petroleum Resources
DTS	Distributed Temperature Sensing
E_a	Activation energy
ECD	Equivalent Circulating Density
EMAT	Electro-Magnetic Acoustic Transducers
EOR	Enhanced Oil Recovery
EPDM	Ethylene Propylene Diene Monomer
ESP	Electrical Submersible Pump
F	Failure force
FAB	Formation as Barrier
F_{fc}	Filter cake – Formation Friction Force
FFKM	Perfluoroelastomer
FKM	Fluoroelastomer
FPSO	Floating, Production, Storage, and Offloading
F_R	Reservoir Force
F_{sb}	Shear Bond Force
H	Reservoir Depth
h_{MSD}	Minimum Setting Depth Of Plug
HMV	Hydraulic Master Valve
HPHT	High-Pressure High-Temperature
HSE	Health, Safety and Environment
HSE	The British Health and Safety Executive
HWU	Hydraulic Workover Unit
ICD	Inflow Control Device
IFT	Interfacial Tension
IOR	Improved Oil Recovery
KV	Kill Valve
L_p	Plug Length
LPWH	Low-Pressure Wellhead Housing
LVR	Lower Variable Ram
LWIV	Light Well Intervention Vessel
MMF	Mud Mobility Factor
MMH	Mixed Metal Hydroxide
MMV	Mechanical Master Valve
MODU	Mobile Offshore drilling Unit
MPE	Ministry of Petroleum and Energy
MSD	Minimum Setting Depth
MVR	Middle Variable Ram
NA	Not Available

NCS	Norwegian Continental Shelf
nD	NanoDarcy
NIOC	National Iranian Oil Company
NPD	Norwegian Petroleum Directorate
NPT	Non-Productive Time
OBM	Oil-based Mud
OD	Outside Diameter
P&A	Plug and Abandonment
PDF	Probability Distribution Function
PE	Pulse Eco
PEEK	Polyether Ether Ketone
P_{FP}	Final Reservoir Pressure
$P_{frac.}$	Fracture Pressure Gradient
P_g	Gas Gradient Pressure
PMF	Probability Mass Function
PNGRB	The Indian Petroleum and Natural Gas Regulatory Board
POB	Personnel on Board
P_p	Pump Pressure
ppg	Pounds per Gallon
PPS	Polyphenylene Sulfide
PSA	The Norwegian Petroleum Safety Authority
PSD	Particle Size Distribution
PTFE	Polytetrafluoroethylene
PUE	Polyurethane Elastomers
PVDF	Polyvinylidene fluoride
PWC	Perforate, Wash and Cement
PWV	Production Wing Valve
RDT	Radial Differential Temperature
RIH	Run in Hole
ROV	Remotely Operating Vehicle
SCP	Sustained Casing Pressure
SCSSV	Surface Controlled Subsurface Safety Valve
SEM	Scanning Electron Microscopy
sk	Sack
SW	Swab Valve
T_g	Glass transition temperature
TLP	Tension Leg Platform
TOC	Top of Cement
TVD	True Vertical Depth
UNMIG	Ufficio Nazionale Minerario per gli Idrocarburi e le Georisorse (National Office for mining, Hydrocarbons, and Geothermal Resources)
UVR	Upper Variable Ram
VDL	Variable Density Log
WBAC	Well Barrier Acceptance Criteria
WBE	Well Barrier Element

WBM	Water-Based Mud
WBS	Well Barrier Schematic
W_{dp}	Drillpipe Tag Weight
XMT	Christmas Tree
YP	Yield Point

Chapter 1
Introduction

Every beginning has an end. This book covers the beginning of the end of well life. When a well reaches the end of its life, it must permanently be plugged and abandoned. Plug and abandonment can easily contribute to 25% of the total cost of drilling exploration wells offshore Norway. The cost of running a plug and abandonment operation on some offshore production wells may have a cost impact similar to the cost of the original drilling operation. Therefore, cost efficient plug and abandonment technology is a necessity without compromising the scope of the operation. The occasion that dictates the end of a well life could be integrity issues, subsidence induced well failure, depleted reservoir, water/gas coning, negative cash flow, or finished data gathering from exploration. In addition, there are other circumstances that force the wellbore(s) to be permanently plugged and abandoned. For instance, a platform in the Gulf of Suez, Egypt, was struck by a cargo vessel on December 1989. Due to the massive damage, the nine wells were forced to be plugged and abandoned and a field re-development had to be performed [1]. A question rises; what is the purpose of a plug and abandonment operation? Why are not wells left behind as they are? One answer is establishment of barriers for preventing flow of hazardous fluids to surroundings. The surroundings can be the marine environment, groundwater, ground or atmosphere. The objective of plug and abandonment operations is to restore the cap-rock functionality, securing the well-integrity permanently. In order to succeed, an appropriate permanent barrier shall be placed across a suitable formation through the utilization of relevant equipment to fulfill the local requirements.

Now a comprehensive definition of *Plug and Abandonment* (P&A) could be given as a collection of tasks and actions taken to isolate and protect the environment and all fresh water zones and surroundings from a source of potential inflow. The source of potential inflow is a formation with permeability and it may be either a water or a hydrocarbon bearing zone. The outline of a P&A operation varies a little; whether the well is offshore or onshore, or if the well is going to be abandoned permanently or temporarily, although the main goal is to secure all formations which have the potential to leak. Therefore, we begin the discussion of plug and abandonment with some basic definitions.

M. Khalifeh and A. Saasen, *Introduction to Permanent Plug and Abandonment of Wells*, Ocean Engineering & Oceanography 12,
https://doi.org/10.1007/978-3-030-39970-2_1

1.1 Abandonment Types

Once the downhole activities or production is discontinued, the well status needs to be clarified. Generally, three different statuses may be defined; suspension, temporarily abandoned or permanently abandoned [2]. When a well is subjected to construction or intervention, the operation may need to be suspended without removing the well control equipment. In this scenario, the well status is called *suspension*. The operation could be suspended due to waiting on weather, workover on another well, waiting on equipment, rig skidded to do short-term work on another well or batch drilling (top section of hole only), or to accommodate pipe lay activities in the field.

Temporarily abandoned is a status where the well has been abandoned and the well control equipment is removed with the intention of later re-entry or permanent abandonment. Another phrase for temporarily abandoned could be *long-term suspension*. Temporary abandonment could be through a long shutdown, waiting on a workover, waiting on field development, re-development, etc. Temporarily abandoned status begins when the main reservoir has been fully isolated from the wellbore and may last from days up to several years. Different regulatory authorities have their own requirements with respect to the maximum period of temporary abandonment. A temporarily abandoned well may be with or without monitoring a system which depends upon the requirements of the regulatory authority, and well location.

Permanently abandoned is a status where the well or part of the well, has been permanently plugged and abandoned with the intention of never being re-used or re-entered.

1.2 Asset Retirement Obligation

Asset Retirement Obligation (ARO) addresses legal obligations and associated costs related to future retirement of long-lived assets. According to ARO, operators are obliged to demonstrate that sufficient assets have been allocated to cover the cost of future P&A operations [3, 4]. An ARO liability includes downhole abandonment, surface abandonment, facility site abandonment, infrastructure dismantling, and site decommissioning [5]. One of the main reasons to bring the ARO mechanism into action is the reported failure to properly abandon wells and facilities which create serious issues for environment, safety, and security. Dry wells or improperly abandoned wells or fields that are left behind require huge public funds to be allocated; however, operators were supposed to be in charge. The ARO, however, does not apply to unplanned clean-up costs such as cleaning up of an accident.

1.3 Prepared for Permanent Plug and Abandonment

When a well reaches the end of its life-cycle, it must be permanently plugged and abandoned. In addition, there are many other reasons for a well to be partially or fully plugged. A safe production operation is primarily about maintaining well integrity and sufficient barriers throughout the well life-cycle. It is common practice to perform risk assessments for all wells. Once risks are identified, wells are assigned color codes. Based on the color codes, whenever well integrity is not maintained or is compromised, the well should be economically repaired or alternatively be permanently plugged. A wellbore that has not encountered hydrocarbons of a commercially viable quantity is usually plugged. These types of wellbores are called either *dry-holes* or *dusters* even though they may contain water. Generally, most dry holes are exploratory wells. Regardless of was an exploration success or not, a common procedure for exploratory wells is to permanently plug and abandon them after data gathering is complete. This is due to their inappropriate well design for production and the costs and risks associate with modifying their design (e.g. uncertainties in the sealing capabilities of the intermediate and production casings, unknown cement tops and damaged formation nearby casing shoes).

Occasionally a sidetrack needs to be drilled to bypass an unusable section of the original wellbore or to explore a nearby geologic feature. Prior to beginning such a sidetrack the borehole below the sidetrack should be permanently plugged.

Slot recovery, re-development and well integrity issues are some other reasons that may initiate a permanent plug and abandonment operation. Slot recovery is a process of recovering an existing drilling or template slot to reach a new target. Slot recovery may be done due to limited rig skidding capacity, an irretrievable fish in a slot, not hitting the target with the original well, or a limited number of slots on a drilling platform or template.

1.3.1 Plug and Abandonment Challenges

Every well is unique and the associated challenges with it as well. The main challenges which have been reported associated with the P&A of wells, can be categorized as high temperatures, unconsolidated formations, changes in formation strength as a result of depletion, uncertain ultimate reservoir pressure after abandonment, formation permeability, tectonic stresses exerted by formation (e.g. shear stress and subsidence), sustained casing pressure (SCP), lack of data from old drilled wells, deep section milling, and verifying the casing cement behind the second casing string. These are the main challenges that industry moat deal with; however, all of these may not be applicable to a specific well.

1.4 Past, Present, and Future of Plugged and Abandoned Wells on the NCS

Since the first discovery in the Norwegian sector of the Norwegian Continental Shelf (NCS) in 1966 until June 2015, nearly 5600 wells have been drilled to date. Of these wells, 4037 are development wells and 1542 are exploration wells. Of the exploration wells, 1480 have been permanently plugged and abandoned. Of the development wells, approximately 1400 have been permanently abandoned and 467 have in a temporarily abandoned status. It is estimated that 2637 development wells need to be plugged and abandoned in the near future. In addition, the number of future wells that will be drilled should be added to these statistics [6]. Availability of a database for permanently plugged and abandoned wells could be beneficial for industry, government, and tax payers. This could result in knowledge sharing, optimized planning, and a better understanding of strategies and technology development related to the P&A of wells [7]. Figure 1.1 presents an overview of the status of all the wells drilled in the Norwegian sector of the NCS.

1.5 Digitalization in Plug and Abandonment

Digitalization is a process in which information and knowledge are converted into a digital format [8]. In this way, it is organized into discrete units of data, known as bits. Digitalization has already been implemented by different industries such as the automobile industry. It is not a new concept in the oil and gas industry;

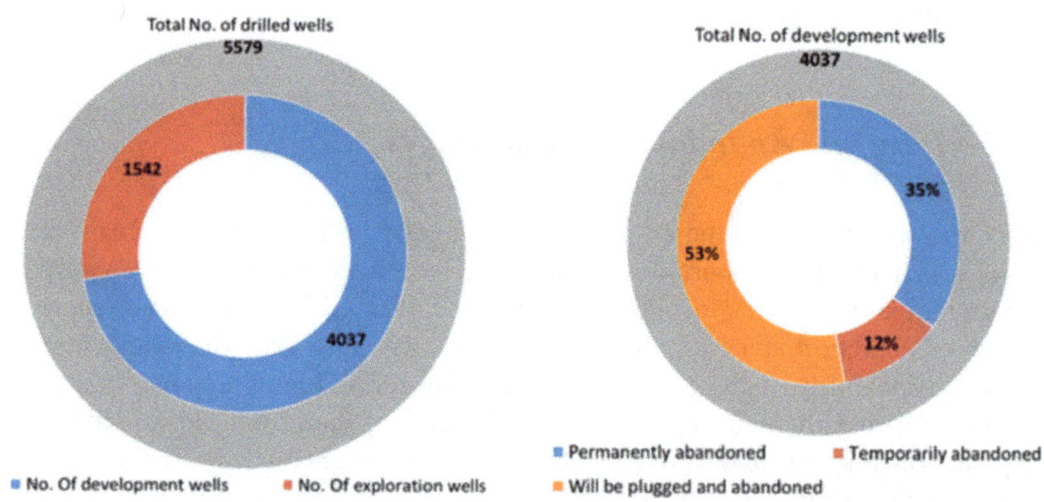

(a) Total number of all drilled wells.

(b) Total number of all drilled development wells.

Fig. 1.1 A status overview of all wells drilled on the Norwegian sector of the NCS

the upstream industry has been relying on digital technologies for many years. Of these, one can refer to seismic data processing from the 1980s and monitoring and optimizing critical production processes from the 1990s [9]. Digitalization has some core benefits including access to data, data management, improving accuracy in engineering by implementing the latest theories and models, optimal planning and operation, minimizing human error or human factors which contribute to failures or incidents, changing human involvement to a supervisory role and finally leading to the automation of the drilling process [10]. But digitalization creates big data volumes and it has associated challenges including data capturing, data storage, data analysis, search, sharing, transfer, visualization, querying, updating and information security [11]. These challenges need to be considered along the way of digitalization in the oil and gas industry. Digitalization of standards may also be considered in different ways; integration of standards and regulations in software programs to "police" the plans and operations or inclusion of standards as help files [12].

Applying digitalization in P&A can be differentiated for old wells and new wells. Perhaps, the most challenging part will be the digitalization of old wells; new wells can be equipped with sensors to monitor the wells and track them from the day of design and construction to abandonment. When considering digitalization of old wells, it is possible to register the well location, well status, well schematic, mechanical failures, HSE issues, previous rig footprints, and archiving the well data. One of the "low-hanging fruit" benefits of digitalization is having an updated overview of well numbers and their status. This has been implemented properly in the oil and gas sector of Norway by the Norwegian Petroleum Directorate.

1.6 The Regulatory Authorities

Regardless of abandonment type, operators must leave wellbores behind which are secured in accordance with local regulations. Figure 1.2 maps some regulatory authorities managing petroleum activities in their own territories. Different regulatory bodies have their own requirements and operators must strictly adhere to local well-abandonment regulations. Local regulations are the minimum requirements and have changed considerably over the years to facilitate P&A operations in a safe manner. Nevertheless, some operators have their own internal requirements and tend to follow them where the regulatory authorities do not provide minimum requirements.

The North Sea could be divided into four sectors; the United Kingdom, the Norwegian, the Danish and the Dutch. The Health and Safety Executive (HSE) is the appropriate department that oversee the petroleum activities in Britain. In the Danish sector, the Danish Energy Agency (DEA) is the regulatory authority. The Dutch Supervision of Mines and the Norwegian Petroleum Directorate (NPD) are the regulatory authorities for the Dutch and Norwegian sectors, respectively. The NPD is the governmental specialist directorate and administrative body for the NCS. The NPD acts as an adviser to the Ministry of Petroleum and Energy (MPE) of Norway. In the Norwegian maritime territory, there is an independent government regulator,

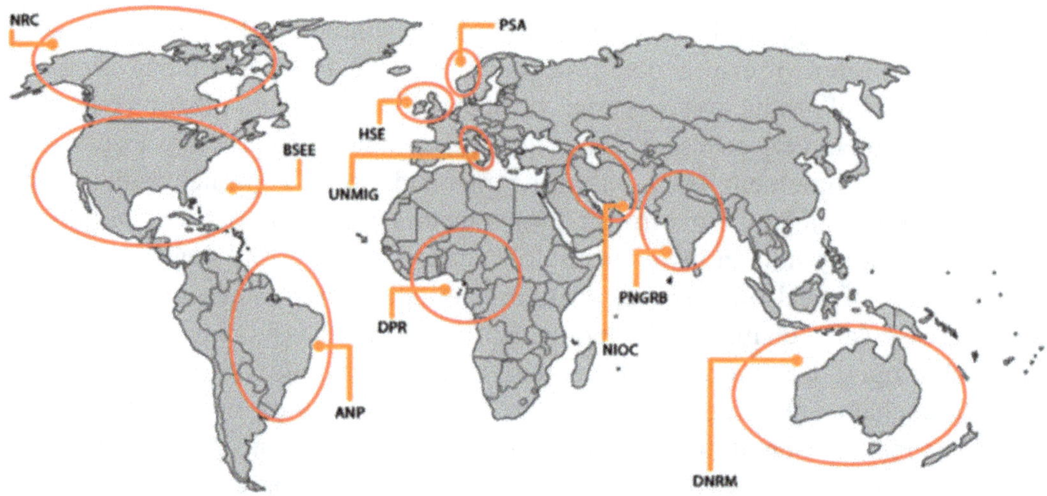

Fig. 1.2 Regulatory authorities, oversee their own petroleum activities

which is known as the Petroleum Safety Authority (PSA) Norway with responsibility for safety and emergency preparedness in the Norwegian petroleum industry. The PSA is the legislative authority for P&A activities and reviews the proposed P&A plans for the NCS. The PSA is the responsible organization for overseeing the P&A operations.

1.7 P&A Barrier Philosophy

There is a generally accepted philosophy for well barriers that the well should be equipped with sufficient well barriers to prevent uncontrolled flow from the potential sources of flow. In addition, it is generally accepted that no single failure of a well barrier component should lead to unacceptable consequences. This means that, in practical terms, the well should be equipped with two independent well barriers; a primary and a secondary barrier. This is also known as "hat-over-hat" principle whereas the secondary barrier acts as a back-up to the primary well barrier, as shown in Fig. 1.3.

The function of the barrier philosophy could be slightly different in circumstances where the P&A operation is ongoing or a well has been permanently plugged and abandonment. For a well P&A operation, some barrier elements need to be in an open position to allow access to the borehole and perform the P&A operation. It is critical that these elements close in circumstances when it is necessary to halt the operation. So the primary and secondary barrier elements may vary based on the pre- or post-abandonment status. The P&A barrier principles will be discussed thoroughly in the next chapter.

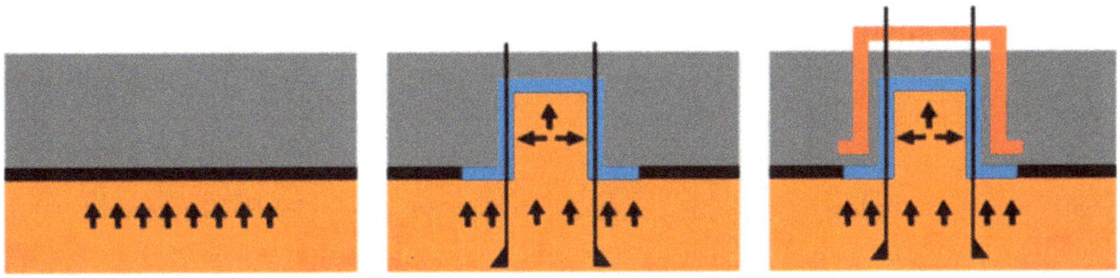

Cap rock holds the pressure | Primary barrier hold the pressure | Secondary barrier is independent of primary and acts as back-up

Fig. 1.3 Two-barrier philosophy shown using the "hat-over-hat" representation

1.8 The Beginning of the End—Decommissioning

All the activities conducted to shut down and remove facilities from service is defined as *decommissioning*. Decommissioning of facilities is highly complex, often even more so than the original installation. Decommissioning is a generic description and it is applicable for both offshore and onshore facilities, and it could be regarded as the beginning of the end of the facilities. Decommissioning can be challenging especially for offshore facilities and particularly in deep waters; a decommissioning process can be monumental and requires detailed considerations by specialized crews [13]. Decisions about when and how to decommission platforms involve complicated issues of environmental protection, safety, technical feasibility and associated costs.

Prior to conducting decommissioning, a *decommissioning plan* needs to be prepared and submitted to the competent authority. A decommissioning plan may consist of two main parts; a disposal plan and an impact assessment. The impact assessment provides an overview of the expected consequences of the disposal such as environmental consequences. According to the Norwegian Act 29, issued in November 1996 No. 72 relating to petroleum activities [14], the cessation of petroleum activities, Sect. 5.1; "*the decommissioning plan shall be submitted at the earliest five years, but at the latest two years prior to the time when the use of a facility is expected to be terminated permanently*".

A decommissioning plan generally includes descriptions of [15, 16]:

- Results of a documentary survey relating to facility design, fabrication, installation, commissioning, etc.;
- Possible risks during and after facility removal;
- Intended methods and strategies to be used during decommissioning, including re-floating of structures;
- Intended analyses which are planned to be carried out;
- Operations planned to be carried out in the event of a possible removal;
- Possible impacts of a removal on adjacent fields and facilities;
- Methods of waste control; and

- Possible monitoring systems which may be designed to secure the area against possible future pollution from permanently abandoned wells or polluted cuttings deposits.

The remaining issues regarding decommissioning are the associated cost and the question of who holds the liability; and according to the OSPAR Convention,[1] the ultimate responsibility of decommissioning remains with the facility owner.

References

1. El Laithy, W.F., and S.M. Ghzaly. 1998. Sidki well abandonment and platform removal case history in the Gulf of Suez. In *SPE international conference on health, safety, and environment in oil and gas exploration and production*. SPE-46589-MS, Caracas, Venezuela: Society of Petroleum Engineers. https://doi.org/10.2118/46589-MS.
2. NORSOK Standard D-010. 2013. *Well integrity in drilling and well operations*. Standards Norway.
3. Merryman, A. 2002. Fair value: What your accountant wants to know. In *SPE annual technical conference and exhibition*. SPE-77508-MS, San Antonio, Texas: Society of Petroleum Engineers. https://doi.org/10.2118/77508-MS.
4. Zhao, L., J. Wells, and B. Dong, et al. 2013. Environmental liabilities in oil and gas industry and life-cycle management. In *International petroleum technology conference*. IPTC-16945-MS, Beijing, China: International Petroleum Technology Conference. https://doi.org/10.2523/IPTC-16945-MS.
5. Austin, D.G. 2007. Strategy for managing environmental liabilities in an onshore oil field. In *SPE Asia Pacific health, safety, and security environment conference and exhibition*. SPE-108784-MS, Bangkok, Thailand: Society of Petroleum Engineers. https://doi.org/10.2118/108784-MS.
6. Khalifeh, M. 2016. Materials for optimized P&A performance: Potential utilization of geopolymers. In *Faculty of science and technology*. University of Stavanger: Norway. http://hdl.handle.net/11250/2396282.
7. Myrseth, V., G.A. Perez-Valdes, and S.J. Bakker, et al. 2016. Norwegian open source P&A database. In *SPE Bergen one day seminar*. SPE-180027-MS, Grieghallen, Bergen, Norway: Society of Petroleum Engineers. https://doi.org/10.2118/180027-MS.
8. Carpenter, C. 2017. Digitalization of oil and gas facilities reduces cost and improves maintenance operations. *Journal of Petroleum Technology* 69 (12): 62–63. https://doi.org/10.2118/1217-0062-JPT.
9. Dekker, M., and A. Thakkar. 2018. Digitalisation—The next frontier for the offshore industry. In *Offshore technology conference*. OTC-28815-MS, Houston, Texas, USA: Offshore Technology Conference. https://doi.org/10.4043/28815-MS.
10. Brechan, B., A. Teigland, and S. Sangesland, et al. 2018. Work process and systematization of a new digital life cycle well integrity Model. In *SPE Norway one day seminar*. SPE-191299-MS, Bergen, Norway: Society of Petroleum Engineers. https://doi.org/10.2118/191299-MS.
11. Murray, J., and K. Eriksson. 2018. Data management and digitalisation: Connecting subsea assets in the digital space. In *Offshore technology conference*. OTC-28997-MS, Houston, Texas, USA: Offshore Technology Conference. https://doi.org/10.4043/28997-MS.

[1]The Convention for the Protection of the marine Environment of the North-East Atlantic (the OSPAR Convention) is the current legislative instrument regulating international co-operation on environmental protection in the North-East Atlantic. It was open for signature at the Ministerial Meeting of the Oslo and Paris Commissions in Paris on 22 September 1992.

12. Brechan, B., S.I. Dale, and S. Sangesland. 2018. New standard for standards. In *Offshore technology conference*. OTC-28988-MS, Houston, Texas, USA: Offshore Technology Conference. https://doi.org/10.4043/28988-MS.
13. Ekins, P., R. Vanner, and J. Firebrace. 2006. Decommissioning of offshore oil and gas facilities: A comparative assessment of different scenarios. *Journal of Environmental Management* 79 (4): 420–438. https://doi.org/10.1016/j.jenvman.2005.08.023.
14. Norwegian Petroleum Directorate. 1966. *Act 29 November 1996 No. 72 relating to petroleum activities*. http://www.npd.no/en/Regulations/Acts/Petroleum-activities-act/. Accessed 2017.
15. Petroleum Safety Authority Norway. 2001. *Regulations relating to health, environment and safety in the petroleum activities* (*The framework regulations*). PTIL: Norway.
16. Shaw, K. 1994. Decommissioning and abandonment: The safety and environmental issues. In *SPE health, safety and environment in oil and gas exploration and production conference*. SPE-27235-MS, Jakarta, Indonesia: Society of Petroleum Engineers. https://doi.org/10.2118/27235-MS.

Chapter 2
General Principles of Well Barriers

The principle of well integrity is primarily occurred with maintaining well control with sufficient barriers. *Well integrity* is defined as "*application of technical, operational and organizational solutions to reduce risk of uncontrolled release of formation fluids and well fluids throughout the lifecycle of a well*" [1]. To control the well, two qualified independent well barrier envelopes should be present at each stage of a well's life. The petroleum industry has employed the principle of a two-barrier philosophy since 1920s [2]. Generally speaking, the overbalance from the drilling fluid is the primary barrier and the blowout preventer (BOP) with casing string comprise the secondary barrier during well construction. Over time, the petroleum industry has entered into more complex and challenging environments, and therefore, the need to clarify and standardize the well barrier integrity has been increasing. In practice, the application of the well barrier philosophy is more complicate due to technical and operational limitations. Figure 2.1 illustrates the two-barrier philosophy of a well throughout its lifecycle, and Table 2.1 presents examples of barrier systems through its lifecycle of the given well.

2.1 Well Annuli

An annulus is any void space between two strings, or a string of casing and formation. When a well is completed, different annuli might be distinguished. In well engineering, the annular space between production tubing and production casing is called *A-annulus*. The annular space between production casing and intermediate casing is called *B-annulus*. The naming procedure is continued until the last annular space, which is between the conductor and formation (see Fig. 2.2) [1]. Generally, these annuli should not have any connection to wellbore fluids. But the annuli are filled with completion fluid or drilling fluid for protection of steel and maintaining the pressure to ensure the integrity of the strings [3].

M. Khalifeh and A. Saasen, *Introduction to Permanent Plug and Abandonment of Wells*, Ocean Engineering & Oceanography 12,
https://doi.org/10.1007/978-3-030-39970-2_2

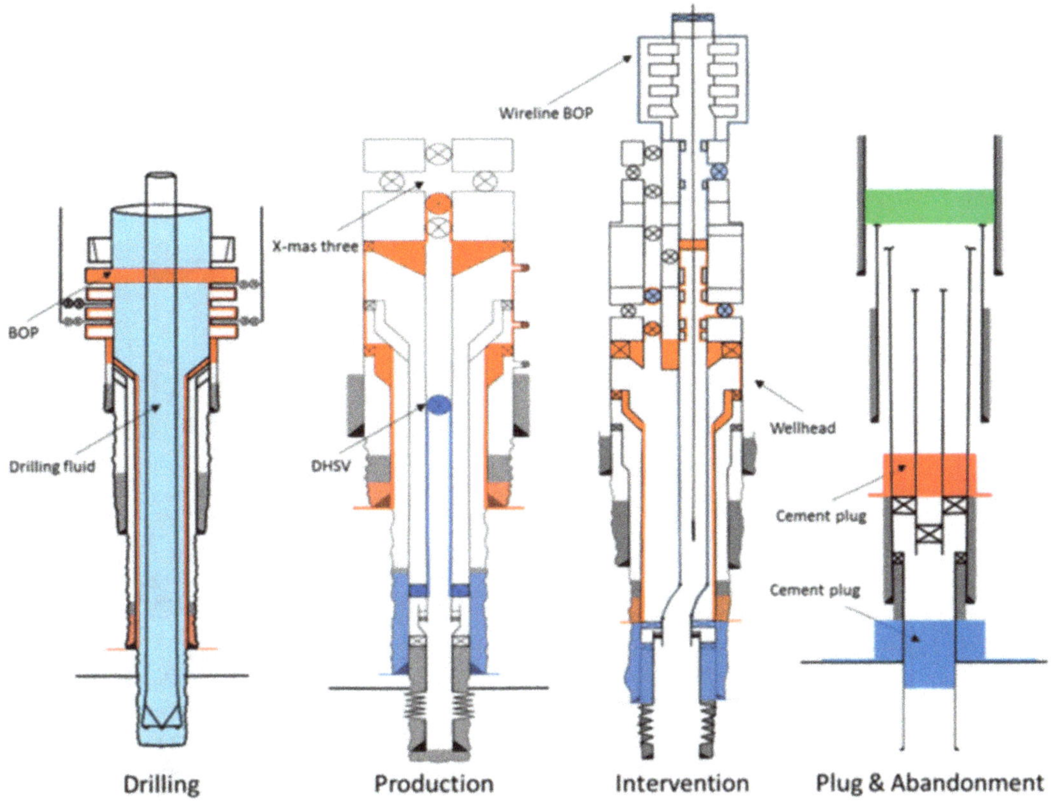

Fig. 2.1 Illustration of the two-barrier philosophy throughout a well's lifecycle [2]

Table 2.1 Examples of barrier systems through the lifecycle of the well given in Fig. 2.1

Example	Primary Barrier	Secondary Barrier
Drilling	Overbalanced mud with filter cake	Casing cement, casing, wellhead, and BOP
Production	Casing cement, casing, packer, tubing, and DHSV (Downhole Safety Valve)	Casing cement, casing, wellhead, tubing hanger, and Christmas tree
Intervention	Casing cement, casing, deep-set plug, and overbalanced mud	Casing cement, casing, wellhead, and BOP
Plug & Abandonment	Casing cement, casing, and cement plug	Casing cement, casing, and cement plug

During coiled tubing well intervention operations, the annular space between the coiled and production tubing should be considered as an annulus and distinguished with a name.

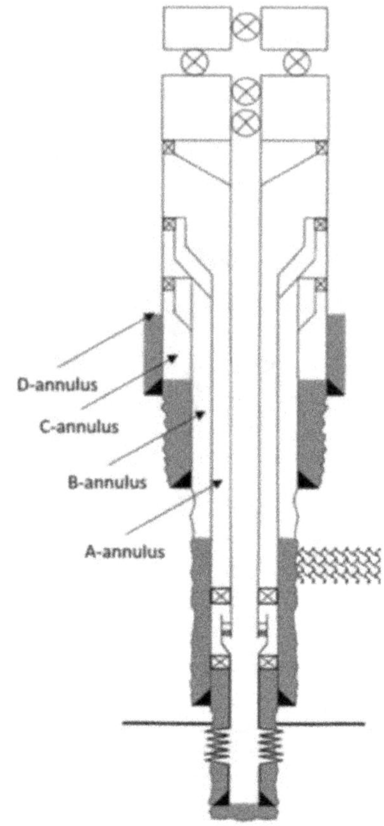

Fig. 2.2 Distinguished annuli in a completed well

2.2 Well Barrier Envelope

2.2.1 Primary and Secondary Well Barriers

To understand the subject of well barrier philosophy, it might be beneficial to start with the following question: What is a barrier? The word barrier has its roots from Middle French barrier, which can be traced back to Anglo-French, from barre bar, in 14th century. Merriam-Webster dictionary defines *barrier* simply as "*something (such as a fence or natural obstacle) that prevents or blocks movement from one place to another*". Different professional disciplines have established their version of the concept, in particular when it comes to operational and organizational barrier elements. Therefore, the term "barrier" is defined in many ways such as human barrier, non-technical barrier, operational barrier, non-physical barrier, or organizational barrier [4]. In the context of well integrity, a barrier is an impenetrable object that prevents the uncontrolled release of fluid. Two-barrier philosophy considers two independent well barrier envelopes; *primary well barrier* and *secondary well barrier*. Primary well barrier is the first enclosure that prevents flow from a potential source of flow. Secondary well barrier is the second enclosure that also prevents flow from the potential source of inflow. The secondary well barrier is a back-up to the primary well barrier and it is not normally in use unless the primary well barrier fails. The principle of the two-barrier philosophy has already been shown in Fig. 2.1; primary well barrier shown as blue line and secondary well barrier as red. For situations where a formation with normal pressure is present, a one-barrier methodology could be acceptable for the abandonment design.

2.2.2 *Environmental Plug*

In the context of the well integrity operating philosophy, one major difference is present between a permanent P&A operation and other activities (e.g. well construction, production, and workover) and that is, the environmental plug. During a permanent P&A operation, in addition to primary and secondary barriers, a supplementary plug is installed close to the surface. It is the shallowest well hindrance that isolates openhole annuli from the external environments that broadly is known as the *environmental plug*. It has also been given, in different literature, other names such as *surface plug*, *openhole to surface well barrier*, and *openhole plug*. These different names have originated due to the definition and functionality of the environmental plug. Some engineers claim that environmental barrier does not provide a well barrier envelope as the surrounding formation cannot hold high pressures and may be bypassed and therefore, it acts as a plug rather than a barrier.

The main function of the environmental plug could be described as to permanently disconnect the open annuli, which are created where casings are cut and retrieved near the seabed, from the external environment. In this manner, three main objectives are achieved; swabbing fluid from sea into the formations through the created annuli is minimized, exposure of surrounding environment to preceding potential hazardous fluids (e.g. drilling fluids) in different annuli is avoided, and potential conduits for leakage from near surface unidentified sources are sealed (Fig. 2.3). However, the obligation to install the environmental barrier is debatable as the cut and retrieval of conductor induces movements that cause the loose sediments to fall down and fill the wellbore. Some authorities do not require the installation of environmental plugs in wells without oil-based fluids in annuli and without zones capable of flow.

2.3 Well Barrier Element

A well barrier envelope consists of different well barrier elements. *Well Barrier Element* (WBE) is a physical element, which in itself may or may not prevent flow but in combination with other WBEs forms a well barrier. Figure 2.4 shows a schematic of primary and secondary barriers and the listed WBEs of a platform well, which is in temporarily abandoned status. The WBEs of a permanent well barrier envelope with its best practices is shown in Fig. 2.5.

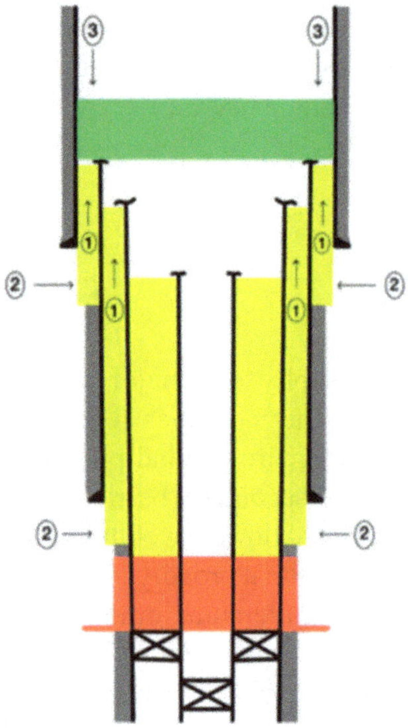

Functions of an Environmental Barrier
1. Avoiding exposure of surrounding environment to preceding hazardous fluids in different annuli.
2. Minimizing the potential of leak from unidentified sources close to surface.
3. Swabbing fluid from sea or near surface fresh water into the formations through the created annuli is minimized.

Fig. 2.3 Functions of an environmental barrier shown here in green color

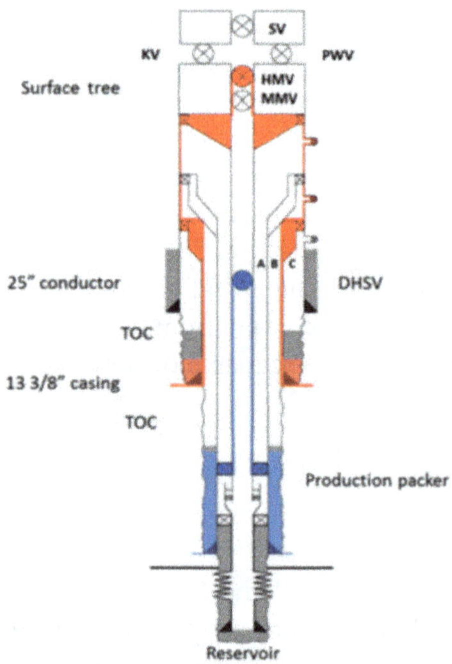

Well Barrier Elements
Primary well barrier
In-situ formation (cap rock)
Casing cement (9 5/8")
Casing (9 5/8")
Production packer
Completion string
Completion string components (chemical injection valves)
Downhole safety valve (and control line)
Secondary well barrier
In-situ formation (13 3/8")
Casing cement (13 3/8")
Casing (13 3/8")
Wellhead (casing hanger with seal assembly)
Wellhead/annulus access valves
Tubing hanger (body seals and neck seal)
Wellhead (WH/XMT connector)
Surface tree

Fig. 2.4 Schematic of well barriers showing well barrier elements for primary and secondary barriers [1]

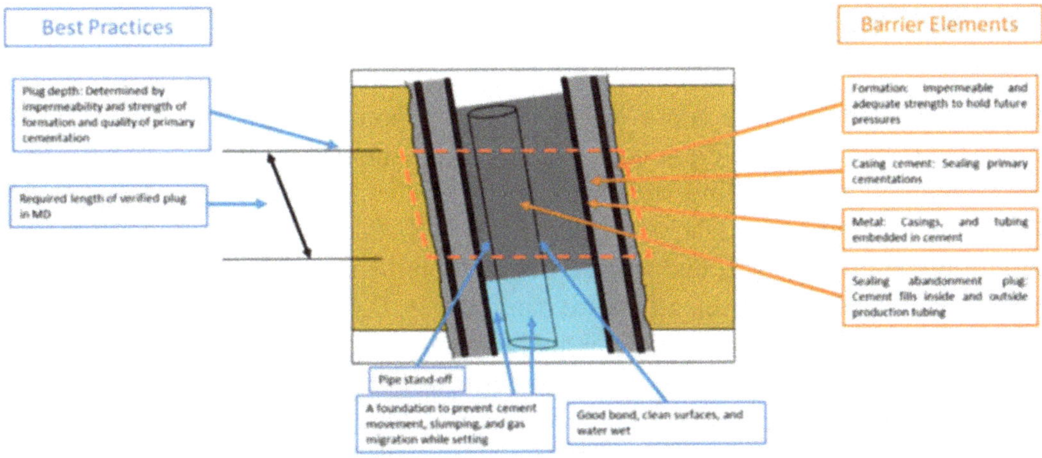

Fig. 2.5 Well barrier elements of a permanent abandonment barrier with its best practices [5]

2.4 Plug

Any object or device which is installed inside the wellbore to block a hole or passageway is called a *plug*. In the context of petroleum engineering, plugs are usually categorized into two main groups; *non-mechanical plugs* and *mechanical plugs*. Non-mechanical plugs will be discussed comprehensively in the subject of permanent plugging materials (see Chap. 4). Mechanical plugs are commonly referred to either bridge plugs or mechanical plugs.

2.4.1 *Bridge/Mechanical Plugs*

Bridge plug (mechanical plug) is a mechanical device installed and used to provide a seal inside the casing or production tubing. Bridge plugs are categorized as permanent, retrievable, or repositionable [6]. A *permanent bridge plug* has no design feature for intact removal from the conduit. For its removal, a substantial destruction process is necessary. However, a *retrievable bridge plug* possesses a design feature that facilitates removal from the conduit intact [7]. A *repositionable bridge plug* includes a design feature that facilitates its relocation inside the conduit (without removal) while re-establishing its intended function.

Throughout the abandonment process, a deep-set bridge plug can be used as a WBE for temporary abandonment. They provide easier and quicker plug retrieval. However, their utilization as a WBE during permanent abandonment should be avoided due to concerns associated with the long-term durability of mechanical plugs. Nevertheless, mechanical plugs can be used to establish a foundation for placing materials (e.g. Portland cement, thermosetting polymers, geopolymers, etc.) to minimize the risk of contamination while setting.

2.5 Well Barrier Illustration

Well barriers and their role in preventing or acting upon leakages from wells may be illustrated in two main different ways; well barrier schematics and barrier diagrams. The concept of documenting well barrier using schematics was introduced to the Norwegian oil and gas industry in 1992 [8]. A *Well Barrier Schematic* (WBS) is a static illustration of a well and its main barrier elements, whereby all the primary and secondary well barrier elements are marked (Fig. 2.4). A *well barrier diagram* is a network illustrating all possible leak paths from the reservoir to the surroundings. The surroundings could be the sea for subsea wells, platform deck for a topside Christmas tree, flowline from a subsea well, ground for onshore wells, etc. Figure 2.6 shows the well barrier diagram for the production well in Fig. 2.4. A well barrier diagram describes the status of barrier elements after a leak occurs. One of the major differences between well barrier diagrams and WBSs, although they have their own specific applications, is the quantification of the barrier diagrams. Well barrier diagrams are widely used to evaluate the likelihood of the consequences illustrated in the diagram.

Hence, in the petroleum industry, well barrier schematics and well barrier diagrams are important tools for reliability and risk assessments of a well in all phases of its lifecycle and for well integrity assessments.

2.6 Prerequisites for Well Abandonment Design

To perform a competent and efficient abandonment practice, the P&A needs to be considered during well design and well construction in order to reduce the associated risks and saving costs. When a well is selected as a candidate for P&A, the abandonment design is initiated. It is usually recommended to start the planning five years ahead of commencement of the P&A operation. This is due to information gathering regarding changes to the well status, the clarity of work scope and accuracy of time estimation. Detailed planning will require the determination of a detailed sequence of all activities and the resources required to perform the job. In the abandonment design stage, it is necessary to study and document the following: well configuration, stratigraphic sequences of each wellbore, cement logs and cementing operation data and documents, formations with suitable WBE properties, and specific well conditions [1].

2.6.1 Well Configuration

It is necessary to know the original and current well configuration. The well configuration includes depths and inclinations, specification of formations that are sources

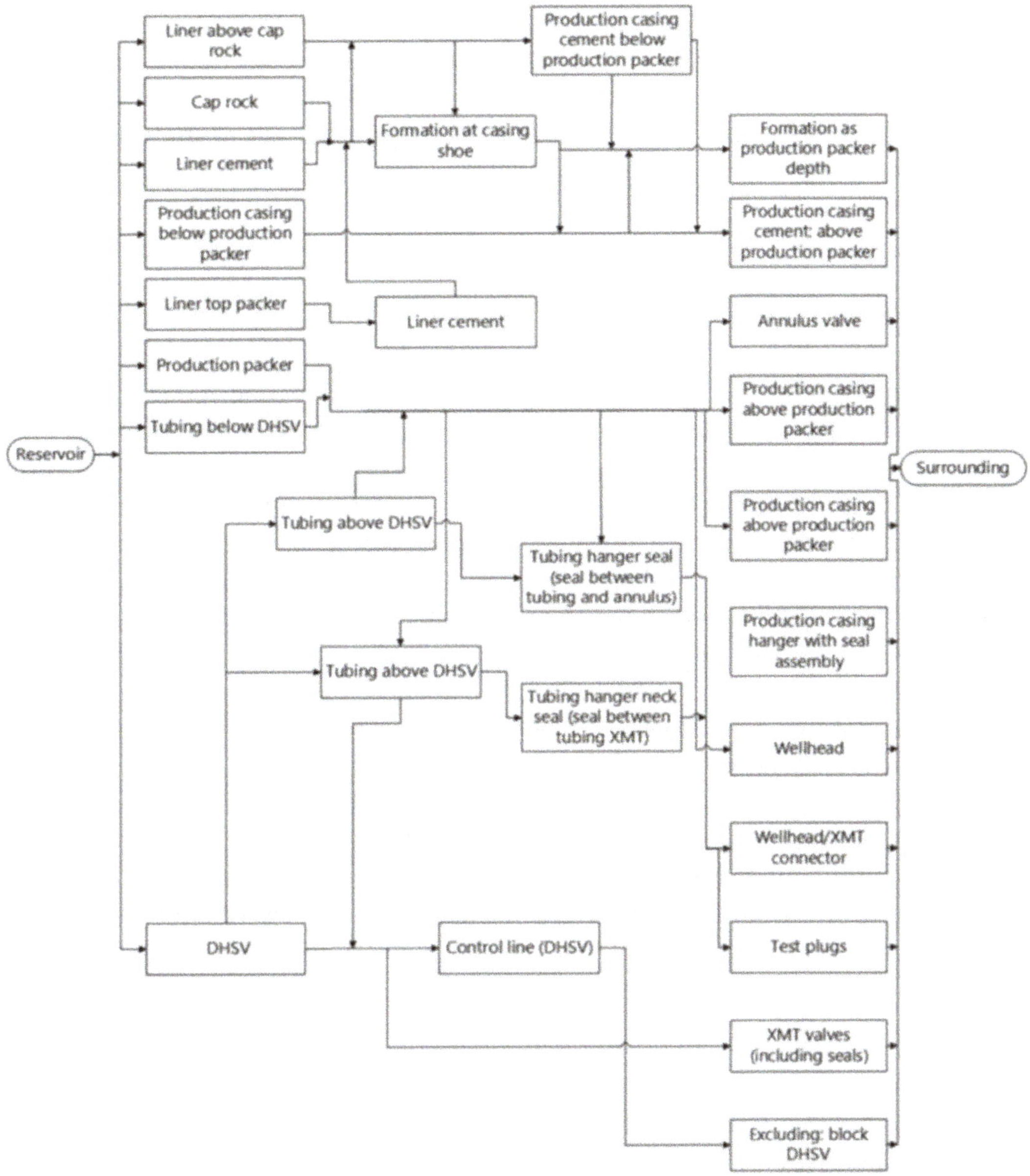

Fig. 2.6 Barrier diagram for the production well in Fig. 2.4

of inflow, casing strings, casing cements and top of cements, casing shoes, and wellbores. In addition, all the active sidetracks, and temporarily and permanently abandoned sidetracks are mapped [1].

2.6.2 Stratigraphic Sequences

Stratigraphic sequence of each wellbore is identified and documented. The stratigraphic report includes reservoir(s) and information about their current and future production potential and their containing fluids. In addition, the initial, current and eternal perspective pressures of each flow potential need to be distinguished and estimated. The identification of flow potentials in the overburden is necessary in order to minimize the risk of leaks or kicks over time. It is also necessary to study and adjust the formation fracture gradients for depleted formations [9].

2.6.3 *Logs and Cementing Operation Data*

Cement logging is one of the most commonly used verification methods that the petroleum industry relies on to qualify casing cement. It is common practice to log and document the casing cement behind the intermediate casing, and production casing strings; however, not the cement behind the surface casing. There are different types of logs that are used for evaluating casing cement, such as cement bond log (CBL), variable density log (VDL), temperature logs, and sonic logs [10]. In addition, displacement efficiency based on the record from the cement operation (e.g. volumes pumped, returns during cementing, differential pressure, slurry rate, density, etc.) is another set of data which is studied to check the quality of casing cement and identify the *Top of Cement* (TOC). Figure 2.7 shows the recording output from a primary cement job. All of these data are considered during P&A design, in addition to the remedial cement jobs performed on the well throughout its lifecycle.

Considering well cementing log data from old wells during P&A design is a concern as the old logging data are less reliable due to their availability and quality. However, there are P&A designs that rely on old CBL-VDL logs reports. Experience shows that casing cement quality of wells constructed 10–15 years ago are still intact when re-evaluated recently.

2.6.4 *Formations with Suitable Well Barrier Element Properties*

Identifying an appropriate formation, for establishing the primary and secondary well barriers across, is a key factor. A suitable formation should possess cap rock properties. It should have sufficient strength to keep the hole in gauge during hole conditioning, hold the exerted hydrostatic pressure of the barrier before it sets (e.g. cement, etc.), and be impermeable or have very low permeability to minimize the risk of integrity loss or providing a conduit for leak around the barrier. Absence of fractures and faults are other properties which in combination with those previously mentioned properties qualify a formation as a suitable candidate for establishing barriers across.

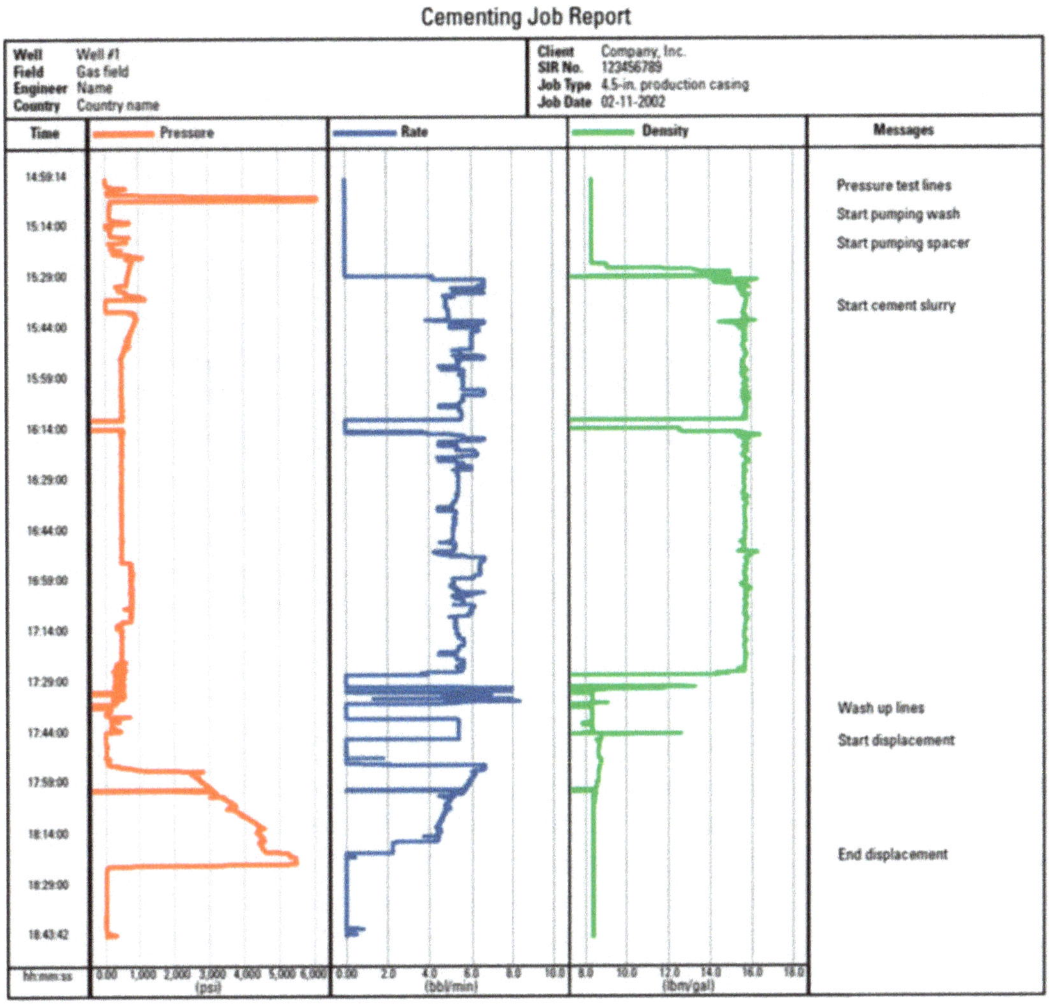

Fig. 2.7 A typical recording output from a primary cement job (Reprint from Well Cementing) [10]

2.6.5 Specific Well Conditions

To establish a permanent barrier for securing a flow potential, it is necessary to get access to the required depth. However, sometimes there are well conditions that dictate contingency plans. Scale build-up, casing wear, collapsed casing, fill, H_2S and CO_2 corrosion, asphaltene deposition, erosion, and hydrates are common specific well conditions to be considered in abandonment design [1].

2.6.5.1 Scale Build-up

Scale is mineral salt deposits or coating that precipitates and adheres to the surface of metal, rock, or other materials [11]. The precipitation is the result of different factors: a chemical reaction with the surface, a change of pressure or temperature, a change

Fig. 2.8 Scale deposit caused a restricted access to downhole (Courtesy of Schlumberger)

in the composition of a solution, or a combination of these factors [12]. In severe circumstances, scale build-up creates a significant restriction, or even completely plugs the production tubing. Typical scales are barium sulfate, calcium sulfate, strontium sulfate, iron sulfate, calcium carbonate, iron oxides, iron carbonate, various phosphates, silicates and oxides, and any compounds that are insoluble or slightly soluble in water. For scale removal, a wide range of mechanical (e.g. milling), chemical (e.g. acid wash, non-acid scale dissolver), and scale inhibitor treatment options are available [13]. Figure 2.8 shows a scale build-up in a production tubing.

Scale build-up is a concern for production tubings due to reducing the effective drift, consequently, limiting wireline activities such as running puncher, cutter, caliper log, etc. As production or injection is through the production tubing, the scale build-up does not occur inside the production casing. Retrieval of the production tubing with occurred scale needs special handling of the scale and its disposal, one feasible solution to minimize the effect of scale on P&A activities could be to leave as much pipe in the well as possible. This approach may lead to a rigless P&A operation as retrieval of production tubing requires high pulling capacity, a rig or a jack.

2.6.5.2 Casing Wear

Casing wear is often a problem in deep and highly deviated wells where doglegs and large tensile loads on the drill string combine to produce high lateral loads where the drill string contacts the casing. It is a complex process involving variables such as temperature, drilling fluid type, percentage of abrasives in the drilling fluid, tool joint hardfacing, revolutions per minute, tool joint diameter, contact load, and many other factors. In the course of P&A operation, casing wear can compromise the integrity of casing and result in blowouts, lost circulation, and other expensive and hazardous problems [14]. Therefore, it is necessary to measure and analyze the casing wear that has occurred over the lifetime of a well (e.g. during construction and intervention operations) and consider it in the abandonment design. The risk of induced casing wear while the P&A operation is performed also needs to be studied in the abandonment design.

2.6.5.3 Collapsed Casing

In all wells, there are natural forces and occasionally induced forces, which may cause casing to collapse. The principle cause of *casing collapse* is compaction of formations and the resultant subsidence of the overlying sediments. Figure 2.9 illustrates casing loads resulting from compaction of reservoir rock. The natural forces are created because of tectonic stresses, subsidence, and formation creep. Subsidence widely occurs in large chalk formations where the depleted chalk reservoir is not able to hold the weight of the overburden; however, the intensity in small chalk reservoirs is not high. To visualize the influence of the size of chalk reservoir on subsidence, consider a beam as representing a reservoir with the overburden acting by the load A (Fig. 2.10). A large reservoir cannot withstand the load of the overburden; however, if the reservoir is small, then it can withstand the load of the overburden without experiencing a compaction effect.

The formation creep is intensified and more subjective in plastic salt zones where a non-uniform formation movement exerts point loading on the casing string and causes collapsed casing. In addition to the natural forces, the induced circumstances

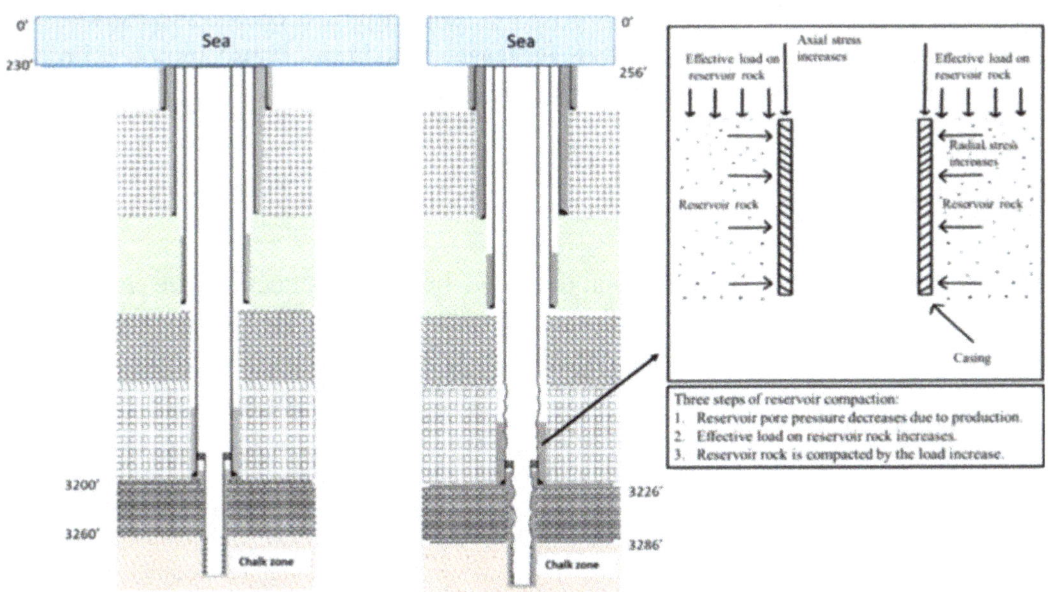

Fig. 2.9 Effect of formation compaction and subsidence on casing string and liner

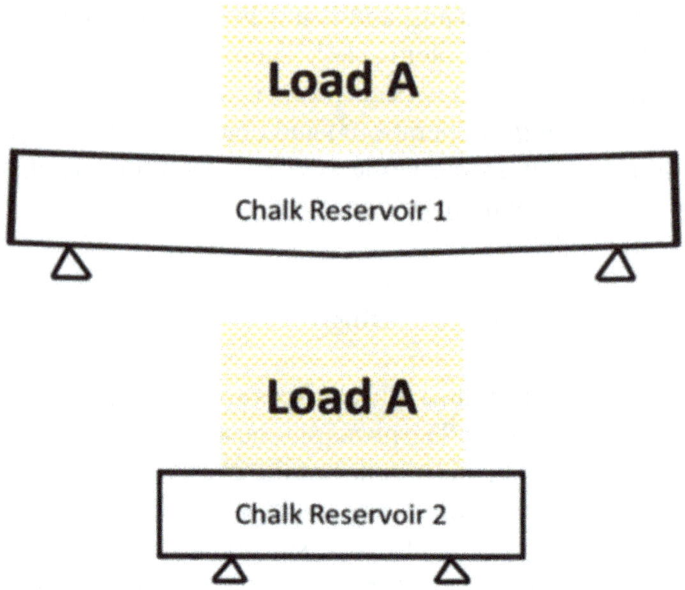

Fig. 2.10 Influence of reservoir size on the compaction load in a chalk basin

such as temperature change, and excessive matrix acidizing, play role as well. Temperature change is one of the bases of the induced forces. Temperature changes encountered during the life of the well are small usually and negligible. However, there are situations where temperature variations are not small. Examples of these large temperature variations that can be encountered include geothermal wells used in extracting steam from volcanic areas of the earth, steam-injection wells used in thermal recovery processes, deep gas wells, and wells completed in abnormally hot areas. Excessive matrix acidizing could result in a lack of lateral support around the casing and consequently lead to buckling as the casing is loaded in compression. Furthermore, when the effects of wear, corrosion, and fatigue are added to the stresses on the casing, the potential for failure increases. Casing collapse imposes limited access to downhole and usually requires section milling.

2.6.5.4 Fill

Drill cuttings, collapse fragments, and settled barite may accumulate around the uncemented casing strings and require more force when pulling the casing string. In most cases during P&A operations, the required force is beyond the *pulling capacity* of the working unit or exceeds the tensile strength of casing. Therefore, collapsed casing situation usually dictates section milling.

2.6.5.5 Corrosion

Hydrogen sulfide corrosion. The general mechanism for hydrogen sulfide (H_2S) attack may be expressed as follows:

H_2S readily dissolves in water and partially dissociates:

$$H_2S(g) + H_2O \leftrightarrow H_2S\,(aq) \tag{2.1}$$

In H_2S corrosion of mild steel, polymorphous iron sulfide is formed:

$$H_2S\,(aq) + Fe^{2+} \rightarrow FeS + 2H^+ \tag{2.2}$$

A research study [15] shows that for H_2S corrosion of mild steel, polymorphous iron sulfides can form iron sulfide (FeS), mackinawite $(Fe, Ni]_{1+x}$ S where x = 0–0.11), cubic ferrous sulfide (FeO_4S), troilite ($Fe_{(1-x)}S$ where x = 0–0.2), pyrite (FeS_2), greigite (Fe_3S_4), marcasite (FeS_2). The formed iron sulfide sets up a galvanic cell in which the steel pipe becomes the anode. This reaction is generally assumed to be responsible for the deep irregular pitting observed in sulfide corrosion.

H_2S corrosion can create cracks in steel pipe in two distinct ways; sulfide stress cracking, and stress corrosion cracking. *Sulfide stress cracking* occurs near room temperature, and it affects the upper parts of wells. This phenomenon occurs during periods of shut-in and cooling down. *Sulfide corrosion cracking* is encountered at high temperatures which occurs at the bottom of wells [16].

Carbon dioxide corrosion. CO_2 corrosion is encountered in both gas wells and oil wells and it is reported in different areas such as Louisiana, the North Sea, Germany, the Netherlands, and Gulf of Guinea. Some of the crucial factors which extend CO_2 corrosion of steel include: temperature, pressure, CO_2 content, salt concentration, basic sediments and water, flowing conditions, etc. [16]. The solubility of CO_2 increases as pressure increases and subsequently the pH decreases. However, as the temperature increases, the solubility of CO_2 decreases and as a result, the pH increases. Certain minerals may act as a buffer preventing the pH reduction. The general mechanism of CO_2 attack in the presence of water may be expressed as [17]:

$$CO_2(g) + H_2O \rightarrow H_2CO_3(aq) \tag{2.3}$$

$$H_2CO_3(aq) + Fe^{2+} \rightarrow FeCO_3 + H_2 \tag{2.4}$$

It has been reported that the CO_2 corrosion of steels is highly localized corrosion, which appears in the form of pits, gutters, or attacked areas of various sizes [16]. Figure 2.11 shows a corroded tubing caused by incompatibility between the tubing type and injection water quality.

Retrieval of a corroded production tubing may cause tubing rupture and it may require a multiple fishing operation. Another scenario is when a kill fluid is pumped through the corroded production tubing, the fluid will be exposed to the production casing before killing the well.

Usually corrosion does not attack production casing as only the production tubing is exposed to production or injection fluids. However, production tubing corrosion may indirectly compromise the well integrity during pressure testing. Consider a

Fig. 2.11 Corroded production tubing due to incompatibility between injection water and tubing type [18]

bridge plug which has been installed in tail pipe of production tubing. The bridge plug is going to be pressure tested and therefore, a fluid is injected through the production tubing. A corroded production tubing with holes exposes the production casing to high pressure and consequently, the casing may burst due to the imposed pressure. Therefore, good knowledge of production tubing condition is necessary prior to starting the P&A operation.

2.6.5.6 Asphaltene Deposition

Asphaltenes are the most aromatic components of crude oil with a high-molecular weight solids, and are insoluble in light alkanes and soluble in aromatic solvents [19]. Several factors such as changes in pressure, temperature, and crude oil composition cause asphaltenes to precipitate from the oil as a black sticky solid material [20]. Traditional methods of removing asphaltene deposits involve mechanical removal, injection of dispersant and solvents, and heat treatment. In P&A operations, in the case of an asphaltene issue, it is a common practice to remove asphaltenes mechanically via scrapers, cutters, coiled tubing deployed jetting tools, or a milling operation [21].

2.6.5.7 Erosion

Erosion is the process of removing material by mechanical action such as particle or droplet impact. The velocity of the particles (e.g. unconsolidated formation) or droplets, which are carried by producing or injecting fluid, provides the energy for erosion of the steel pipe. In addition, fluid flow through the pipe with high velocity creates enough energy for erosion of steel by the fluid. This concept is used for abrasive cutting.

2.6.5.8 Hydrates

Natural gas *hydrates* are solid crystalline compounds in which molecules of natural gas are trapped in water molecules under pressures and temperatures considerably above the freezing point of water. Hydrates tend to form in the near surface environment where the temperature is low such as the wellhead, pipelines, and other processing equipment [22]. In the early phase of P&A, mechanical removal via scrapers, cutters, coiled tubing deployed jetting tools, and milling are common hydrate removal practices.

2.6.5.9 Containment Assurance of the Abandoned Wells or Fields

Subsurface containment assurance is defined as the identification and mitigation of elements that could result in the potential loss of containment of subsurface fluids. The goal of subsurface containment assurance is to ensure no harm is caused to the environment and operated assets, or no impact on well operations due to the leakage of production or injection fluids from their intended zones [23, 24]. One may claim that the subject of subsurface containment assurance fulfills the well integrity requirements, however, in fact it is more comprehensive than the well integrity. It includes well integrity, subsurface integrity, and any aspect of deepwater and surface facilities which are directly relevant to upstream exploration, production, and abandonment operations [25]. As an example the well integrity of abandoned wells or fields might be influenced by nearby injector wells influencing the pressure regime. So possible monitoring of permanently abandoned wells or fields regarding containment assurance may need to be considered.

2.7 Well Abandonment Phases

When the abandonment design is ready, the operator submits the program to the local regulatory body. The authority reviews the program and asks for changes or approves it. Once the program is approved, the operator can commence the P&A operation. Approval of the program does not necessarily load any responsibility on the local authority as all responsibilities during P&A and post-abandonment operations are in the hand of the operator.

Generally, a P&A operation may be divided into three phases; phase 1—reservoir abandonment, phase 2—intermediate abandonment, and phase 3—wellhead and conductor removal. This categorization is regardless of the well location (e.g. offshore, or onshore), well type (e.g. exploratory, producing, injecting, etc.), and the well status (e.g. partially abandoned, shut-in, etc.).

2.7.1 Phase 1: Reservoir Abandonment

Reservoir abandonment starts primarily by inspecting the wellhead and rigging up a wireline unit. The wireline unit is employed to check the access to wellbore by drifting and evaluating the condition of the production tubing by running a caliper log. This preliminary investigation can be regarded as Phase 0—well intervention, and it has a significant influence on time reduction during a P&A operation [26]. In addition, waste handling systems are established for liquid and solid phases. This phase proceeds with an injection test to examine the well integrity. If integrity is maintained, cement slurry is bullheaded to plug the main reservoir and once the cement plug has achieved sufficient strength, its quality is determined by pressure testing. So far, this part of the job is a rigless operation. However, if well integrity is not maintained, a rig needs to be mobilized and a BOP nippled up. Figure 2.12 illustrates a well status where the bullheaded cement is qualified as primary and secondary barriers after conducting the reservoir abandonment phase. Generally speaking, phase 1 is completed when the permanent primary and secondary barriers secure the main reservoir. The production tubing may be retrieved or left in the hole as a part of the well barrier envelope. This phase is completed when the reservoir is fully isolated from the wellbore.

2.7.2 Phase 2: Intermediate Abandonment

Intermediate abandonment phase includes milling, retrieving casing, setting barriers to isolate intermediate hydrocarbon or water-bearing permeable zones, and installing an environmental plug. The production tubing may partly be retrieved if it has not been retrieved in phase 1. Phase 2 is complete when all the flow potentials identified in the overburden are secured.

2.7.3 Phase 3: Wellhead and Conductor Removals

In this phase, the conductor and wellhead are cut below the surface or seabed and retrieved. The reason is to avoid any future incident with other marine activities (e.g. fishing activities). In the Norwegian sector of the NCS, this phase is usually regarded as a marine job and not a drilling operation.

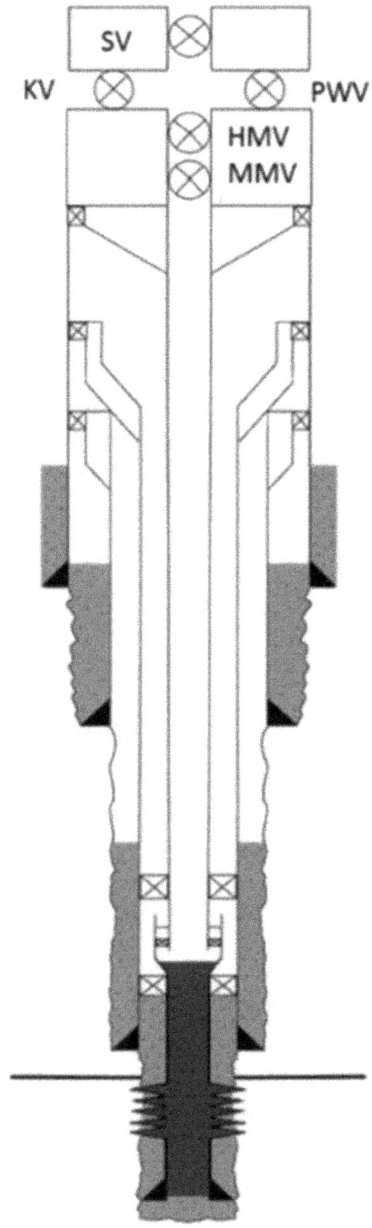

Fig. 2.12 Well status after completing the reservoir abandonment. The assumption prior to bullheading the cement plug is that, the well integrity was maintained and full access to the wellbore was achieved

2.8 Disconnecting the Christmas Tree and Assembling Blowout Preventer

It is a common practice to bring the well to temporarily abandoned status or shut-in prior to commencing the permanent P&A operation or after the reservoir abandonment phase. The reason is to reduce the risk of a kick or release of uncontrolled flow while nippling down the Christmas tree and nippling up the BOP. *Nipple-down* is the activity of disassembling well-control equipment on the wellhead. *Nipple-up* is the process of assembling well-control equipment on the wellhead. So, as we discussed earlier in this chapter, it is necessary to have two independent well barrier envelopes.

This means that when the Christmas tree is disassembled still two intact well barrier envelopes need to be in place. Therefore, it is crucial to understand wellhead and the Christmas tree systems for disconnecting the reservoir from the environment until the BOP is rigged up.

In October 2016, a serious well control incident occurred on a production well in the Troll field, in the North Sea. This incident began after permanently plugging the existing flow paths in the well. Then, a sidetrack was about to be drilled. In connection with pulling the tubing hanger, the completion string with the top drive was suddenly raised six meters without control. Large quantities of fluid and gas flowed out through the rotary table. The blowout lifted the 2.5 tons hydraulic slips and threw some two tons of bushings several meters across the drill floor and the liquid column reached the top of the derrick about 50 m above the drill floor. Fortunately, nobody suffered physical injury, but it could have led to a major incident with loss of several lives. The Norwegian Petroleum Safety Authority investigated the incident and concluded that the direct cause of the incident was the release of a large quantity of trapped reservoir gas underneath of the tubing hanger. Although a BOP wellhead connector test had been performed six hours before the incident, a gas leak that occurred during these six hours, caused the incident [27] (Fig. 2.13).

Another crucial factor that might be considered is wellhead fatigue loads that will be exerted by the BOP stack during P&A operations. Therefore, wellhead systems and their advantages and limitations are reviewed in this chapter.

2.8.1 Wellhead Systems

A wellhead system is the surface termination of a wellbore and it is composed of spools, valves, and assorted adaptors that provide pressure control of a production well. Wellhead systems incorporate facilities for installing casing hangers, tubing hanger, and Christmas tree. The wellhead systems can be categorized depending on the place where the wellhead is installed as *surface wellhead* systems and *subsea wellhead* systems [28, 29]. There are two types of wellheads used for surface applications; spool type and compact type. Other names used for the compact type wellheads are speed head, unitized head, bowl head, multi-bowl head, and unihead. Each of these configurations have their own advantages and challenges in the course of P&A. Table 2.2 lists advantages and challenges for each system. As the subject of wellheads is an extensive area, wellhead systems will be reviewed based on the first classification system.

As wellhead type is related to the number of connections or components and consequently the risk of leakage, it is important to analyze the wellhead condition. In March 2012, on Elgin installation (approximately 200 km east of Aberdeen, Scotland) located in the North Sea experienced a major incident of uncontrolled release of hydrocarbons to atmosphere. In this incident, reservoir gas from the Chalk formation leaked to the A-annulus and in a further step from A-annulus to B-annulus, and then to C-annulus. Due to poor sealing capability of wellhead components and connections,

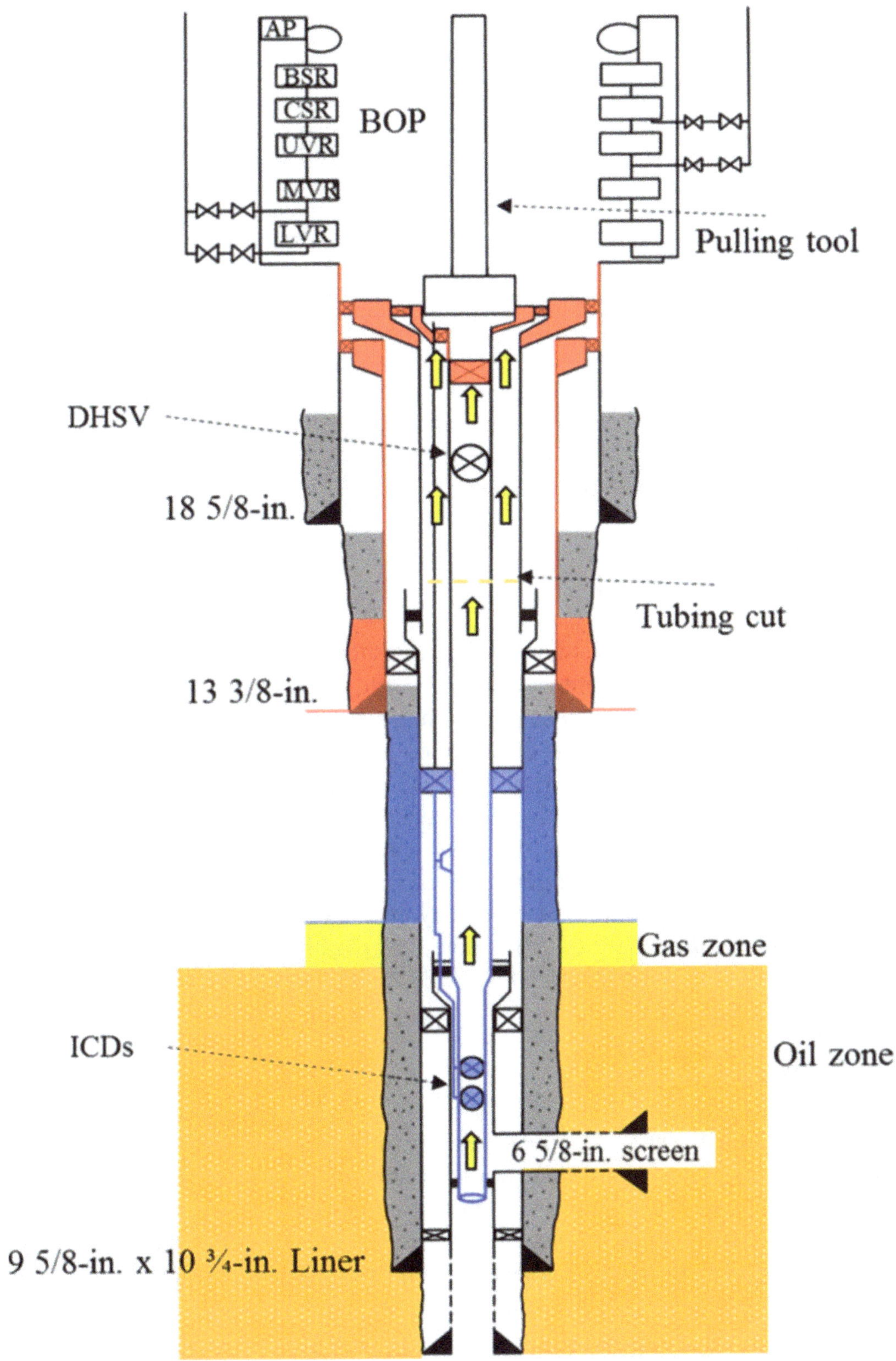

Fig. 2.13 Leak path which caused trapped gas beneath the tubing hanger in Troll field [27]

Table 2.2 Advantages and limitations of wellhead systems with respect to P&A operations

Wellhead type	Advantages	Disadvantages
Spool type	• The relative simplicity of the suspension and sealing systems	• Requires removal and the re-installation and testing of the BOPs for the removal of each casing head spool • They have more connections and consequently more risk of leakage
Compact type	• Less height • Fewer potential leak paths	• Lack of tolerance of damage to the hanger sealing areas

the gas leaked to the conductor, D-annulus [12]. As the conductor annulus, D-annulus, is not connected to any barrier for preventing leaks, the gas leaked to the environment uncontrollably, Fig. 2.14. This incident had no loss of life and well control was achieved by killing the well by pumping kill mud [30].

2.8.1.1 Surface Wellhead

Surface wellheads are used both onshore and offshore. Primary functions of the surface wellhead include pressure isolation, pressure containment, casing and tubing weight suspension, and the Christmas tree housing. shows a surface wellhead model and its main sections; starter head at the bottom, spools for casing hangers, spool for tubing hanger, adaptor, and valves for access to different annuli. There are several factors to be considered in selection of a wellhead during well construction; value, field history, operator preference, lifespan, temperature range, fluid environment, pressure range, mechanical configuration, external loading, and installation or energization method. Some of these factors may endanger the wellhead condition and its performance during the P&A. Therefore, investigating the wellhead quality and running a fatigue analysis are crucial in the P&A design phase (Figs. 2.15 and 2.16).

2.8.1.2 Subsea Wellhead

The main functions of subsea wellheads are the same as surface wellheads. Nevertheless, due to subsea conditions, there are some additional functions such as serving a structural and pressure-containing anchoring point on the seabed for the drilling and completion system, and facilitating guidance, mechanical support, and connection of the systems used to drill and complete the well. A standard subsea wellhead system (Fig. 2.17), typically consists of drilling guide base, low-pressure housing (typically 30-in.), high-pressure wellhead housing (typically 18 ¾-in.), casing hangers, metal-to-metal annulus sealing assembly, bore protectors and wear bushings, and running and test tools. In the course of drilling subsea wells, the Low-Pressure Wellhead Housing (LPWH), conductor, and guide base are run at the same time.

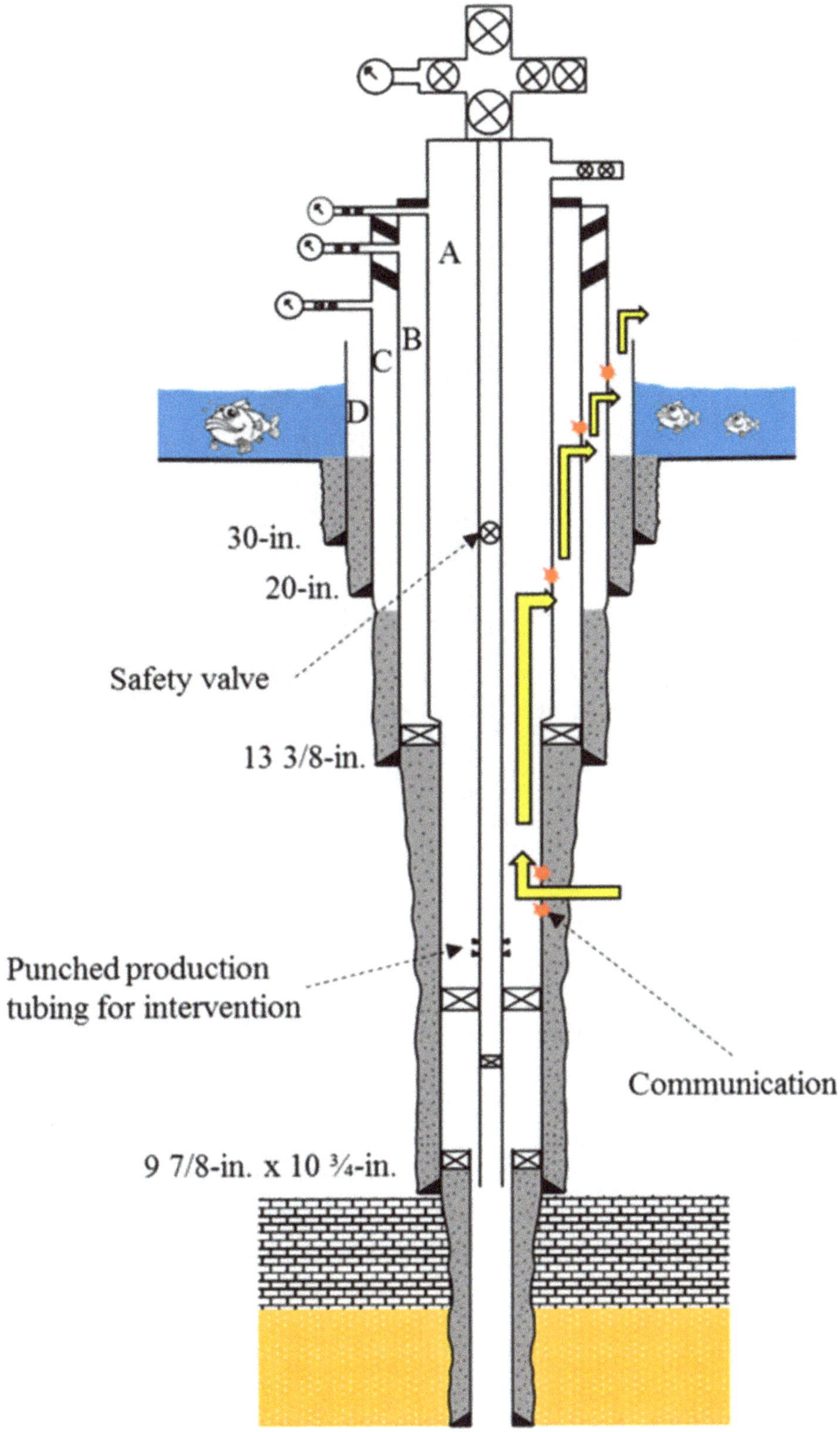

Fig. 2.14 The schematic of the leak path from Elgin, platform well [30]

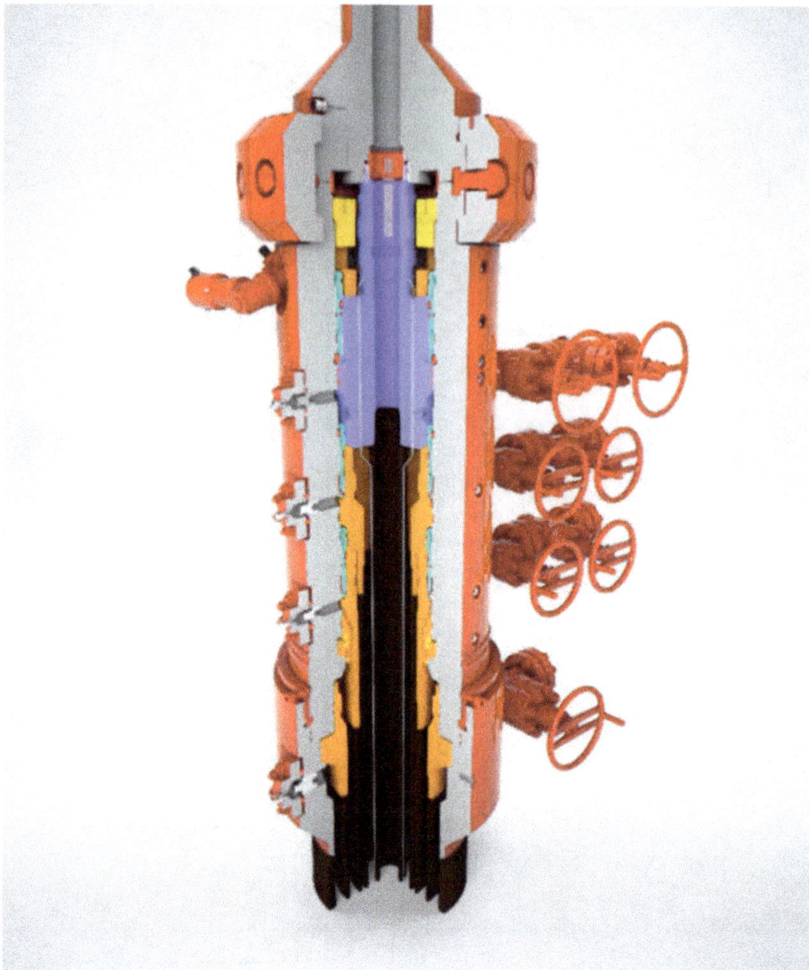

Fig. 2.15 Surface wellhead model with its main different sections. (Courtesy of TechnipFMC)

The subsea wellhead is placed on the seabed and the BOP is installed on top of it. Waves and current forces acting on the marine riser during drilling, production, intervention or P&A operation will cause movements. A subsea wellhead will be exposed to external loads: static and cyclic combinations of bending and tension (compression). The cyclic loads can cause fatigue damage to the well and create well integrity issues. If the subsea wellhead fails then its pressure vessel function will be lost, which can lead to HSE issues [31].

2.8.1.3 Special Consideration for Wellhead Systems

Casing/tubing hanger lockdown—The incident occurring on Troll field [27], could have been prevented if the well was completed with a horizontal Christmas tree rather than a vertical Christmas tree. The horizontal and vertical X-mas tress are discussed in more detailed in this chapter. In offshore wells, casing/tubing hanger is installed and locked down inside a horizontal Christmas tree and the pressure beneath the tubing

Fig. 2.16 Surface wellhead with vertical XMT. (Courtesy of TechnipFMC)

hanger can be measured. However, when a vertical Christmas tree is selected for well completion, the casing/tubing hanger is installed and locked down to the wellhead prior to installation of the tree. For onshore wells, the lockdown and installation of casing/tubing hanger is similar to the scenario for vertical trees.

Fatigue life of a wellhead system—One of the challenges during P&A design and operation, especially for subsea wells, is the fatigue loading exerted on wellheads by the BOP. Some of the older wells, drilled two to three decades ago, have been drilled with BOPs that were smaller and lighter than current designs; therefore, the wellhead design was different and subsequently the wellheads response to induced fatigue. Another challenge linked to the afore-mentioned challenge is BOP pressure rating requirements legislated by some authorities. Consider an old well where its wellhead connector has been pressure rated for 5 kpsi but requirements ask for the utilization of a 10 or 15 kpsi BOP, although a depleted well may require a lower BOP pressure rating. In addition, the challenges associated with fatigue life of wellhead systems is more of a concern in subsea wells due to sea currents. Another concern regarding fatigue is updated regulations. For example: in 1975, a 13 3/8-in. BOP could have an approximate weight of 20 metric tons, however, in 2016, a 18 5/8-in. BOP could weigh up to 400 tons. Therefore, nowadays, the fatigue introduced to the wellhead is much higher compared to old BOPs. In addition, there are wells designed

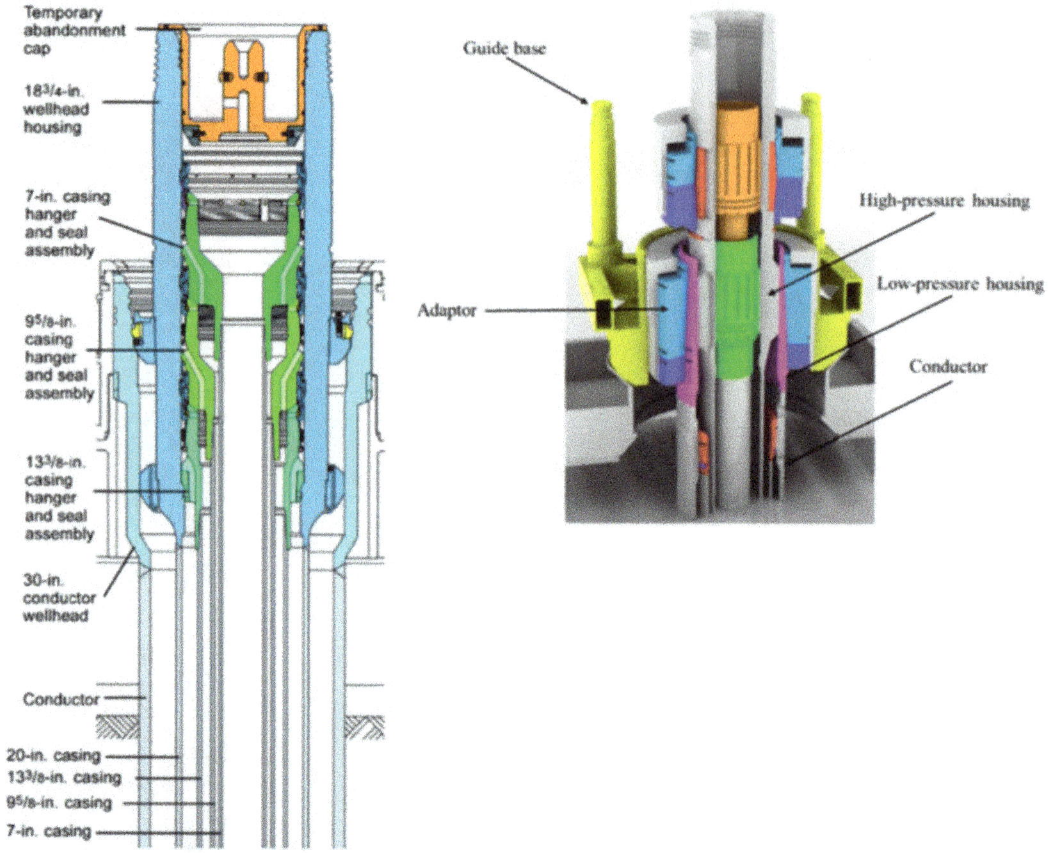

Fig. 2.17 A standard subsea wellhead system and its main sections. (Courtesy of SPE)

and completed for a specific service life with regards to time, however, the service life has extended more (up to a decade) than the design life. These wellheads are a point of failure for wells with strong aquifers where the current reservoir pressure is approximately equal to the initial reservoir pressure.

Recently, a new generation of BOP system being developed is electrical based, which does not require a hydraulic accumulator (Koomey) unit. This system is lighter compared to previous and available BOP systems.

2.8.2 *The Christmas Tree Systems*

The equipment at the top of a well is called "*Christmas tree*". The Christmas tree is assembled of valves, spools, pressure gauges and chokes which is fitted to the wellhead (see Fig. 2.16) of a completed well. The *Christmas tree* (XMT) provides a controllable interface between the well and production facility. It is also called by other names such as cross tree, X-tree, or tree. The functions of a tree are addressed as follows: allowing reservoir fluid to flow from wellbore to the surface facilities in a safe and controlled manner, safe access to the wellbore to perform well intervention procedures, allowing injection of fluids, providing access to a hydraulic line for a *Surface Controlled Subsurface Safety Valve* (SCSSV), providing the electrical interface for instrumentation and the possible electrical wiring for an *Electrical Submersible Pump* (ESP). The XMT is installed on the last casing spool, tubing head adaptor, or high-pressure wellhead housing for a subsea well. They are available in a wide range of sizes and configurations, such as low-or high-pressure capacity and single- or multiple completion capacity. It is a norm to purchase the XMT and wellhead from the same manufacturer due to compatibility. Generally, there are two different approaches to categorize the trees; depending on the place where the XMT is installed or based on the arrangement of the valves and gauges. The first approach divides the trees into two main groups; surface trees (dry trees) for land/platform wells, and subsurface trees (wet trees) for subsea wells. The second approach divides the trees into two main classes based on the configuration of the valves and gauges; vertical trees and horizontal trees. Figure 2.18 illustrates four different configurations of tree

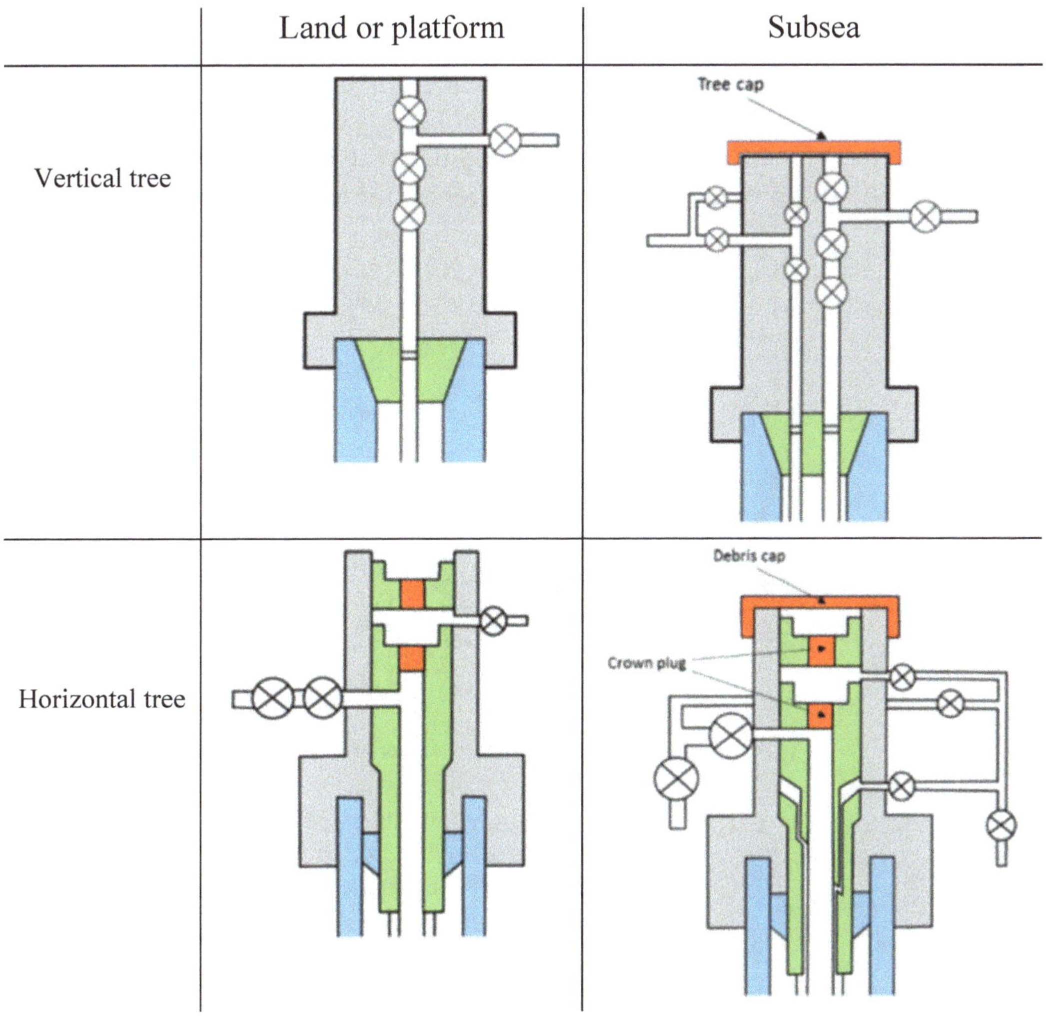

Fig. 2.18 Four different configurations of XMT valves for land/platform and subsea wells [3]

valves. For simplicity and relevancy to P&A, the second approach is preferred and used for discussion in this book.

2.8.2.1 Vertical Christmas Tree

Figure 2.19 shows a drawing of a surface vertical tree. The master valve is located above the tubing head adaptor and its function is to allow a well to flow or to shut-in the well. Typically, there are two master valves; lower master valve, and upper master valve. The two valves are often used because they provide redundancy; if one valve cannot function properly, the other valve is engaged. The upper and lower master valves are shown in Fig. 2.19. Tee type fitting (known as T-block) provides diversion of the vertical flow to the horizontal flowline. Usually a wing valve is located on the side of the tree and used for controlling or isolating production from the well to surface facilities. Based on the tree design, which is the operator requirement, one or two wing valves can be fitted to the tree. As a common practice, usually operators require two wing valves; one for production, known as *Production Wing Valve* (PWV), and another one as backup or as a *Kill Valve* (KV). After the wing valve a small restriction, which is referred to as the choke, is used to control the production rate of the well. On the Christmas tree, the topmost valve is called the *Swab Valve* (SV). The swab valve provides access to the borehole for well intervention operations

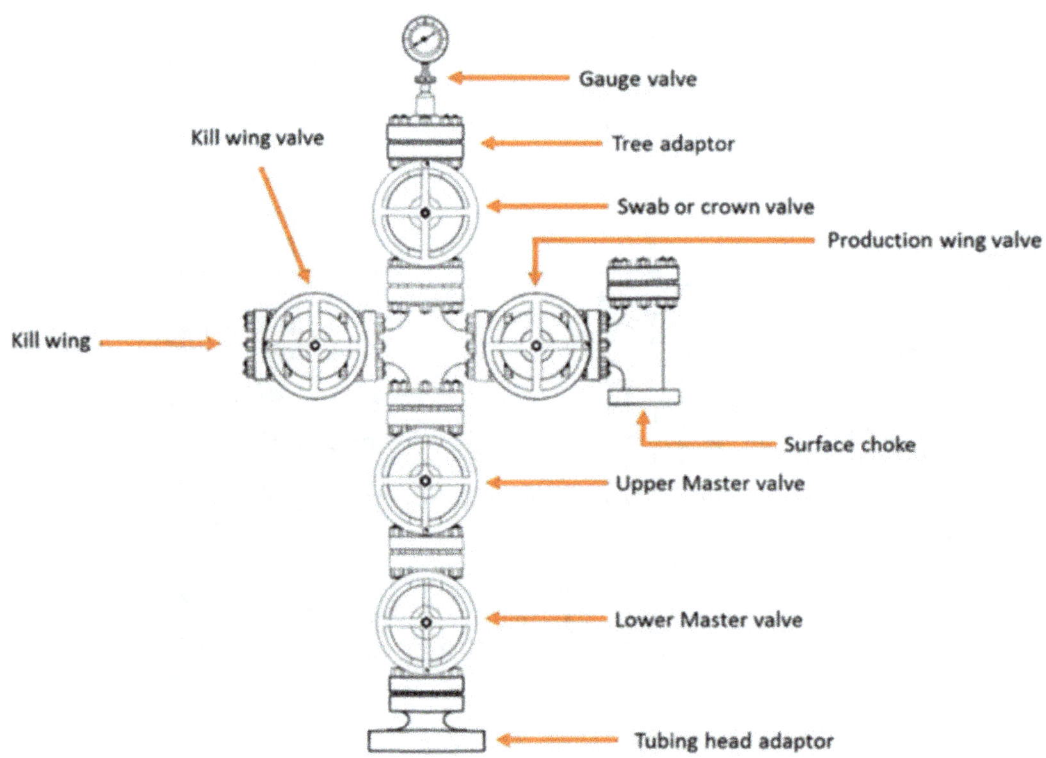

Fig. 2.19 Typical surface vertical Christmas tree

Fig. 2.20 A dual-bore vertical subsea tree which allows to monitor annulus pressure. (Courtesy of Schlumberger)

performed by wireline, slickline, coiled tubing, or a snubbing unit. The inlet of the swab valve is covered by a flange that is called a T-Cap, which allows a wireline lubricator, snubbing unit BOP, or coiled tubing to connect to the well. Finally, a top pressure gauge sits on top of the T-Cap to show the well pressure. The vertical trees can be used for both surface and subsea wells. However, the subsea application needs different interfaces with respect to manipulating the tree valves and access to the A-annulus (Fig. 2.18). It is notable that the configuration of valves on the vertical trees for subsea and surface trees are the same; however, due to the subsea conditions and remotely manipulating the valves, the interface of a subsea well is different (see Fig. 2.20). The vertical subsea tree can be a single-bore or dual-bore (see Fig. 2.18) and this can make a difference to the P&A operation.

Throughout construction and completion of land/platform wells or subsea wells with the vertical XMT, the tubing hanger is installed inside the wellhead (see Fig. 2.15) and then the vertical XMT is installed on top of the wellhead. Therefore, in order to retrieve the production tubing, the vertical XMT must be nippled-down. Accordingly, to maintain well integrity during the tubing retrieval, the BOP must be installed.

2.8.2.2 Horizontal Christmas Tree

The advent of horizontal trees has initially been linked to completion of subsea wells. It is necessary to mention, before any comparison of horizontal and vertical subsea trees is made, the parts and subassemblies are very similar. In horizontal trees, the valves are positioned on the sides of the tree body (Fig. 2.21). The difference between

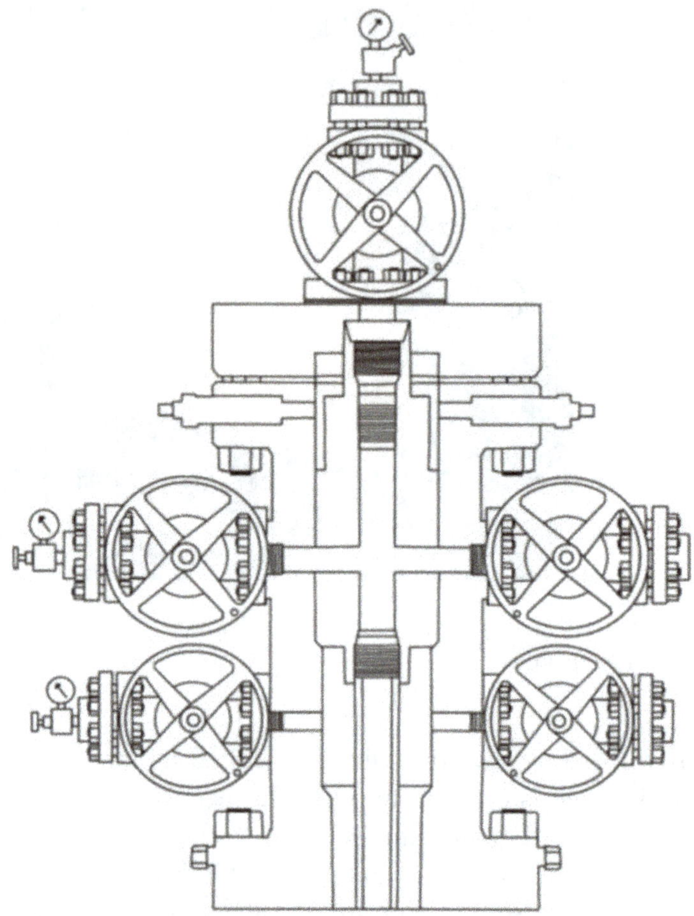

Fig. 2.21 Illustration of a horizontal Christmas tree

horizontal and vertical trees arises largely from the configuration of the valves rather than novel design. One of the main reasons that persuades operating companies toward using horizontal trees are challenges related to subsea well intervention operations. Mostly, a subsea well intervention arises from problems related to the tubing and SCSSV and therefore, ready access to the production tubing and SCSSV, without disassembling the tree, are primary design criteria. Thus, the position of valves on a vertical tree are moved to the sides of the tree and the tubing hanger sits inside the horizontal tree. In this manner, during a subsea well intervention, the BOP is positioned above the horizontal tree and tubing is retrieved without nippling down the tree. Consequently, the workover operation is easier performed, and more efficient. In addition, concerns regarding nippling the tree down/up are minimized. Table 2.3 presents the notable differences between the subsea vertical and horizontal trees. Figure 2.22 depicts a horizontal subsea XMT.

During drilling of a subsea well, the following are run at the same time: LPWH, conductor, and guide base. If the well is planned to be completed with a horizontal XMT, after installation of the subsea wellhead system, the horizontal XMT is installed and subsequently the BOP is installed on top of it and then drilling is resumed. Then in the completion phase, the tubing hanger sits inside the horizontal XMT. In other words, tubing retrieval does not necessitate the removal of the horizontal XMT.

Table 2.3 The most notable differences between subsea trees; vertical and horizontal

Vertical XMT	Horizontal XMT
• Master and swab valves in bore • Tree run after tubing (tree lands on and stabs into the tubing hanger) • Tubing hanger orients via wellhead • External tree cap run after tree landed/tested • Tubing hanger seals normally isolated from well fluid	• No valves in the vertical bore of the well • Tree run before tubing (tubing hanger lands in tree body) • Tubing hanger orients directly from tree (limits tolerance stack-up) • An internal tree cap is used as a secondary pressure barrier above the tubing hanger, two crown plugs are installed by wireline unit • The tubing hanger seals are continuously exposed to well fluids

Fig. 2.22 A horizontal subsea Christmas tree

Where the installation of a BOP is required, it is nippled-up on top of the horizontal tree.

In order to perform an efficient and safer permanent P&A operation, it is necessary to understand the strength and weakness of each type of tree system. Table 2.4 tabulates the advantages and concerns regarding the tree systems.

2.8.3 Assembling BOP

In the course of a P&A operation, regardless of the well type (i.e. land/onshore or offshore) and the XMT type (i.e. horizontal or vertical), at some point, the utilization of a BOP is unavoidable. Therefore, the primary and secondary temporary well barrier envelopes shall be established and maintained to secure the well while disassembling the XMT and assembling the BOP. Theoretically, this procedure may be considered

Table 2.4 Comparison between advantages and concerns of subsea vertical and horizontal tree systems

Tree system	Advantages	Concerns
Vertical	• Lighter than horizontal trees • To nipple-down the tree, there is no need to retrieve production tubing	• Seldom accommodate larger than 5 ½-in. production tubing • Not designed to take the load from BOP • Tubing/casing hanger is locked to the wellhead before tree is installed; consequently, pressure under the tubing hanger cannot be measured
Horizontal	• BOP is installed on top of the tree • The tree has a lower height • Work efficiency is improved (e.g. nippling down/up and testing the tree during well intervention is avoided) • Accommodate up to 7-in. production tubing • Tubing/casing hanger is locked in the tree itself; consequently pressure under the tubing hanger can be measured	• Heavier than vertical trees • Inaccessibility to different annuli except A-annulus

for four different well completion scenarios (Fig. 2.18); a land/platform well completed with the vertical tree, a subsea well completed with the vertical tree, subsea well completed with the horizontal tree, and a land/platform well completed with the horizontal tree. However, the latter scenario is less likely to be accomplished due to some practicality issues such as large weight of the horizontal tree, inaccessibility to different annuli except the A-annulus, etc.

2.8.3.1 Assembling BOP—Land/Platform Well with the Vertical Tree

Consider a platform well which has been completed with the vertical tree (Fig. 2.23). To establish the primary temporary barrier, there should be enough casing cement below and above the production packer in the B-annulus and no reported sustained casing pressure in the A-annulus (see Fig. 2.14 for definition of A-, B-, and C- annuli). If these assumptions are valid, the primary barrier envelope is achieved by installation of a bridge plug in the tail pipe. The bridge plug and primary well barrier envelope are pressure tested for assurance that well integrity is maintained. The primary well barrier elements, illustrated in Fig. 2.23, are listed and marked with a blue line.

For establishing the secondary temporary well barrier envelope, there should be enough casing cement above the production packer in the B-annulus, no sustained casing pressure in the B-annulus, and production casing maintains integrity. Note that the same interval of casing cement, which is used as an element for the primary barrier,

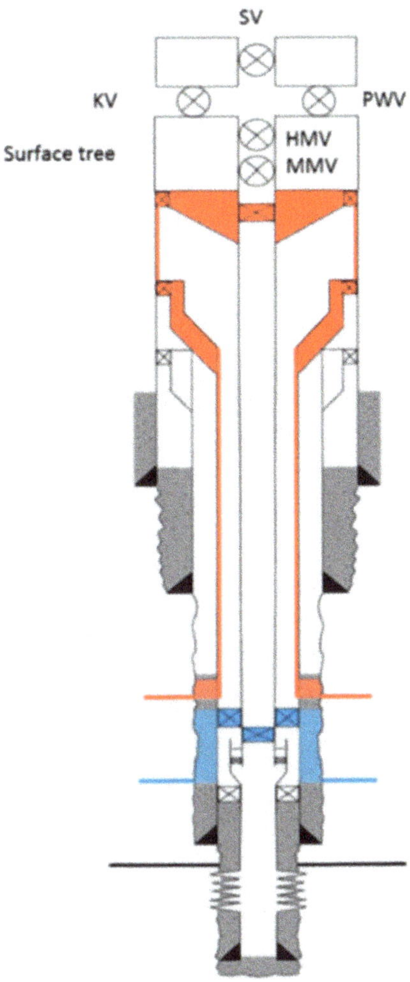

Well Barrier Elements
Primary well barrier
In-situ formation (cap rock)
Casing cement (9 5/8")
Casing (9 5/8")
Production packer
Bridge plug in tail pipe
Secondary well barrier
In-situ formation (9 5/8")
Casing cement (9 5/8")
Casing (9 5/8")
Wellhead (casing hanger with seal assembly)
Wellhead/annulus access valves
Tubing hanger (body seals and neck seal)
Bridge plug inside tubing hanger

Fig. 2.23 List of well barrier elements of a platform well with the vertical tree

cannot be used as an element for the secondary barrier (see Fig. 2.23). If the above-mentioned assumptions are valid, the secondary barrier envelope is achieved by installation of a bridge plug inside the tubing hanger. The bridge plug and secondary well barrier envelope are pressure tested prior to commencing nippling-down the tree. The secondary barrier elements are listed and marked with a redline in Fig. 2.23.

Where a well has been completed with a *Downhole Safety Valve* (DHSV), the DHSV can be used as a well barrier element in the primary well barrier envelope when it is qualified by a function and pressure test.

Example 2.1 A platform well (Fig. 2.24) has been drilled and completed with the vertical tree in 1985. The TOC in the B-annulus is below the permanent packer and the well suffers from sustained casing pressure in the A- and B-annulus. Caliper log shows big holes along the production tubing (shown with triangle on Fig. 2.24). Operator decided to permanently plug and abandon the well. Through the operation, a BOP is necessary to control the well pressure. Make a list of the primary and secondary well barrier elements for nippling-down the tree and nippling-up BOP.

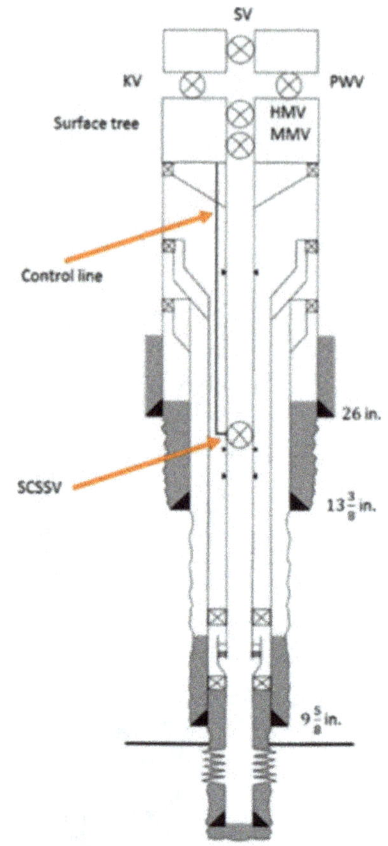

Fig. 2.24 WBS of a platform well with the vertical tree

Solution It is recommended to squeeze cement the perforations and extend it up to the liner packer. Pressure test the cement plug and if it is qualified, bleed off the A-annulus and the B-annulus. If the A-annulus and the B-annulus pressures do not build up, a plug is installed inside the tubing hanger. Then, the primary and secondary well barrier elements, temporary barriers, could be as follows:

Temporary Well Barrier Elements
Primary well barrier
In-situ formation (cap rock)
Casing cement (9 5/8")
Casing (9 5/8")
Liner cement
Liner
Cement placed inside liner
Secondary well barrier
In-situ formation (13 3/8")
Casing cement (13 3/8")
Casing (13 3/8")
Wellhead (casing hanger with seal assembly)
Wellhead/annulus access valves
Tubing hanger (body seals and neck seal)
Plug inside tubing hanger

But if the squeezed cement is not qualified and the A-annulus and B-annulus pressures are building up, then a bridge plug should be installed inside production tubing, below the permanent packer, and A- and B-annuli need to be killed by unconsolidated sand slurries or heavy fluid. Here the assumption is that the SCSSV has successfully passed the pressure test. A plug is installed inside tubing hanger. Then, the primary and secondary well barrier elements, temporary barriers, could be as follows:

Temporary Well Barrier Elements
Primary well barrier
In-situ formation (cap rock)
Kill fluid or unconsolidated sand slurry (B-annulus)
Production casing (9 5/8")
Kill fluid or unconsolidated sand slurry (A-annulus)
Permanent packer
Production tubing
Bridge plug in tailpipe
Secondary well barrier
In-situ formation (13 3/8")
Casing cement (13 3/8")
Casing (13 3/8")
Wellhead (casing hanger with seal assembly)
Wellhead/annulus access valves
Tubing hanger (body seals and neck seal)
Plug inside tubing hanger

2.8.3.2 Assembling BOP—Subsea Well with the Vertical Tree

Vertical subsea trees may be divided into two main groups considering their tubing hanger configuration: single-bore (mono-bore), and dual-bore trees. Figure 2.25 shows configuration of single-bore tubing hanger with an annulus valve and penetrations for control lines and chemical injection lines. Nowadays, single-bore vertical subsea trees are barely used for the completion of subsea wells; however, dual-bore vertical trees are more commonly used for the completion of subsea wells as they provide access to the A-annulus.

Figure 2.26a is an illustration of a subsea well which has been completed with a single-bore vertical subsea tree. It is a conventional procedure to bullhead cement into the main reservoir through the production tubing when the production tubing is in good condition. Then, the cement plug is pressure tested. If it passes the test successfully, it can be used as primary temporary barrier and primary permanent barrier (Fig. 2.26b). However, if it does not pass the pressure test, the primary and secondary well barrier envelopes are established by rigging up a wireline unit. A wireline BOP is positioned on top of the tree, and a bridge plug is installed in the tailpipe. The envelope is tested and if it maintains its integrity, a secondary temporary barrier is established by placing a bridge plug inside the tubing hanger and verified by pressure testing the barrier envelope (Fig. 2.26c).

Dual-bore vertical subsea trees provide direct access to the production and annulus bores via a completion riser. As there is more than one bore inside the tubing hanger,

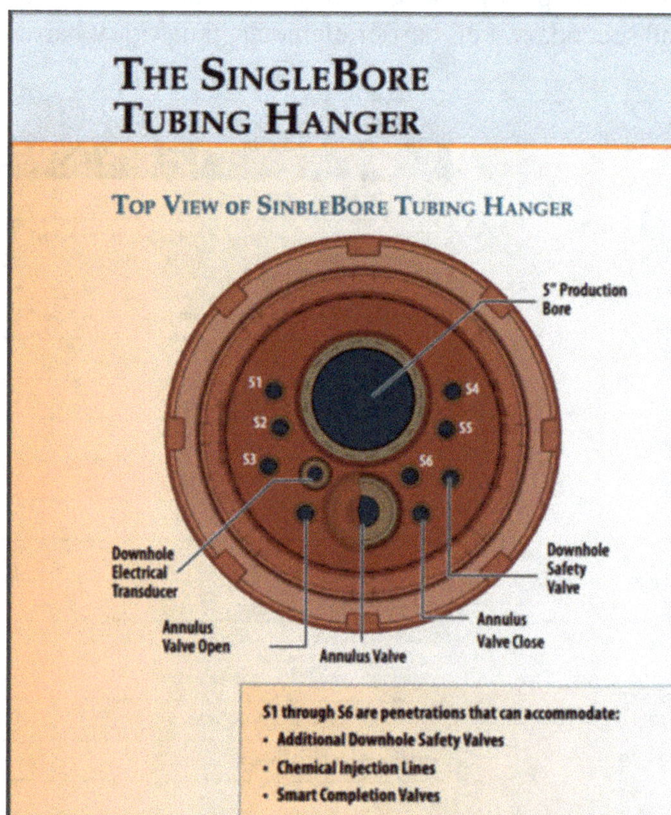

Fig. 2.25 Configurations of tubing hanger for subsea well completed with the vertical trees. (Courtesy of Dril-Quip)

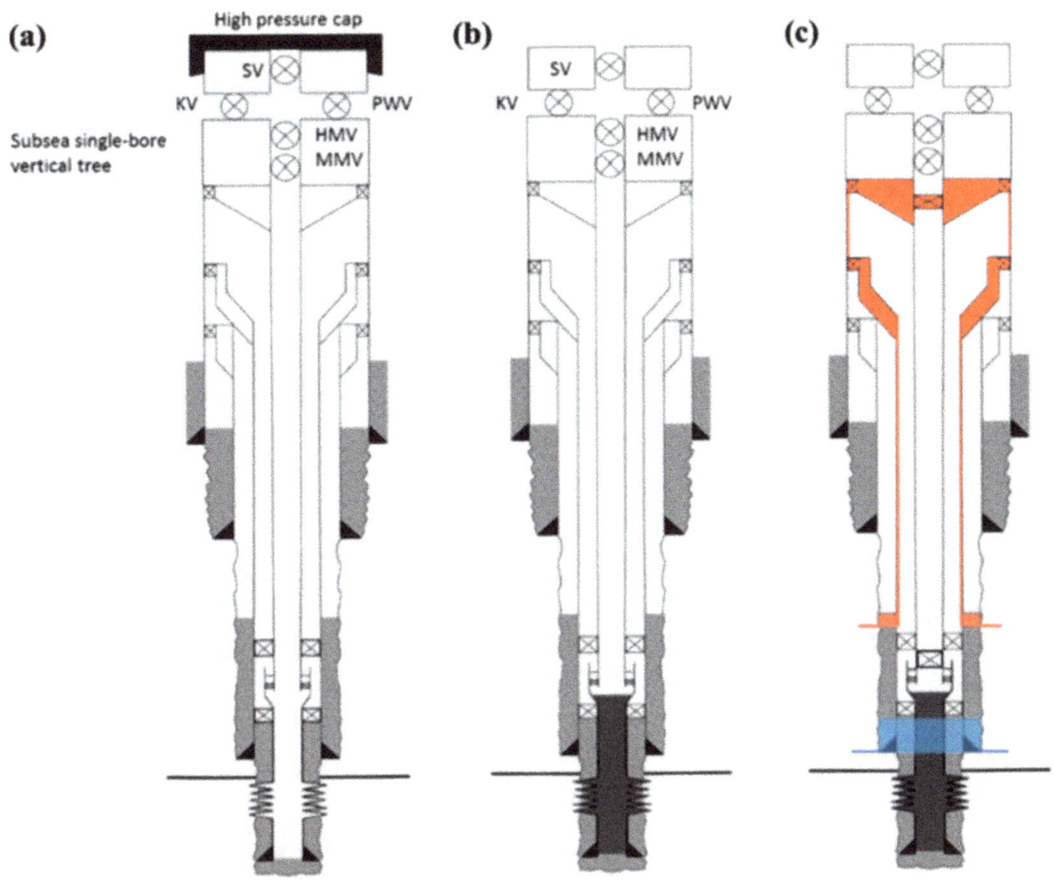

Fig. 2.26 Subsea well completed with the single-bore vertical tree

two bridge plugs are required to establish the secondary barrier and consequently, a dual-bore riser is required. A wireline BOP is installed on top of the tree and bridge plugs are installed and eventually, the envelopes are pressure tested. It is noteworthy to know that a dual-bore tubing hanger does not affect establishment of the primary temporary barrier (see Fig. 2.27).

2.8.3.3 Assembling BOP—Subsea Well with the Horizontal Tree

Often, it is assumed that nippling up the BOP for wells completed with horizontal trees is not as complex as for wells completed with vertical trees; however, that is not true. Maybe the reason for this belief is rooted in the reality that the BOP is installed on top of the horizontal tree and there is no need to nipple-down the tree and consequently, there is no need to establish the primary and secondary temporary barriers. Indeed, to get access to the wellbore, the high-pressure tree cap needs to be removed (see Fig. 2.28), and pressure underneath of the tree cap needs to be controlled. Therefore, a well control equipment is necessary and will be employed.

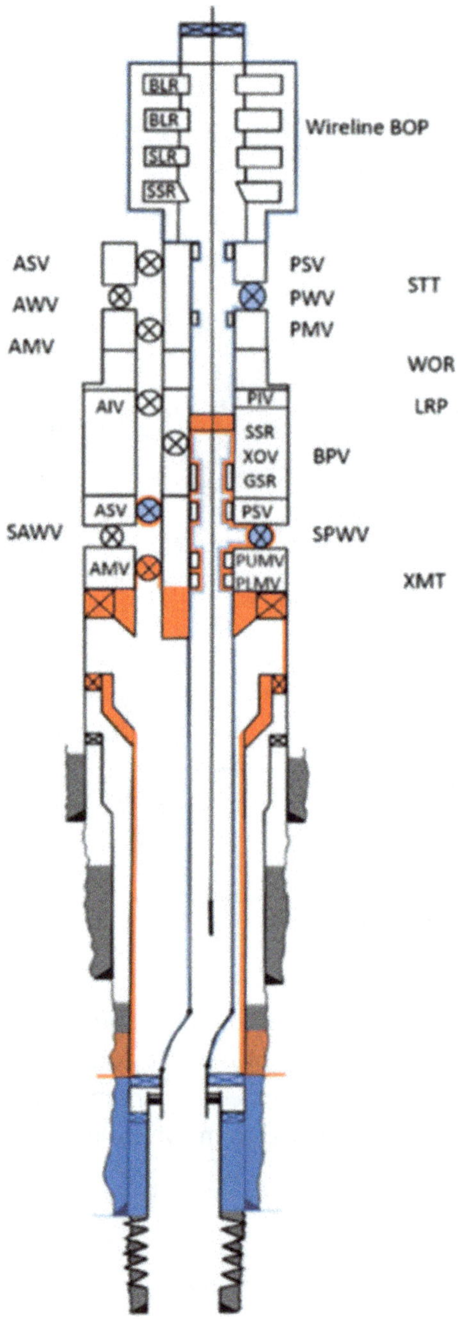

Well Barrier Elements
Primary well barrier
In-situ formation
Casing cement
Casing
Production packer
Completion string
Tubing hanger
Subsea tree
Lower riser packer
High pressure riser
Surface test tree
Wireline BOP
Wireline lubricator
Wireline stuffing box/grease injection head
Secondary well barrier
In-situ formation
Casing cement
Casing
Wellhead (casing hanger with seal assembly)
Tubing hanger (body seals and neck seal)
Subsea tree
Lower riser package

Fig. 2.27 Subsea well with the dual-bore vertical tree

2.9 Special Considerations in Abandonment Design

2.9.1 Control Lines

A control line is a small-diameter hydraulic line used to channel fluid from surface to operate downhole completion equipment such as the SCSSV. Wellhead is designed in such way that provides penetrations for control lines to go through. Figure 2.29 shows

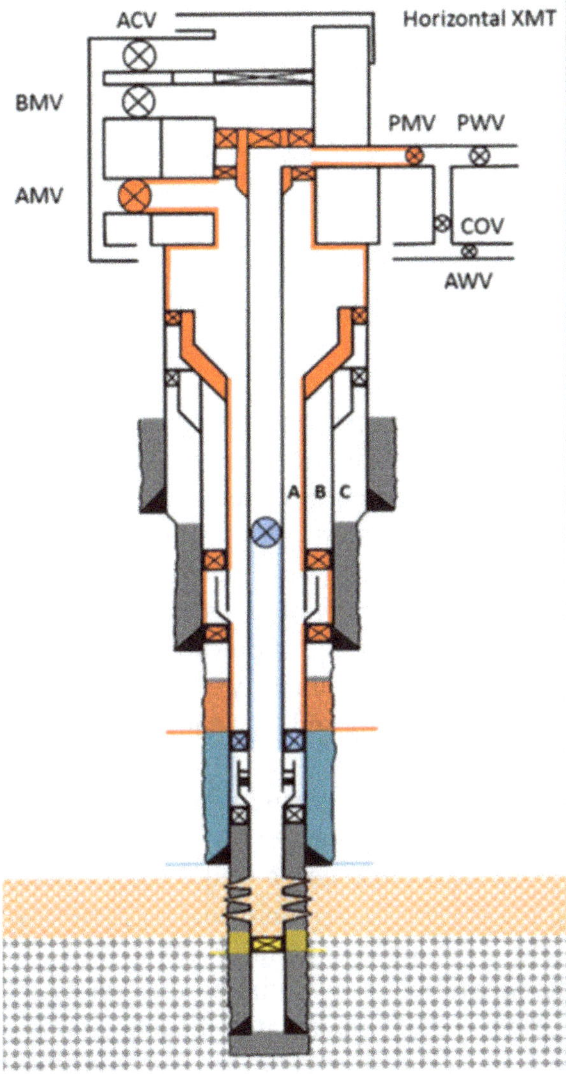

Well Barrier Elements
Crossflow well barrier
In-situ formation
Casing cement (liner cement behind plug)
Casing (liner behind plug)
Deep set plug
Primary well barrier
In-situ formation (at casing shoe)
Production casing cement (shoe to production packer)
Production casing (below production packer)
Production packer
Completion string
DHSV/control line
Secondary well barrier
In-situ formation
Production casing cement (above production packer)
Production casing (liner above production packer)
Liner hanger packer
Intermediate casing (between liner packer and tie-back packer)
Tie-back packer
Production casing (above tie-back packer)
Production casing hanger with seal assembly
Tubing hanger with seals
Wellhead (WH/tree connector)
Subsea tree

Fig. 2.28 Subsea production well with a horizontal tree

Fig. 2.29 Control lines with and without electrical line

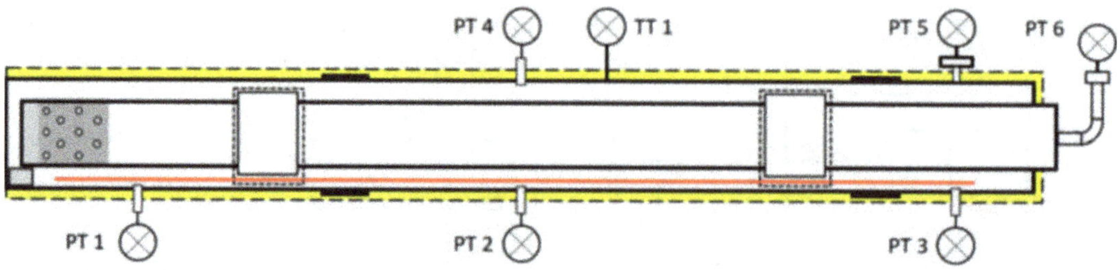

Fig. 2.30 Full-scale test set-up used for investigating the sealing ability of cement plug when tubing and control lines are left in hole [32]

two different types of control lines; a control line equipped with four small-diameter hydraulic lines, and a control line equipped with two small-diameter hydraulic lines and one conductor line (electric line to the right). Two major concerns exist regarding control lines that persuade engineers to plan for control line retrieval; flow potential of hydraulic line(s) and quality of the bonding between plugging material and surface of the control line. During cement plug placement, a variety of fluids are pumped and each has its own function. These fluids adhere to the surface of downhole equipment (e.g. control line) and its removal is challenging due to wettability of the control line surface. Therefore, the bonding between the cement plug and the control line surface is a concern.

Nowadays most wells are completed as smart wells, meaning control lines and electric lines are part of the well completion to control *Inflow Control Devices* (ICDs). There are different opinions regarding the risk of leakage or well integrity issues concerning control lines as part of the well barrier envelope. Aas et al. [32] performed a full-scale test on the sealability of annulus cement when tubing is left in hole with control lines. In this study, a 7-in. production tubing and a 9-in. production casing were used in the experiments. Figure 2.30 shows the schematic of the full-scale test assembly used in the study. In this study, a 16.0 (ppg) conventional class G Portland cement displaced a 10.0 (ppg) brine, and then the cement was cured for 7 days. The 9-in. production casing was insulated and the temperature development due to cement hydration was recorded. The maximum temperature was recorded as 75 °C after 1 day of curing. The test assembly was 40 (ft) long and it was inclined 85° while pumping cement and during curing. Cement was allowed to fill in the control line. Then, the sealing ability of the cement plug was investigated by pressure testing. Water was pumped at 725 psi to the A-annulus and 1450 psi to inside of the tubing. Aas et al. [32] observed no leakage through the established barrier.

One of the challenges regarding leaving control lines in hole during P&A is related to placing a plug inside the control lines at the desired depth interval. Control lines have a small diameter (1/8 to 1-in. OD) and the fluid inside ranges from water-based hydraulic fluid to a mixture of this fluid with reservoir fluid including formation water. When plugging material is pumped through control lines, it is contaminated in such a way that pressure testing will show a failure. Different plugging materials (i.e. cements, resins, and silicone materials) have been tested for sealing the annuli of control lines [33]. The compatibility of fluid inside control lines and plugging material appear to be the key to success and there is an interest for more research on this subject.

2.9.2 Well Design

One of the best and most cost effective solutions for P&A is to consider the P&A scenarios during the well design phase. The consideration of the following parameters during well design may strongly influence permanent P&A operations: proper primary cement jobs, depth of TOC, qualification and documentation of primary cement jobs, depth of control lines, and identification of pressure sources in the overburden.

Primary cement job—Consider a well with two different well design scenarios, Fig. 2.31. In the first scenario, the well has been designed and constructed in such a way that there is high enough qualified cement across a suitable formation, in the B-annulus, and the primary cement has been qualified and documented, Fig. 2.31a. In the second scenario, the same well has been designed in such a way that there is neither cement in the B-annulus across a suitable formation nor has the cement been qualified and documented, Fig. 2.31b. For the first scenario, if the well does not experience SCP, then the tubing may be retrieved (i.e. if control line is present at the plugging depth) and then a cement plug is placed inside the production casing and across the casing cement. In the second scenario, as there is an uncemented casing, access to the suitable formation requires section milling or other techniques, and then the cement plug can be placed across the formation. The first scenario may take 1 day per plug while the second scenario may take several days per plug due to the required access to the formation.

When there are more than one well to be plugged and abandoned in a field and all of them fulfill the circumstances of the given well in Fig. 2.31a, a common practice may be accepted, which is based on experience. According to the practice, production tubings of two or three wells are retrieved, fully or partially, and their casing cements in the B-annulus are logged. If the casing cements are qualified, then an assumption is made for all the wells in the field. As the wells do not experience any sustained casing pressure, the assumption is that, since the casing cements of the selected wells have been qualified during P&A operations by logging, the casing cement of the other wells are intact and qualified as well. Therefore, tubings are left in hole and the A-annulus and the inside of the production tubings are filled with cement. Furthermore, the new cement plug is tested and if qualified, it is verified and documented. It is necessary to remember that in this scenario, there is no control line in the well barrier envelope.

Pressure in overburden—Identification of pressure sources in the overburden during P&A operations is a challenge with a high uncertainty. Therefore, it is recommended to identify and document all pressure sources in the overburden during well construction. Experience shows that unidentified formations with flow potential, in the overburden, can create challenges to qualify permanent barriers below the unidentified influx formation.

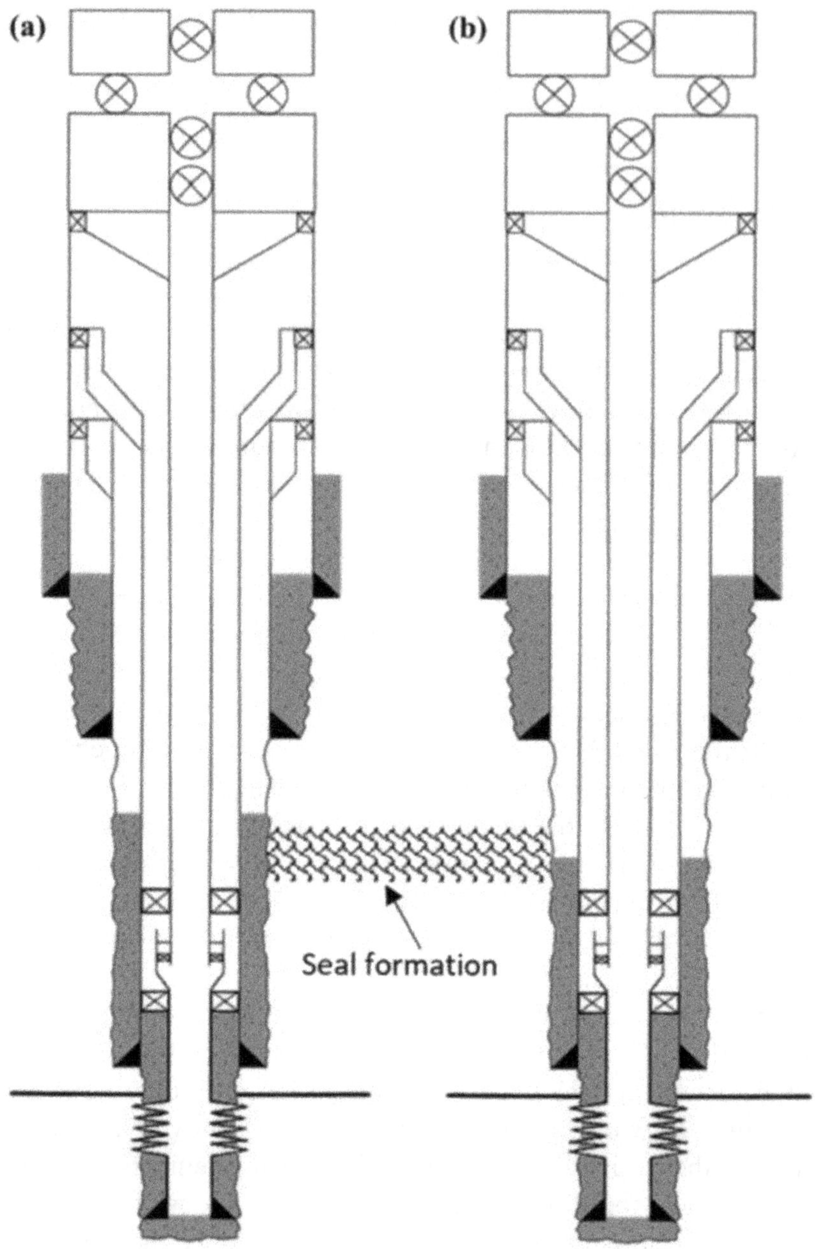

Fig. 2.31 A production well in two different scenarios: **a** high enough qualified cement in the B-annulus, **b** not enough cement in the B-annulus

Well objectives—Wells are drilled with different objectives; exploration, delineation, appraisal, development, production, or injection (Table 2.5). When investigating new areas, the interpretation of seismic data helps find areas where there are probable hydrocarbon reserves. The goal of an exploration well is to confirm or reject this hypothesis. In addition, the exploration well gathers the necessary information for a better understanding of the area and its potential future production. As a result of exploration, when the exploration well penetrates an accumulation of petroleum,

Table 2.5 Well types and their objectives

Well Type	Objectives
• Exploration	• P&A or keeper[a] [34]
• Delineation	• Size of reservoir
• Appraisal	• Reservoir characteristics
• Development/production	• Reservoir drainage
• Injection	• Pressure maintenance • Cutting reinjection • Disposal of unwanted fluids

[a]An exploration well intended for completion

a delineation well is drilled for estimating the size of the reservoir and its commercial value. Appraisal wells are drilled for investigating the reservoir characteristics. Development wells are drilled for reservoir drainage and production of petroleum. Injection wells are drilled with different objectives; pressure maintenance, disposal of fluids, and cutting reinjection. Pressure maintenance injection wells are drilled for *Enhanced Oil Recovery* (EOR) and *Improved Oil Recovery* (IOR) processes to increase the recovery factor of reservoirs. Disposal wells are drilled to reinject unwanted fluids (e.g. connate water) produced with petroleum to a non-hydrocarbon bearing formation. Cutting reinjection wells are drilled to dispose of drilling cuttings and contaminated mud in a formation while drilling.

Drilling culture has been influenced by different time regimes and the above-mentioned types of wells have been designed based on the needs over time, and subsequently their design criteria has been changing with regards to the knowledge of engineers. Figure 2.32 shows a timeline for different design criteria eras in well construction. Perhaps from past to 1970s could be called as classic era of well design, 1980s as horizontal drilling and slot recovery era, 1990s as well integrity era, 2000s as rotating liner and the last decade could be marked as era of P&A in well design and it has been accelerated due to Macondo incident. There is an indisputable subject which has been focused continuously over time, and that is cement and its properties.

There have been eras where wells were drilled and completed without comprehensively considering their future P&A. Therefore, each type of these wells which have been drilled over decades, have their unique specific well design and need to permanently be plugged and abandoned using a best practice.

Pore pressure and fracture pressure profiles—In a pressure depleted reservoir, the reservoir pore pressure will be lower than the initial pressure and the fracture pressure will also be reduced. Consequently, the margin between pore pressure and

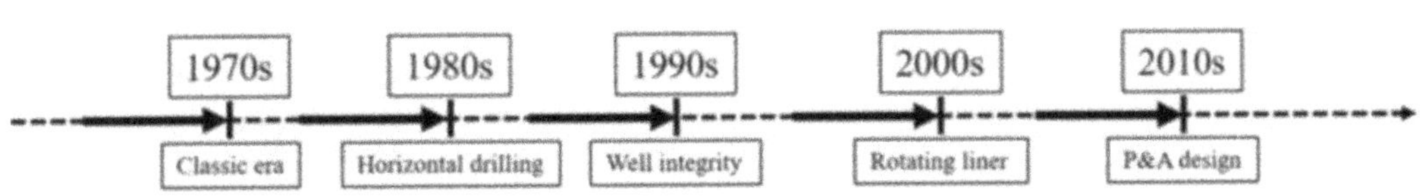

Fig. 2.32 Timeline showing different eras for focusing on different subject

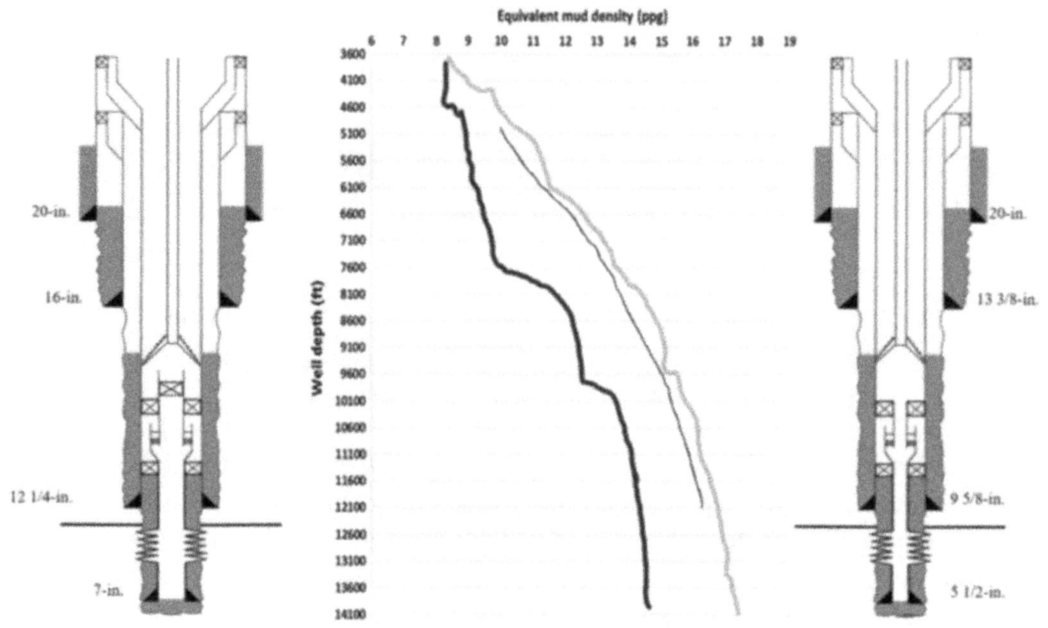

Fig. 2.33 Effect of annular space on ECD and margin between pore pressures and fracture gradient profiles

fracture pressure gradients will be smaller. In P&A operations, having a narrow margin between pore pressures and fracture gradient profiles can be a challenge and limiting factor, particularly in deep-water offshore wells. This narrow margin results in limitations with respect to the selection of hole cleaning fluid systems, swarf/cutting removal, cement design and its placement, etc. Compare a scenario where a 5-in. drillpipe as a working string is performing section milling in a 9 5/8-in. casing with another scenario where the same drillpipe is conducting section milling on the same well at the same depth but milling a 12 ¼-in. casing, Fig. 2.33. As a result of different pressure drop in the annulus, the *Equivalent Circulating Density* (ECD) will change and the exposed formation may be fractured.

Example 2.2 Consider a 5-in. drillpipe is performing section milling of a 9 5/8-in. casing at 8000 (ft) True Vertical Depth (TVD) with a mud weight of 10.8 (ppg) whereas the annular pressure loss is 460 psi. Assume that the pore pressure and fracture pressure at 8000 (ft) are 10.2 (ppg) and 11.7 (ppg), respectively.

a. Does the downhole pressure fracture the formation?
b. Consider the same situation where the casing is 12 ¼-in. and the annular pressure loss is 340 (psi). Does the downhole pressure fracture the formation in this scenario?

Solution

$$ECD\,(ppg) = mud\ weight\,(ppg) + \frac{annular\ frictional\ pressure\,loss\,(psi)}{0.052 \times TVD\,(ft)}$$

(a) $ECD = 10.8 + \frac{460}{0.052 \times 8000} = 11.9\ ppg$
The ECD is higher than the fracture gradient and induces fractures.

(b) $ECD = 10.8 + \frac{340}{0.052 \times 8000} = 11.6\ ppg$
The ECD is slightly lower than fracture gradient.

Casing seats—It is the set point of the end of casing and generally it is based on consideration of pore-pressure gradients and fracture gradients of formations to be drilled. Casing seat is normally placed in an impermeable and stable formation. Casing seat depth (also known as casing setting depth) can influence P&A operational time; therefore, it should be considered and selected properly during well design with regards to future P&A of the well. The example shown in Fig. 2.34 illustrates the situation where a lost-circulation zone (DPZ 5) leads to a very low height of TOC for the production casing. As the intermediate casing string has the right casing-seat

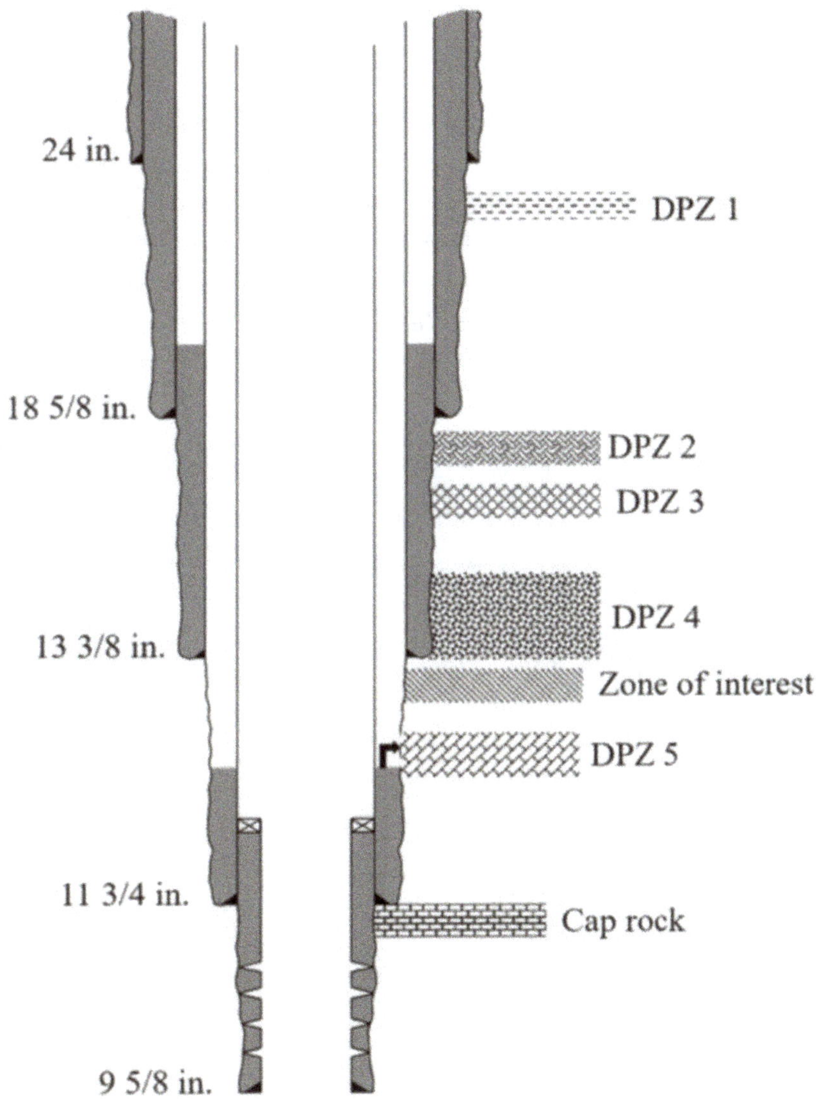

Fig. 2.34 Relationship among casing seat and permanent barrier establishment

depth, an interval above the lost-circulation zone is provided to establish primary and secondary barriers for the reservoir. But if the casing seat for the intermediate casing was deeper, then barrier establishment for the reservoir could be more challenging.

2.9.3 Well Schematic

A well schematic is suggested to be produced and updated on a daily basis, during P&A operations. It shows the actual phase of each well with its status. In this manner, it becomes clear to a wider audience if the P&A operation is falling behind, or is ahead of schedule, and if necessary, extra resources can be applied.

2.9.4 Horizontal Wells

A well with an inclination of generally larger than 85° is called a horizontal well. The horizontal section of a horizontal well is in the pay zone and is not normally an interesting interval for P&A. The build and tangent sections are important, Fig. 2.35. It is recommended to establish the permanent barrier as close as possible to the cap rock and across a suitable formation. Therefore, the build and tangent sections are interesting intervals; however, the high angle of these sections imposes some serious challenges. Performing wireline operations, hole cleaning, and cement plug

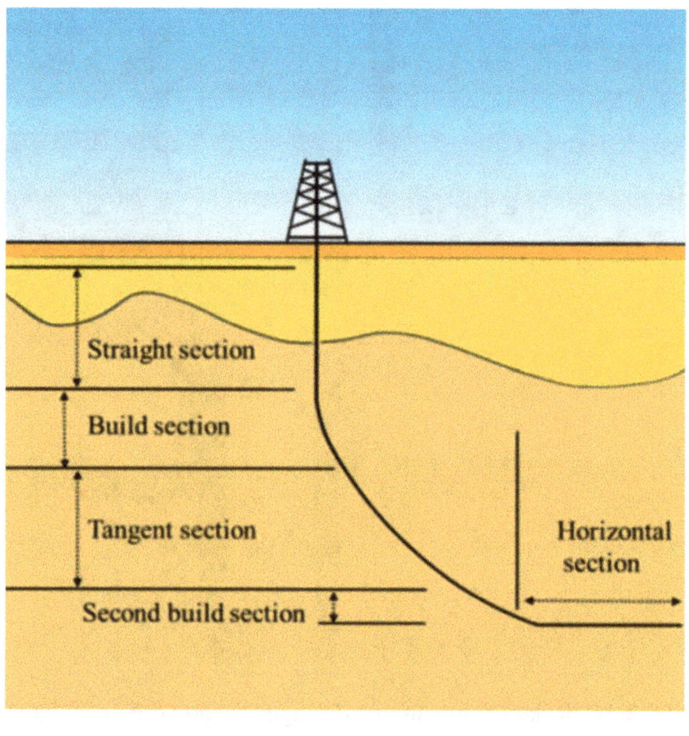

Fig. 2.35 Different sections of a horizontal well

placement at high angles (usually more than 65°) are some of the challenges to be considered, impacting the operation time and associated risks.

2.9.5 High-Pressure High-Temperature Wells

Over the years and across companies, definitions of *High-Pressure High-Temperature* (HPHT) wells have varied and no industry-wide standard defines HPHT conditions. The *American Petroleum Institute* (API) attempted to define HPHT terminology by publishing guidelines for equipment used in HPHT operations. According to the API Technical Report 1PER15K-1, a well having pressure higher than 15000 psi is defined as a HP well; and a well having temperatures higher than 350 °F is defined as a HT well [35]. However, NORSOK Standard D-010 [1] defines a well as a HP well when the shut-in pressure is exceeding 10000 psi and a well as a HT well when the static bottomhole temperature is higher than 300 °F. Figure 2.36 shows a proposed system for the classification of HPHT conditions for well-service-tool components. As HPHT conditions impose a unique situation, particular considerations should be taken during the design and operational phase of P&A.

Usually HP conditions dictate the use of larger BOPs. A larger BOP means limited space and handling capacity for offshore wells, more fatigue stresses on the wellhead, and more time consumed function testing the BOP. In addition, HP wells impose the need for a large increase in mud weight to control formation pressure; however, hydrostatic pressure may then approach fracture pressure.

For high temperature conditions, the impact of thermal expansion of drilling fluid on kick tolerance, effect of high temperature on equipment performance, and limited

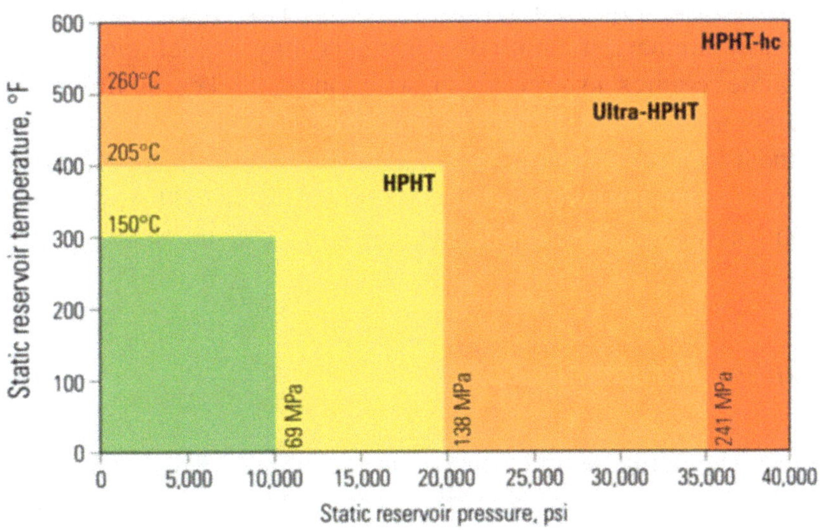

Fig. 2.36 A proposed HPHT classification system [36]

range of suitable plugging materials need to be considered during abandonment design.

More challenging conditions exist where high pressure and high temperatures are present together. Achieving adequate hydrostatic pressure while avoiding fracturing the formation is a challenge for the HPHT mud engineer. Wellbore instability, in milled sections, in high temperature and highly-depleted reservoirs is another challenge to be considered. Establishment of a cross sectional barrier (see Fig. 2.43) may require section milling and where a pressure depleted, and high temperature reservoir is the subject, there is the risk of wellbore instability due to changes of thermal elasticity of porous medium. Subsequently, leading to narrowing the mud density window. The reduced fracture gradient and drilling fluid loss are challenges which are rooted in pore pressure depletion, reduction of wellbore temperature, and drilling fluid osmosis in plugging and abandoning high temperature and highly-depleted reservoirs [37].

In HPHT and deep-water conditions, considerable temperature variations are seen in the transition from circulating conditions to geothermal gradient conditions during static periods. Generally, the milling interval of a wellbore experiences cooling as cold mud is pumped down the drill string. The upper part of the wellbore experiences heating as warm drilling fluid is circulated up, especially during hole cleaning. When the circulation is stopped, the temperature profile will, by heat conduction, return to geothermal conditions and subsequently, heat the mud in the milling interval and cool the mud in the upper part of the wellbore. Heating and cooling of mud in lower part and upper part of the wellbore in static conditions counteract each other. If the heating process dominates the wellbore condition, then mud experiences a thermal expansion and a gain in the mud pits will be observed. Consequently, hydrostatic pressure of the mud column drops and well control will be a concern.

Cement plug placement of HPHT wells requires a higher cement density and the situation imposes higher ECDs and therefore, low pumping rates are preferred. Subsequently, due to a low pumping rate which may change the flow regime (i.e. from turbulent to laminar), the mud displacement may be inefficient. In addition, a low pumping rate requires a slurry design with longer thickening times which needs more chemicals as retarders at high temperature conditions as high temperature accelerates the hydration process. As a rule of thumb, the more retarders used the more side effects on properties of the cement [38, 39]. So, HPHT wells require special consideration during abandonment design.

2.9.6 *Shallow Permeable Zones*

Identification and sealing production potentials in the overburden formation are assessed during abandonment. Shallow sources include shallow gas, coalbed methane, and water bearing zones. Although protection of the surface environment from any contamination caused by an abandoned well is the goal of P&A, protection of drinking water sources (surface water and ground water) increases the importance of permanent P&A. So, identification of shallow potentials during the initial life-cycle of wells (drilling) and P&A is necessary. These potentials need to be sealed properly, assessed, and finally documented.

2.9.7 Multilateral Wells

Where a well has more than one branch radiating from the main borehole, the well is called *multilateral well*. Multilateral wells are an evolution of horizontal wells. Permanent abandonment of multilateral wells requires a detailed study to design the required number of permanent plugs per borehole. If possible some of the boreholes may be regarded as one borehole to reduce the number of plugs to be installed, Fig. 2.37.

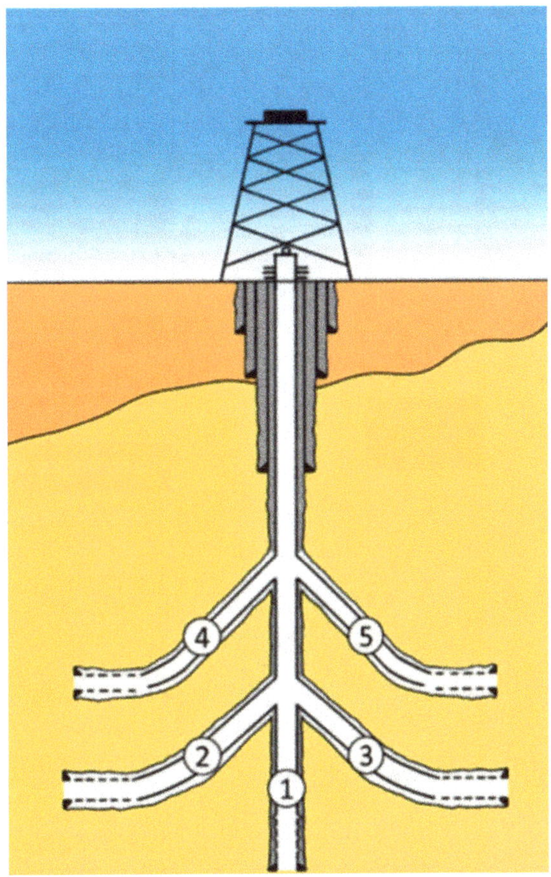

Fig. 2.37 Cased and cemented main bore with openhole multilaterals

2.9.8 Slot Recovery Sidetracks

Some regulators require, prior to slot recovery or sidetracking, the original wellbore to be permanently plugged and abandoned, Fig. 2.38. The permanent barrier can be a crossflow barrier or primary and secondary permanent barriers depending on the pressure regimes, formation strengths and available window between the formations. However, if at the time of sidetracking a permanent abandonment of the original borehole is not feasible, the primary and secondary temporary barriers need to be designed and established for the intended period. During the abandonment design, all of these original boreholes need to be mapped.

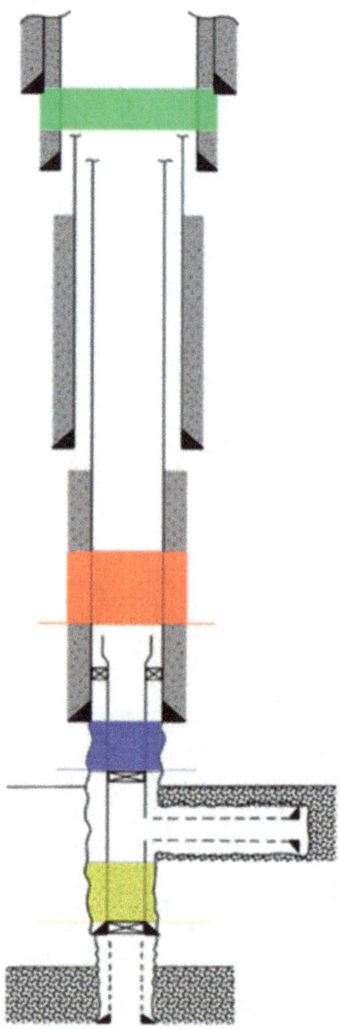

Well Barrier Elements
Crossflow well barrier/Primary & secondary barriers
In-situ formation
Casing cement
Casing
Cement plug
Primary well barrier
In-situ formation
Casing cement
Casing
Cement plug
Secondary well barrier
In-situ formation
Casing cement
Production casing
Cement plug
Environmental well barrier
Cement plug
Casing cement

Fig. 2.38 Permanent abandonment, multi-bore with slotted liners

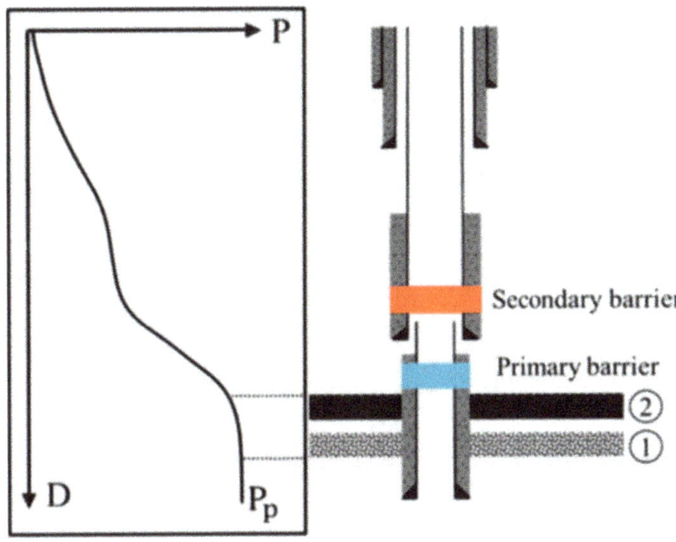

Fig. 2.39 Permanent plug and abandonment of multiple reservoirs with the same pressure regime

2.9.9 *Multiple Reservoirs*

Multiple reservoir zones located within the same pressure regime can be regarded as one reservoir. In such a scenario, primary and secondary permanent barriers are established above the upper potential, Fig. 2.39. If potentials are in the same pressure regime but crossflow is not acceptable, then a barrier can be established between the potentials, Fig. 2.40. In this scenario, the crossflow barrier may be regarded as a primary barrier for reservoir 2 and the primary permanent barrier for reservoir 3 is regarded as secondary permanent barrier for reservoir 2, Fig. 2.40. In this case, the reservoir 2 has three barriers including the crossflow barrier. A combination of permanent plugs for different potentials, which are in the same pressure regime and crossflow is acceptable, saves time and reduces costs. However, risk analysis needs to be performed to investigate the long-term consequences.

When the potentials have different pressure regimes and crossflow is not acceptable, each potential is secured with two barriers, primary and secondary barriers, Fig. 2.40.

2.9.10 *Slotted Liner*

The *screen* or *slotted liner* is a mechanical device which may contain gravel-pack sand in the annular space between it and the casing wall or openhole, Fig. 2.41.

Reservoirs or zones completed with a slotted liner or screen should be plugged above the slotted liner or screen. One of the reasons is that screen or slotted liner acts as filter for plugging materials. Consider a cement slurry which is pumped across a screen, the screen acts as a filter and drains the water such that hydration of cement

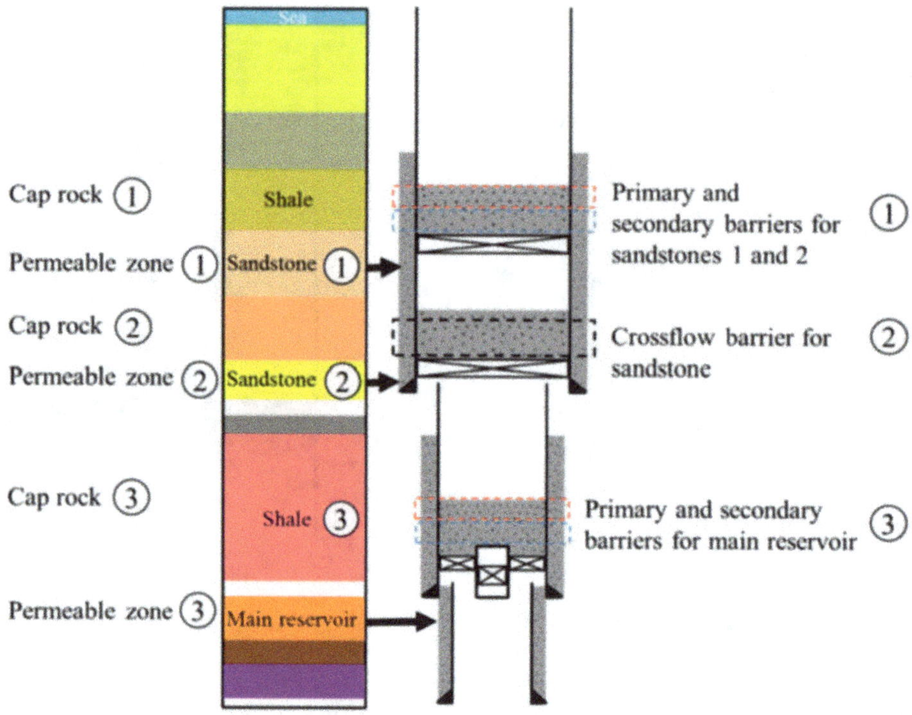

Fig. 2.40 Two flow potentials; same pressure regime but crossflow is not permitted, different pressure regimes with no permitted crossflow

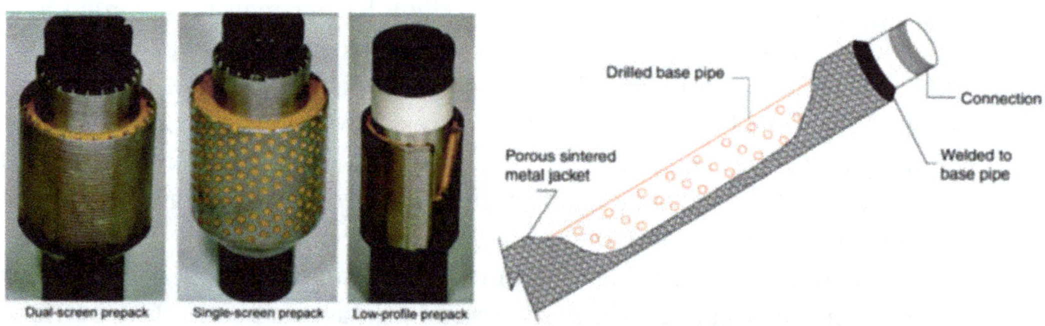

Fig. 2.41 Types of prepacked screens. (Courtesy of Baker Hughes)

would not be completed and consequently, poor chemical and physical properties of the cement plug would be expected.

2.9.11 Inflow Control Device

An inflow Control Device (ICD) is a surface controlled device which is installed as a part of well completion to help to optimize production by reducing water inflow contribution. Usually, during completion multiple ICDs are installed along the well,

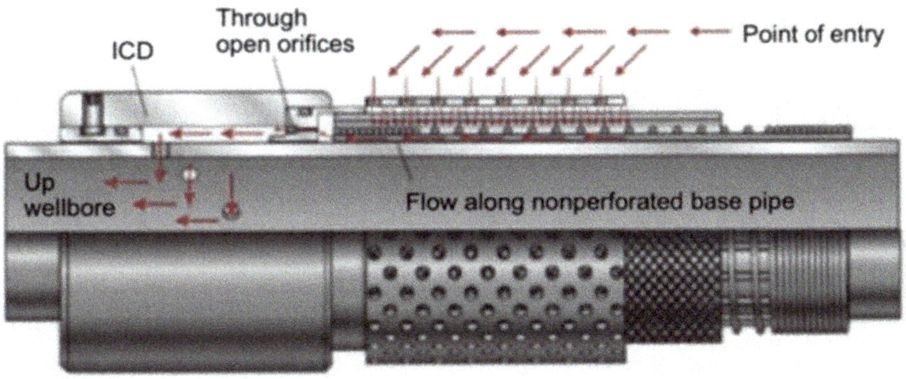

Fig. 2.42 Flow path of a typical ICD [40]

in horizontal sections. ICDs are frequently used with sand screens and in open-hole completions. ICDs allow controlled inflow but prevent outward flow, Fig. 2.42. Therefore, pumping cement slurry through ICDs is a challenge.

2.9.12 *Tubing Left in Hole*

Retrieval of production tubing is a time consuming and costly operation. Therefore, leaving the tubing in hole is often a desired option. If the production tubing is in a good condition, it can be used as work string for cementing. However, mechanical strength of a tubing for being used as work string should be analyzed [32].

2.9.13 *Hydrocarbons in the Overburden*

It is important to identify all sources of inflow in the overburden, run a risk analysis, and if necessary secure them by establishing of barriers. Due to uncertainty of old data, it is necessary to use data obtained from recent wells drilled in the same reservoir or field, to identify shallow sources of inflow. Retrieving the production tubing and logging behind production casing and intermediate casing may be necessary as these sources can contaminate ground water, soil or the marine environment. Hydrocarbons in the overburden may exist naturally or form due to well integrity issues.

2.10 Requirements for Designing Permanent Barriers

2.10.1 Well Cross Sectional Barrier

A permanent well barrier shall extend across the full cross section of the well while sealing all annuli both vertically and horizontally, Fig. 2.43. It is placed across a suitable formation; an impermeable formation with sufficient strength to hold the maximum anticipated pressure from the source of inflow. The barrier is known by different names such as *cross-sectional barrier*, or *formation-to-formation barrier*.

2.10.2 Plug Setting Depth—Formation Integrity

The adjacent formation to the permanent plug shall be capable of holding the maximum anticipated pressure from the source of inflow. The maximum anticipated pressure is either the original (initial) reservoir pressure for reservoirs with strong aquifers or an estimated final pressure that is below original reservoir pressure. The estimated final reservoir pressure is defined in a time interval and obtained through simulations. The maximum anticipated pressure is important for calculating the *Minimum Setting Depth of Plug* (MSD). The minimum setting depth of plug is the shallowest depth where the formation withstands the maximum anticipated pressure without being fractured. As the secondary plug is a backup to the primary plug, the MSD is the shallowest depth where the top of the secondary permanent plug can be placed. It is a common practice to place the plug as close as possible to the source of inflow. The MSD is estimated by using either pressure-gradient curves or a fluid gradient concept. The gradient curve method is a quick and reliable estimation for finding the MSD but the fluid gradient concept can also be used when several leak-off data are available.

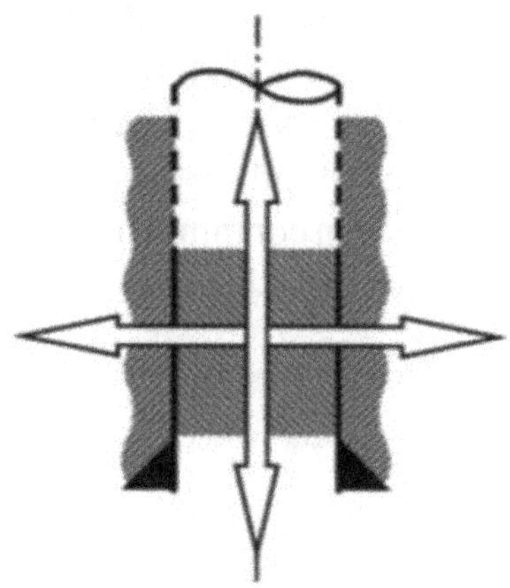

Fig. 2.43 Cross sectional barrier seals all the annuli. (Reprint from NORSOK D-010)

2.10.2.1 Minimum Setting Depth—Gradient Curves

In this method, initial pore pressure, fracture pressure, minimum horizontal stress, and overburden pressure curves are plotted. Then, a gas gradient line is drawn from the reservoir pressure towards the surface. For drawing the gas gradient line, it is necessary to know the final reservoir pressure and then subtract the hydrostatic effect of the gas column. The intersection of the curve for a closed well filled with gas and the minimum horizontal stress curve is the MSD, (see Fig. 2.44). The selection of final pressure is a crucial decision whereby selection of a lower value shifts the gas column curve to the left and consequently the MSD will be closer to the surface. However, when the reservoir builds up pressure, the adjacent formation to the plug

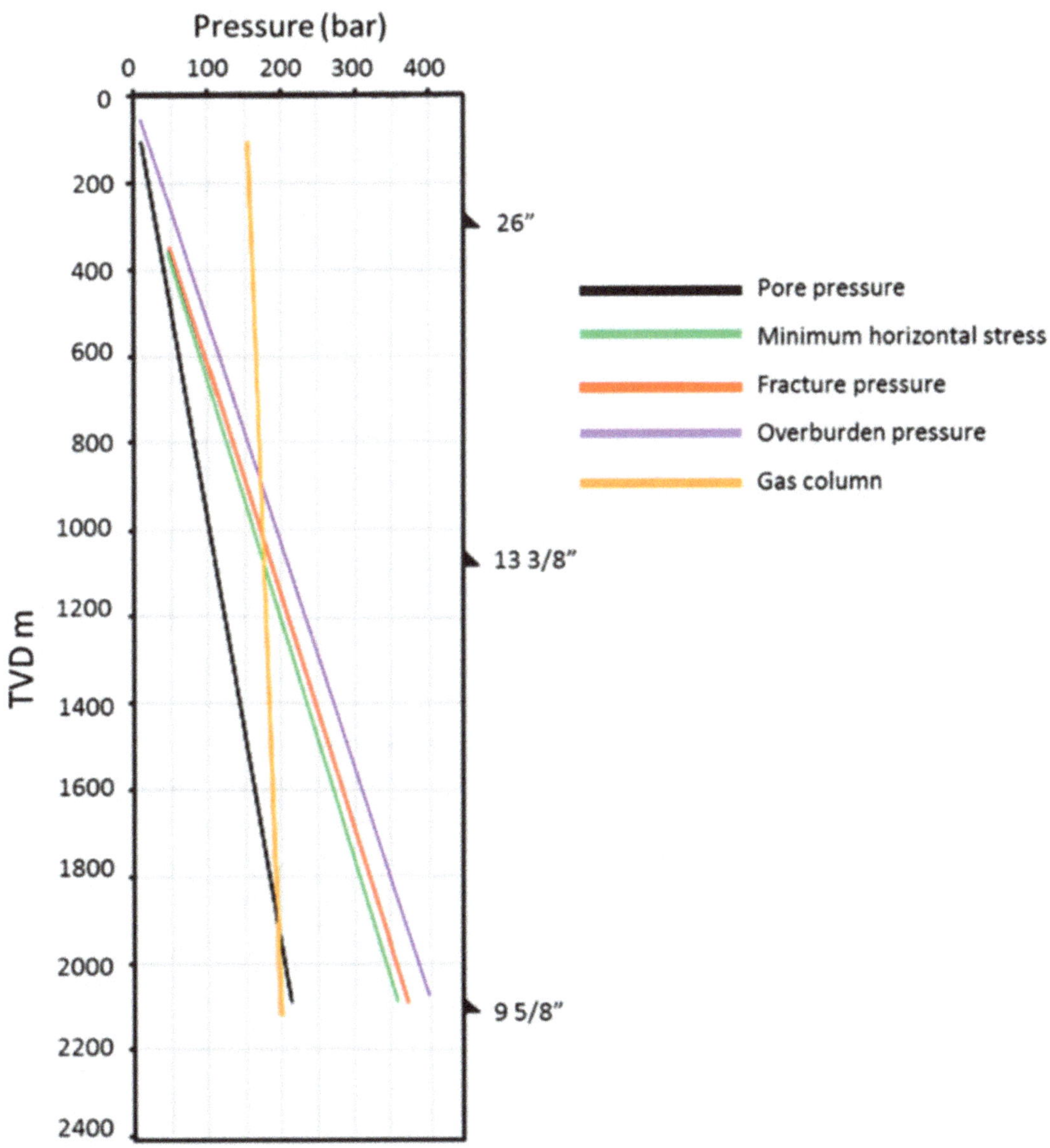

Fig. 2.44 Pore pressure and fracture pressure gradient curves of a platform well

is not able to hold the pressure and subsequently fractures. Selection of a higher reservoir pressure shifts the gas column to the right and MSD will be further away from the surface. As a result, the window for finding the appropriate interval for plug to be placed is limited.

Example 2.3 The following gradient curves, Fig. 2.45, have been reported for a well. Based on the given information estimate the minimum setting depth for the plug and propose an interval for the plug placement.

Solution The gas gradient curve should be plotted and extended to overburden pressure. The intercept point between the gas gradient curve and the minimum horizontal stress curve is the MSD of secondary plug. It means that the top of secondary barrier can be up to the depth of interception. However, it is recommended to install the permanent barriers, primary and secondary, as close as possible to the source of inflow.

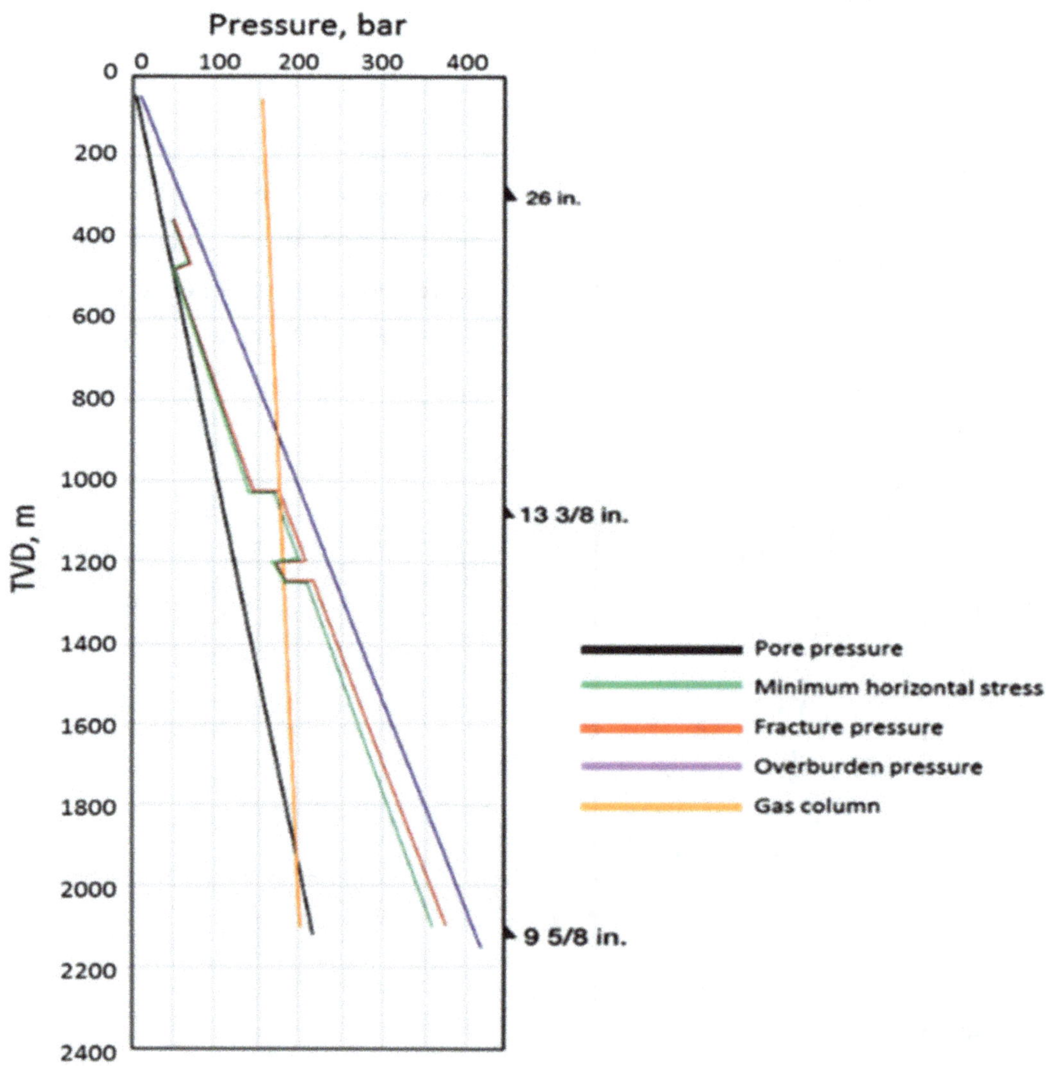

Fig. 2.45 Pore pressure—Fracture pressure gradient curves

In this way, in case of barrier failure, if permanent barriers are not qualified, there will be a window left to install new barriers.

2.10.2.2 Minimum Setting Depth—Fluid Gradient

In this method, the intersection between the fracture pressure and gas column is obtained mathematically. The MSD is the unknown parameter while the final reservoir pressure, fracture gradient, fluid gradient, and TVD of the reservoir are known parameters. The MSD is given by:

$$P_{FP} - P_g \times (H - h_{MSD}) \leq \frac{12}{231} \times P_{frac.} \times h_{MSD} \tag{2.5}$$

$$h_{MSD} \geq \frac{P_{FP} - P_g \times H}{\left(\frac{12}{231} \times P_{frac.}\right) - P_g} \tag{2.6}$$

whereas P_{FP} is the final reservoir pressure (psi), P_g is the gas gradient pressure (psi/ft), H is the TVD of reservoir (ft), h_{MSD} is the minimum setting depth (ft), and $P_{frac.}$ is the fracture pressure gradient (ppg).

Example 2.4 A platform well was drilled in the NCS in 1999 and the initial reservoir pressure was 2915 psi. The well is an oil producer with a fluid gradient of 0.32 (psi/ft). The reservoir is supported with a strong aquifer and the current reservoir pressure is 2755 psi. An average fracture gradient from leak off test is estimated to be 10.0 (ppg). The production casing shoe was placed at 6050 (ft) TVD and the cap rock thickness is 200 (ft) TVD. Calculate the minimum setting depth for primary and secondary barriers. Assume a gas gradient of 0.1 (psi/ft).

Solution This question shows how the gas presence can create a small window for placing the barrier.

The question mentions that the reservoir is supported by an active aquifer, which means that the reservoir pressure can build up to initial reservoir pressure. If we assume that the reservoir is under saturated, then oil is present and oil gradient is used for calculations. By using Eq. (2.6):

$$h_{MSD} \geq \frac{P_{FP} - P_g(H)}{\left(\frac{12}{231} \times P_{frac.}\right) - P_g}$$

$$h_{MSD} \geq \frac{2915 - 0.32 \times 6250}{\left(\frac{12}{231} \times 10\right) - 0.32}$$

$$h_{MSD} \geq 4575\,(ft)\,TVD$$

Now consider that the reservoir is a saturated reservoir, which means gas is present in the wellbore.

$$h_{MSD} \geq \frac{2915 - 0.10 \times 6250}{\left(\frac{12}{231} \times 10\right) - 0.10}$$

$$h_{MSD} \geq 5452\,(ft)\,TVD$$

It means that the window to install both primary and secondary barriers is smaller.

References

1. NORSOK Standard D-010. 2013. *Well integrity in drilling and well operations*. Standards Norway.
2. Anders, J., B. Mofley, and S. Nicol, et al. 2015. Implementation of well barrier schematic workflows. In *SPE digital energy conference and exhibition*. SPE-173433-MS, The Woodlands, Texas, USA: Society of Petroleum Engineers. https://doi.org/10.2118/173433-MS.
3. Bellarby, J. 2009. *Well completion design*. Elsevier.
4. Aggelen, A.V. 2016. Functional barrier model—A structured approach to barrier analysis. In *SPE international conference and exhibition on health, safety, security, environment, and social responsibility*. SPE-179214-MS. Stavanger, Norway: Society of Petroleum Engineers. https://doi.org/10.2118/179214-MS.
5. Oil & Gas UK. 2012. *Guidelines for the suspension and abandonment of wells*. Oil & Gas UK, United Kingdom.
6. ISO Standard. 2008. *Petroleum and natural gas industries: Downhole equipment: packers and bridge plugs*. Norge Standard.
7. Gee, N., S. Brown, and C. McHardy. 1993. The development and application of a slickline retrievable bridge plug. In *Offshore Europe*. SPE-26742-MS. Aberdeen, United Kingdom: Society of Petroleum Engineers. https://doi.org/10.2118/26742-MS.
8. Fjagesund, T. 2015. Technology update: Using schematics for managing well barriers. *Journal of Petroleum Technology*, SPE-0915-0034-JPT. https://doi.org/10.2118/0915-0034-JPT.
9. Alberty, M.W., and M.R. McLean. 2001. Fracture gradients in depleted reservoirs—Drilling wells in late reservoir life. In *SPE/IADC drilling conference*. SPE-67740-MS. Amsterdam, Netherlands: Society of Petroleum Engineers. https://doi.org/10.2118/67740-MS.
10. Nelson, E.B., and D. Guillot. 2006. *Well Cementing*. second ed. Sugar Land, Texas: Schlumberger. ISBN-13: 978-097885300-6.
11. Smith, P.S., C.C. Clement, Jr., and A.M. Rojas. 2000. Combined scale removal and scale inhibition treatments. In *International symposium on oilfield scale*. SPE-60222-MS, Aberdeen, United Kingdom: Society of Petroleum Engineers. https://doi.org/10.2118/60222-MS.
12. Yan, F., N. Bhandari, and F. Zhang, et al. 2016. Scale formation and control under turbulent conditions. In *SPE International oilfield scale conference and exhibition*. SPE-179863-MS, Aberdeen, Scotland, UK: Society of Petroleum Engineers. https://doi.org/10.2118/179863-MS.
13. Sanchez, Y., E.C. Neira, and D. Reyes, et al. 2009. Nonacid solution for mineral scale removal in downhole conditions. In *Offshore technology conference (OTC)*. OTC-20167-MS, Texas: Offshore Technology Conference. http://dx.doi.org/10.4043/20167-MS.
14. Wu, J., and M. Zhang. 2005. Casing burst strength after casing wear. In *SPE production operations symposium*. SPE-94304-MS, Oklahoma City, Oklahoma: Society of Petroleum Engineers. https://doi.org/10.2118/94304-MS.
15. Ning, J., Y. Zheng, and D. Young, et al. 2013. A thermodynamic study of hydrogen sulfide corrosion of mild steel. In *NACE international*. NACE-2462, Florida, USA: NACE International.
16. Crolet, J.-L. 1983. *Acid corrosion in wells* (CO_2, H_2S): *Metallurgical aspects*. SPE-10045-PA. http://dx.doi.org/10.2118/10045-PA.

17. Browning, D.R. 1984. CO_2 Corrosion in the Anadarko Basin. In *SPE deep drilling and production symposium*. SPE-12608-MS, Amarillo, Texas: Society of Petroleum Engineers. http://dx.doi.org/10.2118/12608-MS.
18. Torbergsen, H.E., and H.B., Haga, and S. Sangesland, et al. 2012. *An introduction to well integrity*. Norskoljeoggass.
19. Asian, S., and A. Firoozabadi. 2014. Deposition, stabilization, and removal of asphaltenes in flow-lines. In *Abu Dhabi international petroleum exhibition and conference*. SPE-172049-MS, Abu Dhabi, UAE: Society of Petroleum Engineers. http://dx.doi.org/10.2118/172049-MS.
20. Salahshoor, K., S. Zakeri, S. Mahdavi, et al. 2013. Asphaltene deposition prediction using adaptive neuro-fuzzy models based on laboratory measurements. *Fluid Phase Equilibria* 337: 89–99. https://doi.org/10.1016/j.fluid.2012.09.031.
21. Frost, K.A., R.D. Daussin, and M.S. Van Domelen. 2008. New, highly effective asphaltene removal system with favorable HSE characteristics. In *SPE international symposium and exhibition on formation damage control*. SPE-112420-MS, Lafayette, Louisiana, USA: Society of Petroleum Engineers. http://dx.doi.org/10.2118/112420-MS.
22. Guo, B., and A. Ghalambor. 2005. Chapter 12—Special problems. In *Natural gas engineering handbook* (Second Edition). 263–316. Gulf Publishing Company. 978-1-933762-41-8.
23. Bybee, K. 2007. A risk-based approach to waste-containment assurance. *Journal of Petroleum Technology* 59(02). https://doi.org/10.2118/0207-0057-JPT.
24. Guo, Q., T. Geehan, and K.W. Ullyott. 2006. Formation damage and its impact on cuttings injection well performance: A risk-based approach on waste-containment assurance. In *SPE international symposium and exhibition on formation damage control*. SPE-98202-MS, Lafayette, Louisiana, USA: Society of Petroleum Engineers. https://doi.org/10.2118/98202-MS.
25. Schultz, R.A., L.E. Summers, K.W. Lynch, et al. 2014. Subsurface containment assurance program—Key element overview and best practice examples. In *Offshore technology conference-Asia*. OTC-24851-MS, Kuala Lumpur, Malaysia: Offshore Technology Conference. https://doi.org/10.4043/24851-MS.
26. Moeinikia, F., K.K. Fjelde, A. Saasen, et al. 2015. A probabilistic methodology to evaluate the cost efficiency of rigless technology for subsea multiwell abandonment. *SPE Production & Operations* 30 (04): 270–282. https://doi.org/10.2118/167923-PA.
27. Gergerechi, A. 2017. *Report of the investigation of the well control incident in well 31/2-G-4 BY1H/BY2H on the Troll field with the Songa Endurance drilling unit*. Petroleum Safety Authority. http://www.ptil.no/getfile.php/1343478/Tilsyn%20p%C3%A5%20nettet/Granskinger/2016_1154_eng%20Granskingsrapport%20Songa%20Endurance%281%29.pdf.
28. API Specification 6A. 1996. *Specification for wellhead and Christmas tree equipment*. American Petroleum Institute: Washington, DC.
29. Kaculi, J.T., and B.J. Witwer. 2014. Subsea wellhead system verification analysis and validation testing. In Offshore technology conference. OTC-25163-MS, Houston, Texas: Offshore Technology Conference. https://doi.org/10.4043/25163-MS.
30. Henderson, D., and D. Hainsworth. 2014. Elgin G4 gas release: What happened and the lessons to prevent recurrence. In The SPE international conference on health, safety, and environment. SPE-168478-MS, Long Beach, California, USA: Society of Petroleum Engineers. https://doi.org/10.2118/168478-MS.
31. Reinås, L. 2012. Wellhead fatigue analysis: Surface casing cement boundary condition for subsea wellhead fatigue analytical models. In *Department of petroleum engineering*. University of Stavanger: Stavanger, Norway.
32. Aas, B., J. Sørbø, and S. Stokka. et al. 2016. Cement placement with tubing left in hole during plug and abandonment operations. In *IADC/SPE drilling conference and exhibition*. SPE-178840-MS, Texas, USA: Society of Petroleum Engineers. https://doi.org/10.2118/178840-ms.
33. Iversen, M. and J.A. Ege. 2000. Plugging of a downhole control line. In *Offshore technology conference*. OTC-11929-MS, Houston, Texas: Offshore Technology Conference. https://doi.org/10.4043/11929-ms.

34. The university of Texas at Austin. 1997. *A dictionary for the petroleum industry*. University of Texas at Austin Petroleum: USA.
35. API TR 1PER15K-1. 2013. *Protocol for verification and validation of high-pressure high-temperature equipment*. Washington.
36. DeBruijin, G., C. Skeates, and R. Greenaway, et al. 2008. High-Pressure, high-temperature technologies. In *Oilfield review*. France: Schlumberger
37. Baohua, Y., Y. Chuanliang, T. Qiang, et al. 2013. Wellbore stability in high temperature and highly-depleted reservoir. *Electronic Journal of Geotechnical Engineering* 18: 909–922.
38. Frittella, F., M. Babbo, and A.I. Muffo. 2009. Best practices and lesson learned from 15 years of experience of cementing HPHT wells in Italy. In *Middle east drilling technology conference & exhibition*. SPE-125175-MS, Manama, Bahrain: Society of Petroleum Engineers. https://doi.org/10.2118/125175-MS.
39. Gjonnes, M., and I.G. Myhre. 2005. High angle HPHT wells. In *SPE Latin American and Caribbean petroleum engineering conference*. SPE-95478-MS, Rio de Janeiro, Brazil: Society of Petroleum Engineers. https://doi.org/10.2118/95478-MS.
40. Denney, D. 2010. Analysis of inflow-control devices. *Journal of Petroleum Technology* 62 (05): 52–54. https://doi.org/10.2118/0510-0052-JPT.

Chapter 3
Specification for Permanent Plugging Materials

Portland cement is currently the prime barrier material used in the petroleum industry for zonal isolation and permanent well abandonment. In addition, the industry considers alternative plugging materials. Therefore, it is necessary to consider functional requirements, operating conditions and qualification procedures for any newly developed alternative plugging material.

3.1 Material Requirements for Permanent Barriers

In order to qualify the well barrier for its intended use, some requirements are necessary to be defined. These requirements are called *Well Barrier Acceptance Criteria* (WBAC) and include functional, and verification requirements of the well barrier [1]. The main functional characteristics of permanent barrier materials are addressed as [1, 2].

1. Very low permeability or impermeable,
2. Long-term durability at downhole conditions,
3. Non-shrinking,
4. Ductile or non-brittle,
5. Resistance to downhole fluids and gases, and
6. Sufficient bonding to casing and formation.

3.2 Functional Requirements of Permanent Well Barrier Elements

A permanent well barrier element has to fulfill a number of functional requirements including sealing capability, bonding properties, downhole placeability, durability, and reparability. These requirements are discussed in this chapter in more detail.

M. Khalifeh and A. Saasen, *Introduction to Permanent Plug and Abandonment of Wells*, Ocean Engineering & Oceanography 12,
https://doi.org/10.1007/978-3-030-39970-2_3

3.2.1 Sealing Capability

The main function of a permanent barrier is to seal a potential and prevent the movement of fluids. The sealability of a material is a function of its permeability and bond strength. However, definition of impermeability is a controversial subject as most, if not all, materials have some degree of permeability for some elements. For example, cap rocks have some permeability within the range of 10^{-3}–10^{-6} millidarcy. It means that in the context of permanent well barrier materials, it is inevitable that a fluid within the well will ultimately migrate through a barrier, even though at a low rate. Table 3.1 presents permeability of some materials.

In order for a leak to occur, fluids must be able to enter into a barrier and breakthrough must happen. Then the definition and investigation of permeability gets its meaning. A fundamental requirement for an effective seal is that the *entry pressure* of the sealing material shall be higher than the capillary forces of fluids in the formation beneath. The *seal entry pressure*, the seal capacity, is the capillary pressure at which fluid pressure exceeds the capillary entry threshold and therefore, fluid leaks into the pore space of the barrier material. This is dependent on both barrier and fluid parameters. Barrier parameters include the size distribution of connected pore throats. Fluid parameters include the present fluids (e.g. water, oil or gas), fluid density, and *Interfacial Tension* (IFT) of the fluids.

The capillary entry pressure (dynes/cm^2) is defined by Eq. (3.1) [5].

$$P_c = \frac{2\sigma(\cos\theta)}{r} \tag{3.1}$$

where σ is the interfacial tension (dynes/cm), θ is the contact angle of the water with the pore surface (degrees), and r is the pore radius (microns). Capillary entry pressure, also known as seal capacity, could potentially be defined as a means of resisting permeation of WBE by fluids. Among the contributing factors in capillary entry pressure, the contact angle and pore radii are prone to modification with time. In the case of water as the entering fluid, the capillary entry pressure is only exceeded when the contact angle between WBE and water is greater than 90°.

Table 3.1 Permeability of some materials [3]

Material	Permeability (millidarcy)
Portland cement (neat class G) [4][a]	10^{-2}
Shale	10^{-3}–10^{-5}
Granite	10^{-3}–10^{-4}
Halite	10^{-7}–10^{-9}
Anhydrites	10^{-5}–10^{-7}

[a]Although neat class G Portland cement has such a permeability, use of cement permeability reduction additives reduce the permeability significantly

3.2.1.1 Methods for Measuring Capillary Pressure

Due to the complex structure of pores, it is impossible to use Eq. (3.1) to calculate the capillary pressure of porous media. Therefore, laboratory measurements have been developed and are the most reliable methods for capillary pressure measurements. Methods for measuring capillary pressure are categorized as:

- Mercury methods
- Porous-plate method, and
- Centrifuge method.

Mercury method—For experimental convenience, it is a common practice to use air-mercury system for capillary pressure measurements. In the mercury method, the specimen is dried and placed inside a cell and then the cell is vacuumed. Subsequently, mercury is injected into the cell and the volume of mercury that enters the specimen at increasing pressures is measured. To apply the proper overburden pressure, a cylindrical specimen could be placed inside a confining sleeve and then the overburden could be applied. Figure 3.1 shows a simplified schematic of the mercury setup.

By measuring the displacement pressures for the assessment of WBE, seal capacity could be assessed from capillary pressure curves. However, there are some disadvantages associated with this method: it is a destructive test, it is performed on dried specimens which does not include fluid-surface interactions, and it may cause collapse of accumulations of grain surface coating minerals. In addition, the HSE issue related to mercury is a challenge. The advantages of the technique are that it is rapid and irregular specimens can be used in the case with no overburden pressure effect [6].

Porous-plate method—This technique can yield very accurate capillary pressure relationships. In this technique, a cylindrical specimen is saturated with water. A flat end of the specimen is then pressed against a flat porous plate, to make a good contact, in a cell filled with gas. The porous plate is also saturated with water. To improve the contact between the porous plate and the specimen, usually a moist tissue is placed between them. Subsequently, the gas pressure above the specimen is

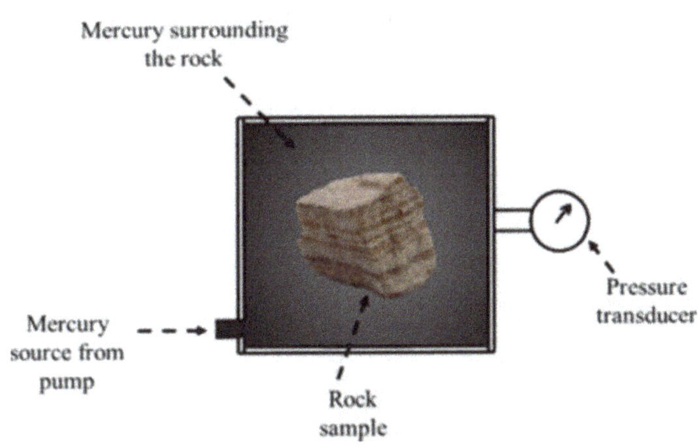

Fig. 3.1 Mercury is metered into a vacuumed specimen

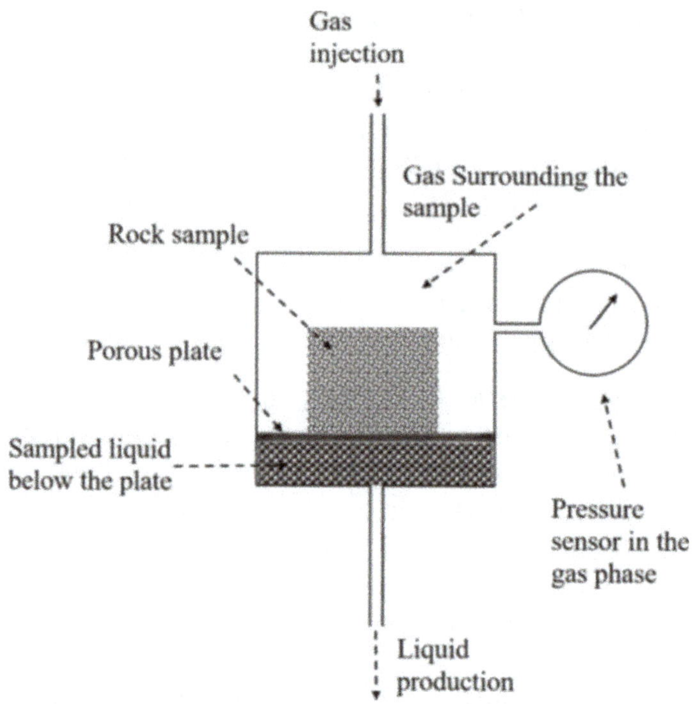

Fig. 3.2 Schematic of the porous-plate setup. The differential pressure between the gas and the water must not exceed the threshold pressure of the porous plate

increased in small steps to force the gas to displace the water from the specimen. In this procedure, the high displacement pressure of the porous plate allows brine from specimen to pass through but prevents the flow of gas. The specimen is removed at intervals and weighed until weight equilibrium is achieved. Considerable time may be needed, often a week or more, for each displacement step to reach equilibrium [7]. A diagrammatic sketch of this equipment is shown in Fig. 3.2.

Centrifuge method—Centrifuge measurements are more favorable as they take less time compared to porous-plate measurements; however, the measurements are not as quick as mercury measurements. In this method, a cylindrical specimen is saturated with water and then it is placed inside a centrifuge. Subsequently, it is spun in steps of increasing spin rate. The centrifugal forces force the water out of the specimen, replacing water with gas. There is a collector system for the drained water. The average saturation of water in the specimen, at each spin-rate, may be calculated from the volume of accumulated water and the porous volume of the specimen [8]. The capillary pressure distribution, at each spin-rate step, is given by [9].

$$P_c(r) = \frac{1}{2}\Delta\rho\omega^2\left(r_e^2 - r^2\right) \tag{3.2}$$

where r_e is the radius from center of rotation to the upper face of the specimen, r is the radial distance to any point in the sample, ω is the rotational velocity (rad/s), and $\Delta\rho$ is the density difference between displaced and displacing fluids.

Table 3.2 Fluid flow properties for a reservoir and cap rock [10]

Property	Reservoir	Cap rock
Porosity (–)	0.125	0.05
Permeability (md)	2.028	1.11×10^{-3}
Irreducible water saturation (–)	0.3	0.66
Entry capillary pressure (psi)	0	39
Maximum capillary pressure (psi)	145	924

3.2.1.2 Permeability

Permeability is a property of the material, representing the ability of the material to transfer fluids. The WBE's permeability is the property controlling the movement and leak rate of formation fluids. As the rock permeability was first defined mathematically by Henry Darcy in 1856 [5], the equation that defines permeability in terms of measureable quantities is called Darcy's Law and is given in Eq. (3.3). Darcy's Law shows that permeability, k, is directly proportional to flow rate, q, length of the medium, L, and fluid viscosity, μ, and inversely proportional to cross-sectional area, A, and the differential pressure across the medium, Δp.

$$k = -\frac{q \cdot \mu \cdot L}{A \cdot \Delta p} \quad [\mathrm{m}^2] \tag{3.3}$$

When a fluid with one centipoise viscosity and a pressure gradient of one atmosphere per centimeter of length flows with a flow rate of one cubic centimeter per second across a cross-sectional area of one square centimeter, the permeability is unity. For the units described above, k has been arbitrarily assigned a unit called Darcy in honor of Henry Darcy. One Darcy is a relatively high permeability as the permeabilities of most reservoir rocks are less than one Darcy. Therefore, the term millidarcy is normally used, where one millidarcy is equal to one-thousandth of one Darcy [5]. Table 3.2 presents example of fluid flow properties of a carbonate reservoir rock and a shale cap rock.

3.2.1.3 Acceptance Criteria for Fluid Flow Through Plugging Material

The goal of permanent P&A is to restore the cap rock in the wellbore or its functionality by placement of a plugging material across a suitable formation. Although the definition of a competent plugging material might be a matter of discussion, it is reasonable to consider properties of cap rock as the acceptance criteria for selection or design of any plugging materials. This adaptation is also valid for permeation characteristics of any plugging material as all existing materials, have some degree of permeability.

3.2.2 *Bonding*

Plugging materials should remain intact in place and block the migration of fluids. Therefore, sufficient bond strength and hydraulic bond strength of plugging materials with formations and steel are required. For zonal isolation purposes, hydraulic bonding is normally more important than bond strength.

Bond strength failure, debonding, may eventuate from two different loading scenarios; shear load, and tensile load. These loads can be induced by thermal cycling, hydraulic forces, volume changes of material, tectonic stresses, or a combination of these [11–13]. The volume change could be due to shrinkage and may occur either during curing or after setting due to changes in downhole conditions. Shrinkage of the plugging materials may impose sufficient tensile stresses on bonding between the plugging material and steel or formation to compromise the bonding. Another scenario that may result in tensile failure of bonding is the expansion of casing where the plugging material is placed inside the casing. When a reservoir starts to build up pressure underneath the plugging material, it may cause expansion of casing and consequently, *debonding* may occur. Debonding due to expansion of casing is predominantly for large casing sizes [2].

Hydraulic bond strength failure may eventuate from shrinkage or expansion of plugging material or expansion of casing caused by reservoir pressure build-up and/or due to interaction in the interface of casing and plugging material [14].

Bonding properties, bond strength and hydraulic bonding, are studied to improve knowledge and determine bonding ability of plugging materials. In 1962, Evans and Carter [15] published the result of their extensive study on shear bond and hydraulic bond strength of oilwell cement covering the effect of closed-in pressure, new mill varnish, uncoated pipe (wire-brushed, rusty, and sandblasted), dry pipe surface, and pipe surface being wet with either water-based or oil-based mud. Determining bond strength and hydraulic bonding of plugging materials considering effects of the above-mentioned factors are necessary.

3.2.2.1 Shear Bond Strength to Pipe

Shear bond strength defines the bond that mechanically supports pipe in the hole, and it is determined by measuring the force required to initiate pipe movement inside a sealing material (Fig. 3.3). The force is applied parallel to the contact surface [16]. This force when divided by contact surface area between the plugging material and casing, yields the shear bond [17].

$$\text{Shear bond strength} = \frac{\text{Force}}{\text{Contact area}} \tag{3.4}$$

The shear bond strength to pipe can be measured for two different scenarios; plugging materials placed inside the casing and plugging material placed outside the casing. The shear bond strength induced by a push-out test for cement outside casing is calculated by:

$$\tau_{av} = \frac{F}{\pi \cdot D_o \cdot L_c} \tag{3.5}$$

where F is the failure load which is applied on pipe, D_o is the pipe outside diameter, and L_c is the cement length. The shear bond strength of cement inside casing is given by:

$$\tau_{av} = \frac{F}{\pi \cdot D_i \cdot L_c} \tag{3.6}$$

where F is the failure load which is applied on pipe, D_i is the pipe outside diameter, and L_c is the cement length inside the pipe.

In one attempt, Evans and Carter studied shear-bond strength of cement to pipe [15, 17]. Variation between brands of API class A cements (see Chap. 4), curing temperature, condition of pipe, mud-wet and dry pipe, and closed-in pressure were factors studied by them. According to their results, there are correlations between compressive strength and shear bond on dry pipe. Closed-in pressure during setting of cement is detrimental to shear bond of cement to pipe after pressure is released. Shear bond strength is increased when cement is squeezed and wall pipe is water-wet. Mill-coated finish surface is detrimental to shear bond strength. It is important to mention that Evans and Carter applied both hydraulic and shear loads and therefore, their true measured hydraulic-bond strengths are uncertain. Table 3.3 presents shear bond strengths measured by Evans and Carter.

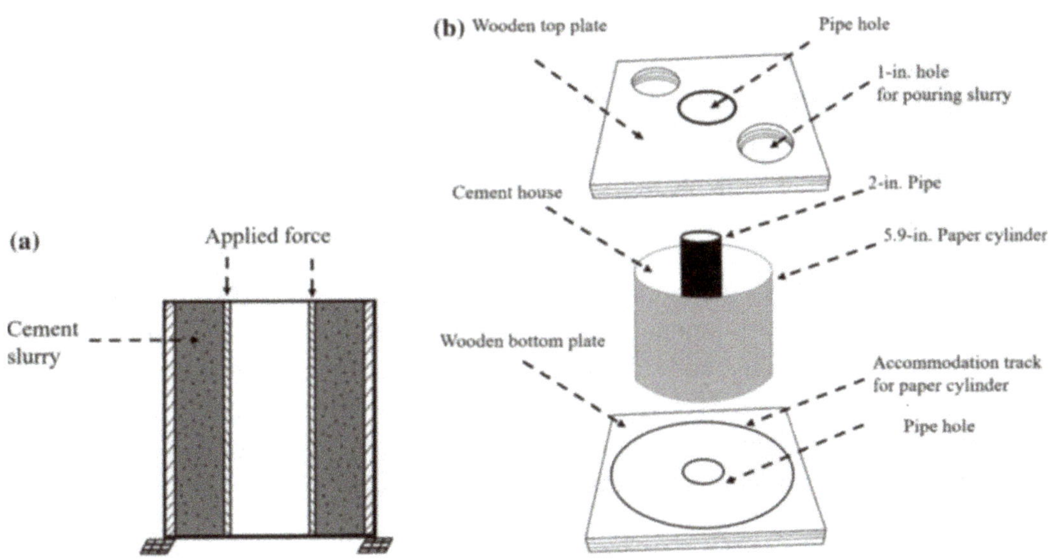

Fig. 3.3 Shear bond strength of cement to pipe **a** setup used by Evans and Carter [15], **b** setup used by Khalifeh et al. [18]

Table 3.3 Examples of hydraulic- and shear-bonding properties of new and used pipe [15]

Casing type	Time (Days)	Hydraulic bond (psi)	Shear bond (psi)
New	8 h	–	10
Used (rusted)	8 h	–	53
New	1	300	79
New (sandblasted)	1	500	123
Used (slightly rusty)	2	500–700	182
Used (wire brushed)	2	500–700	335
New (sandblasted)	2	500–700	395
Used (rusted)	2	500–700	422
New Pipe			
Water-based mud	2	175–225	46
Dry	2	375–425	284
Used (slightly rusty)			
Oil-based mud	2	–	75
Water-based mud	2	–	174
Dry	2	–	182
Latex cement			
New	1	500	105
Used (slightly rusty)	1	360	58

- API Class A Cement
- Curing temperature: 80 °F
- Casing size: 2-in. inside 4-in.
- Cement-sheath thickness: 0.812-in.

Cement-pipe and cement-formation bond strength investigation shows that the bond strength depends on the nature of the contact surfaces and the cement hydration characteristics [16]. For a permanent plug, it is necessary to determine the appropriate bond strength for supporting the plug inside either openhole or casing and test it when the plug has set.

The shear bond strength to pipe is improved by use of expandable cement or bonding agents such as latex and surfactants. Expanding properties of expandable cements prevent the development of microannuli at the interface between casing and formation and cement plug, and subsequently, ensure good bonding with casing [19]. The inclusion of latex additives to cement slurry lowers the surface tension between the slurry and casing and helps cement adhere to casing while setting. Surfactants treat the oil-wet surfaces by removing oil and allow better bonding contact [20].

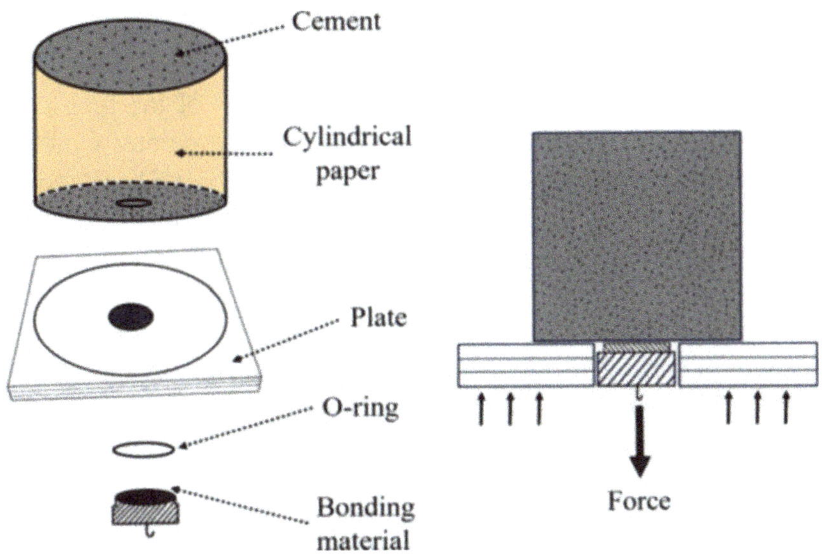

Fig. 3.4 Experimental setup for measuring tensile bond strength of cementitious materials to steel

3.2.2.2 Tensile Bond Strength to Pipe

Tensile bond strength is defined as the force which acts normal to the contact surface [16]. Tensile forces are applied perpendicularly to the contact surface of the supported specimen. Few publications are available for tensile bond strength to pipe and this area needs more attention [21]. Figure 3.4 shows an experimental setup for measuring cement-steel tensile bond strength.

3.2.2.3 Hydraulic Bond Strength to Pipe

Hydraulic bond is defined as the bond between pipe and cement, which helps to prevent the flow of fluids [15]. Hydraulic bond is determined by applying pressure at the pipe-cement interface until leakage occurs at either end of the specimen, Fig. 3.5.

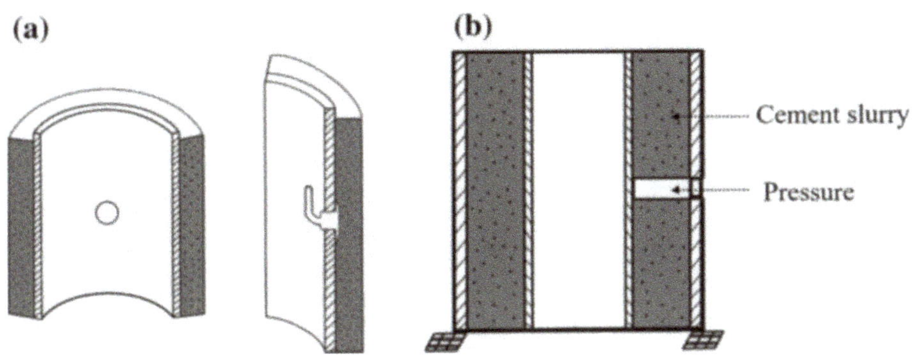

Fig. 3.5 Test setup for measuring hydraulic bond test; **a** the used setup by Scott and Brace [22], **b** the used setup by Evans and Carter [15]

The hydraulic pressure when leakage appears is defined as the bond failure pressure. Studies conducted by different researchers [15, 17] show that pressures at which hydraulic bonding failure occur depends on the viscosity of the pressurizing fluid. Hence, the choice of pressurizing fluid is an important parameter which influences the breakthrough time and failure pressure. Therefore, gas bond tests and liquid bond tests should be considered for hydraulic bonding measurements. The gases could be compressed air, nitrogen, CO_2, methane, etc. and the liquids could be crude oil and brine.

In 1966, Scott and Brace [22] studied the hydraulic bond strength at the casing-cement interface for various conditions of the external surface of the casing, effect of mud film on the casing surface, effect of temperature on the resin-sand coated pipe, and effect of corrosive atmosphere on the resin-sand coating as important parameters. Table 3.4 shows the relative hydraulic bonding strengths of casing-cement interface with various surfaces measured by Scott and Brace.

Scott and Brace [22] found that excellent hydraulic bonding strengths are maintained at temperatures in the range to 350–400 °F. In addition, poor hydraulic bonding resulted from untreated pipe where mud film remained and surfaces which were mill-varnished. However, resin-sand coating greatly improves the casing-cement bond. Figure 3.5 shows two different test setups used by different researchers for measuring hydraulic bond strength of cement to pipe.

In another effort, Evans and Carter [15] studied hydraulic bonding strengths of casing-cement (API class A cement) while measuring shear-bond strengths. Table 3.3 presents their results from these hydraulic- and shear-bonding tests. Evans and Carter investigated the effect of surface finish, drilling fluid, pipe size and length, cement-curing conditions, temperature and pressure on pipe, cement types, and the effect of

Table 3.4 Effect of mud film on hydraulic bonding strength of casing-cement interface [22]

Surface condition	Surface coating	Hydraulic bonding (psi)
Dry	Mill varnish	<20
Mud film	Mill varnish	<20
Dry	Rusty	350–450
Mud film	Rusty	20–50
Dry	Acid-etched	250–400
Mud film	Acid-etched	40–50
Dry	Sandblasted	500–600
Mud film	Sandblasted	50–60
Dry	Epoxy coated, 6–12 mesh sand	700–950
Mud film	Epoxy coated, 6–12 mesh sand	500–600

- Curing time: 24 h
- Curing temperature: 120 °F
- Cement type: not available
- Casing size: 4 ½-in. OD

squeezing. Their investigation concluded that the maximum reduction in hydraulic bonding is caused by a fluid layer at the cement-pipe interface; and hydraulic bonding strengths at cement-pipe interface are governed by surface finish of pipe, type of mud wetting, and degree of mud removal. In addition, they concluded that there is no fixed correlation between compressive strength and hydraulic bond strength to pipe. Low hydraulic bonding strengths at the cement-pipe interface are also a function of the pipe resiliency [15].

3.2.2.4 Shear Bond Strength to Formation

Shear bond strength of materials to formation depends on the nature of the contact surfaces and the reaction characteristics of the materials. Shear bond strength to formation maintains an intact barrier in place. Fluids only adhere to solids when the fluid wets the solid material and therefore, bonding of cement to formation is only possible if cement slurry filtrate is able to wet the wellbore wall. Roughness of formation surface, mineralogy of formation, degree of cement hydration, water-cement ratio, drilling mud and mud cake, downhole pressure and temperature, and types of cement additives are important factors to be considered for measuring shear bond strength of cement to formation [16]. Figure 3.6 presents a schematic of the setup used for formation-cement shear bond strength measurements.

A study performed by Becker and Peterson on cement-formation bond shows that the strength of the developed bond between cement and formation is mainly due to wettability and the degree of cement hydration. In addition, contamination of cement slurry with drilling mud, oil, or gas strongly deteriorate the shear bond strength to

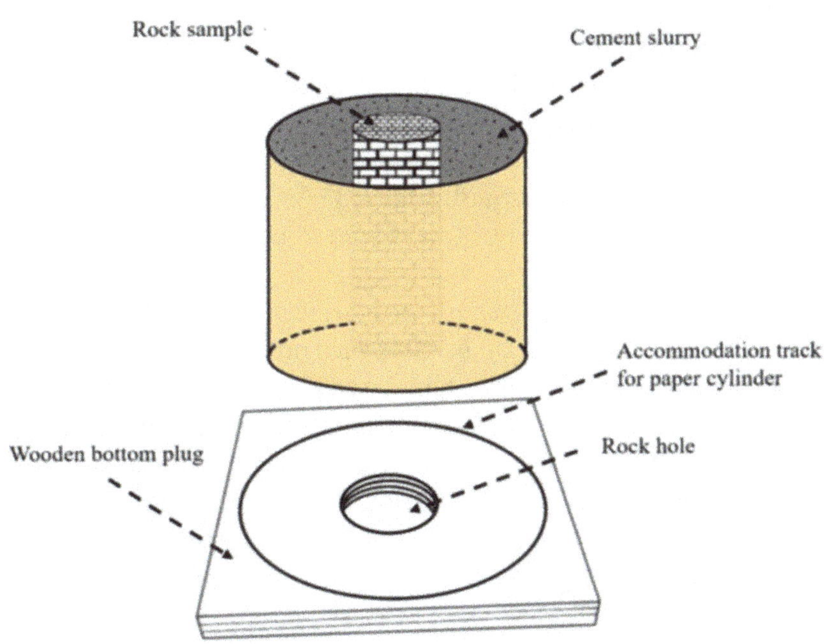

Fig. 3.6 Experimental setup used for formation-cement shear bond strength measurements

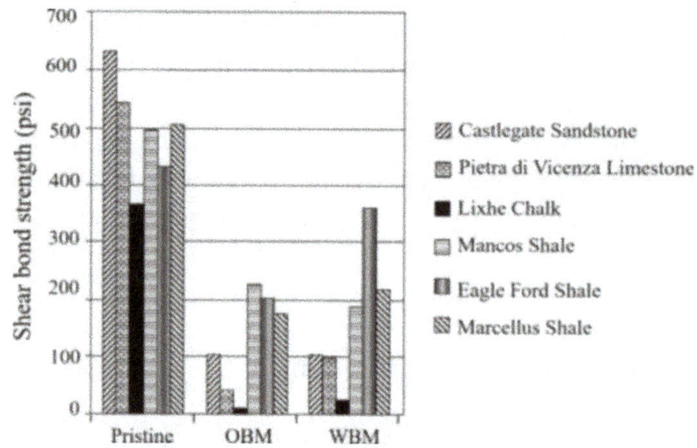

Fig. 3.7 Shear bond strength of cement with various rock types [23]

the formation. Therefore, contamination must be eliminated and mud or oil films on the formation surface should be removed totally. They also found that borehole temperatures up to 250 °F accelerate the development of shear bond strength. Higher temperatures can deteriorate the shear bond strength. However, the addition of silica flour can prevent the bond deterioration [16].

Opedal et al. [23] studied the role of formation type in the development of shear bond strength. A substantial reduction in shear bonding between cement and high porosity rocks (sandstone, limestone, and chalk) was observed in the presence of drilling fluid compared to the situation when low porosity shale was used as a formation. Existence of *Water-based Mud* (WBM) at the interface of rock and cement gives slightly better bonding than presence of *Oil-based Mud* (OBM) [15, 23]. Figure 3.7 presents the shear bond strength of some types of rock with and without drilling fluid film at the cement-formation interface.

When considering minimum shear bond strength (F_{sb}) required by a plugging material to avoid barrier movement, the resultant of two forces needs to be considered; reservoir pressure which pushes the plug upward and barrier weight and hydrostatic pressure above it, which act downward, Fig. 3.8. As a depleted reservoir may start to build up pressure after abandonment, it is safe to use initial reservoir pressure as the final reservoir pressure (F_R).

Cement-formation shear bond strength measurement—One of the main challenges in bond strength evaluation is lack of any standard procedure on how to perform the experiments. However, over the years, many researchers have followed the procedure implemented in the 60's by Evans and Carter [15, 17, 21]. In this method, a cylindrical rock sample is placed in the middle of a mold and then cement slurry is poured in the space between the rock sample and the mold, Fig. 3.6. The mold is covered with a plastic cover to avoid water evaporation during curing. Depending on the size of the test cell, the mold can be cured inside an autoclave to simulate downhole pressure and temperature.

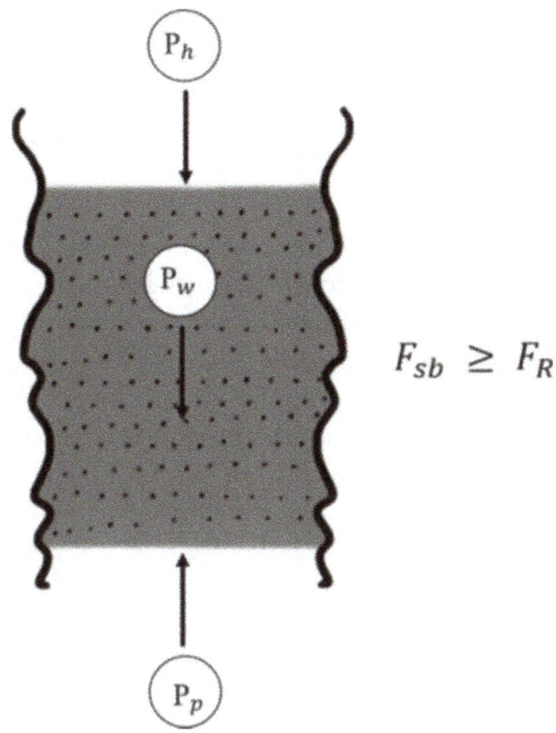

Fig. 3.8 Different forces acting on a barrier to dislocate it

It should be noted that the contribution of frictional force in real situations, at downhole conditions, is not identical to laboratory testing as washouts and other anomalies are not simulated in laboratory testing.

3.2.2.5 Tensile Bond Strength to Formation

Investigation of tensile bond strength of formation with any plugging material is an area which has not been studied so far. In this case, the applied force is perpendicular to contact surface; pulling the formation or the pipe away from the plugging material. The tensile bond strength helps to stop debonding created by lateral tectonic stresses.

3.2.2.6 Hydraulic Bond Strength to Formation

Hydraulic bond is defined as the bond between cement and formation, which helps prevent the flow of fluids [15]. The illusion of having similar cement-formation and cement-pipe hydraulic bonding strengths is one the reasons that few researchers have considered the cement-formation hydraulic bond strength measurement [15, 17]. Cement-formation hydraulic bond strength depends on the formation's mineralogy. Experimental works have shown that when cement is squeezed against a dry core, a higher hydraulic bond strength is attained [15]. Obtained hydraulic bond strength between cement and limestone shows higher pressure compared to obtained results for cement-sandstone, in identical circumstances. The failure path is also dependent on formation mineralogy. When limestone is used as a formation, the failure path is at the formation-cement interface; however when sandstone is used as formation, the failure path is within the core rather than the interface. Presence of drilling fluid in the interface usually lowers the hydraulic bond strength, regardless of the formation type, compared to dry core situations. Therefore, drilling fluid displacement is important to be considered. Different types of mudcakes at the cement-formation interface influence the hydraulic bond strength differently. When the mudcake is fresh and soft, the failure pressure which cause leakage is lower than a situation where the mudcake is old and rigid. In fact, a rigid mudcake does not make a higher bond strength but as it is old and rigid, the mudcake has a higher resistance to flow. This phenomenon occurs in both sandstone and limestone rocks [15]. Generally, when cement is placed against a filter cake, the failure plane is within the filter cake and the flow path is at the filter cake-formation interface [24]. Usually, spacers and chemical washes are pumped ahead of the cement slurry for fluid separation and hole cleaning. Curing pressure has also an influence on hydraulic bond strength. As the curing pressure is increased, the hydraulic bond strength is increased.

Cement-formation hydraulic bond strength measurement—This type of test is accomplished by placing a formation core inside casing and pouring cement on top of it. Cement is allowed to set at the target pressure and temperature. An embedded pressure port on top of the setup provides the pressure for simulating downhole pressures, and the setup can be placed in a heating cabinet for simulating downhole temperatures. A pressure port embedded below the setup guides the applied hydraulic pressure across a predrilled hole along the formation rock to the cement-formation interface, as illustrated in Fig. 3.9.

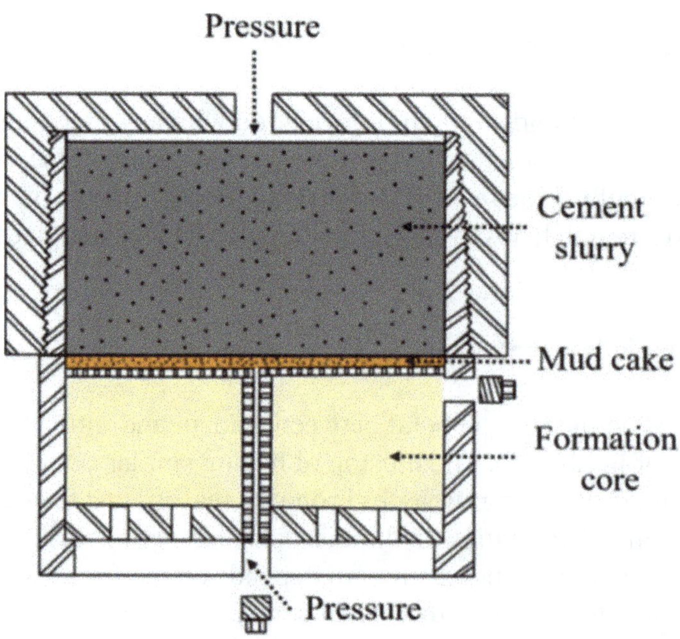

Fig. 3.9 Hydraulic bonding to formation [15]

As there are no standard methods available on how to perform the bond strength measurements, different researchers select different loading rates which consequently influences the reliability of the data. Therefore, it is necessary to consider a realistic loading rate which may occur at downhole conditions.

3.2.3 Placeability of Permanent Barrier Material

The permanent barrier materials are going to be placed downhole and must therefore displace existing fluids. Hence, optimizing displacement and placeability processes of permanent barrier materials must be prioritized. Usually to minimize the instability at the interface between cement and drilling fluid, a spacer fluid is pumped ahead of cement slurry to separate it from drilling fluid. To remove drilling fluid and filter cake by use of a spacer, the force resulting from interaction between viscosity, friction, and buoyancy forces is the critical factor. In addition, the impact of rheological properties of fluids (i.e. yield stress and gel strength), physical and chemical effects must be considered.

To remove filter cake, the friction pressure introduced by displacing fluid (ΔP_{f2}) must be higher than the adhesion force between filter cake and formation (F_{fc}), Fig. 3.10. The filter cake removal is affected by rock permeability, pressure drop across formation and filter cake, properties of displacing fluid and filter cake, and velocity of displacing fluid [25, 26].

Although turbulent flow regime is noted as a solution for the removal of the drilling fluid and filter cake, achieving a turbulent flow regime for cement is challenging due to

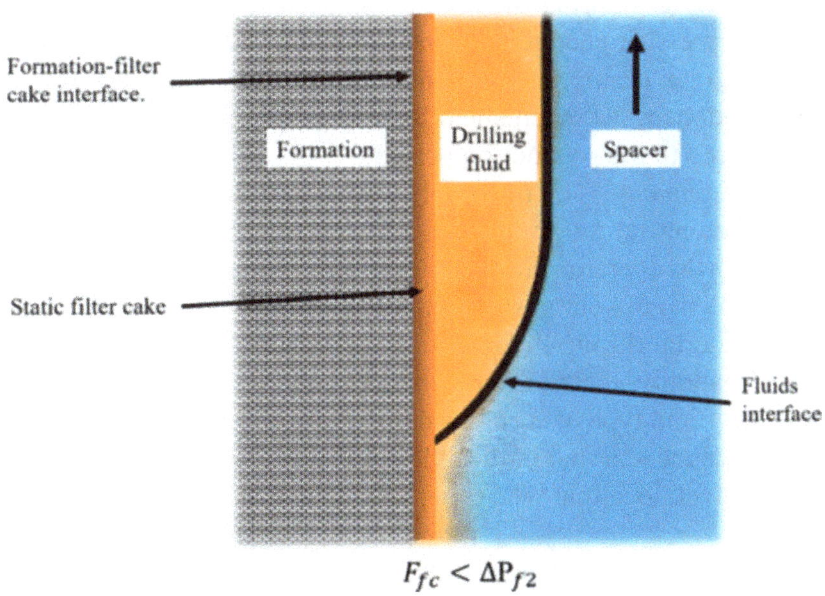

Fig. 3.10 Spacer displaces filter cake when the differential friction pressure of fluids overcome the friction force between filter cake and formation

restrictions on fluid velocities. The rheological properties and chemical composition of spacers should be designed in such a way that a turbulent flow regime is achieved and as it should be compatible with the cement slurry, the spacer should have minimal effect on the properties of the cement [27]. It should be noted that a turbulent flow regime may cause a higher ECD and consequently, a higher risk associated with fracturing the in situ formation in open hole.

Downhole conditions dictate the plug placement techniques utilized but once the barrier material is placed, the placement operation needs to be verified. These operations are explained in Chaps. 7 and 9.

3.2.4 Durability

Durability means that the plugging material keeps its initial quality with respect to mechanical integrity and hydraulic conductivity. To assess durability of a WBE, aging tests are carried out in presence of fluids which represent the wellbore fluid at different life periods. If modification of macroscopic properties of plugging material occurs as a result of chemical evolution however it does not impair the mechanical properties of the material, then it is acceptable and not considered as harmful.

Available standards and/or guidelines address durability of sealing materials for production and abandonment periods together; however due to differences between the two periods of well life, the availability of comprehensive standards (testing procedure) or guidelines considering durability of plugging materials are necessary. The two major differences between production period and abandonment period are addressed as: mechanical loading scenario, and downhole environment [28].

Mechanical loading—During the production period, the stress variations caused by thermal and/or pressure changes are exerted on the wellbore and consequently on wellbore elements (e.g. casing, cement, and formation). When a well is depleted and permanently abandoned, the mechanical loadings still exist but the stress variations are slower compared to the production period.

Downhole environment—The nature of fluids in contact with plugging or sealing materials are different during production life and post-abandonment period. Consider a well which is operating under a sour gas re-injection process as the field produces high contents of sour gases. Hence, the exposure time and rate is different for the sealing material during the injection time of the well compared to its post-abandonment. Generally, the nature of chemical species and their thermodynamic state vary with time and type of well during production and post-abandonment [29]. Therefore, it is necessary to standardize the durability of plugging materials used for P&A considering well location, well type, and the thermodynamic state of chemical species.

Durability of a potential well barrier element is evaluated by considering its long-term behavior when exposed to different chemicals at downhole condition and at different time intervals, besides microstructure, volume, weight and permeability changes. In addition, the role and behavior of interfaces between plugging material and formation/steel caused by material deterioration, and role and behavior of interfaces between plugging materials and formation/steel caused by different mechanical loads and rates require consideration during the durability analysis.

3.2.4.1 Exposure Time

Well barrier elements selected in well barrier envelopes are supposed to maintain their integrity for a long time. NORSOK D-010 [1] suggests an eternal perspective for durability of well barrier elements. Nonetheless, the definition of long time is a matter of interpretation as an established definition has not been published. Some researchers have selected 1-, 3-, 6-, and 12-month time intervals as the exposure time in their studies [28, 30, 31]. However, it is recommended to continue for much longer periods, even up to 5 years [2]. Long-term testing can be useful for better understanding the properties of materials and material qualification for utilization as permanent plugging material.

3.2.4.2 Downhole Condition

Any material used in well barrier envelopes must be selected carefully to withstand downhole conditions. Bottomhole conditions include temperature, pressure, and formation fluids. In addition, geographical location of wells may also be a guideline for the selection of chemicals.

3.2.4.3 Chemicals

Well barrier elements of a permanently abandoned well experience chemical attacks from different chemical substances over a long period of time. The chemical substances include crude oil, brine, hydrogen sulfide, hydrocarbon gas, and carbon dioxide. Obviously, in a single well, a well barrier element may not be exposed to all of these chemicals after abandonment, hence, the chemical selection should be based on the chemicals present in the target reservoir. For instance, sour wells are common in the Republic of Azerbaijan and Russia. Therefore, the durability of WBEs in sour wells is a priority in these countries.

Crude oil—For performing aging tests, selected crude oils should represent the reservoir fluid. It is necessary to represent the chemical composition of crude oil and its density.

Brine—It is a common practice to prepare an artificial seawater as representative of brine. The most commonly used industrial standard for the synthesis of artificial seawater is ASTM D1141-98 [32].

Carbon dioxide—Materials may be exposed to CO_2 in gas state or dissolved in brine or crude oil (liquid state). The exposure scenario should mimic the formation fluids. When CO_2 is dissolved in water, it is partly hydrated and subsequently carbonic acid is formed [33].

$$CO_2 + H_2O \leftrightarrow H_2CO_3 \tag{3.7}$$

The formed carbonic acid dissociated in two steps:

$$H_2CO_3 \leftrightarrow H^+ + HCO_3^- \tag{3.8}$$

and:

$$HCO_3^- \leftrightarrow H^+ + CO_3^{2-} \tag{3.9}$$

Formation of CO_3^{2-} changes the pH value of brine and the resulting pH is a function of the CO_2 partial pressure. CO_2 can corrode metal and deteriorate cement. In the presence of CO_2 dissolved in water, metal is unstable and as a result of the chemical reactions between carbonic acid and steel, Fe^{2+} is released:

$$Fe + 2H_2CO_3 \rightarrow Fe^{2+} + 2HCO_3^- + H_2 \tag{3.10}$$

When concentrations of Fe^{2+} and CO_3^{2-} ions exceed the solubility limit, the following reaction occurs and $FeCO_3$ precipitates:

$$Fe^{2+} + CO_3^{2-} \rightarrow FeCO_3(s) \tag{3.11}$$

The precipitated compound occupies a different volume compared to the initial compounds and it causes the casing to decompose.

High pH stabilizes the steel surface and prevents its corrosion but as explained, the pH of the medium is lowered due to dissolution of CO_2 in brine. Lower pH makes the steel surface unstable and rust is formed (see Eq. (3.11)). Formation of rust causes expansion and extensively deteriorates the cement [34]. Therefore, selection of steel as a WBE for permanent P&A might be a concern in long term.

CO_2 deteriorates cement through two different mechanisms: carbonation and leaching. As presented in Eqs. (3.7)–(3.9), the presence of CO_2 in brine produces CO_3^{2-}, the formed ion reacts with Ca^{2+} and yields:

$$Ca^{2+} + CO_3^{2-} \rightarrow CaCO_3(s) \tag{3.12}$$

The source of Ca^{2+} is supplied in two ways: the dissolution of $Ca(OH)_2$ broadly known as CH, and decomposition of hydrated silicate and aluminate phases or broadly known as calcium silicate hydrate (C-S-H) gel. The $Ca(OH)_2$ becomes unstable at a pH below 12.6 and Ca^{2+} is leached out and if the pH becomes less than 8, the strength giving C-S-H phases are destabilized and Ca^{2+} is leached out [35]. Taylor [34] explains the reactions as following:

$$Ca(OH)_2 \rightarrow Ca^{2+} + 2OH^- \quad (3.13)$$

$$xCaO \cdot SiO_2(aq) + zH_2O \rightarrow 2yCa^{2+} + 2yOH^- + (x - y)CaO \cdot SiO_2(aq) \quad (3.14)$$

Taylor [34] showed that through decalcification of hydrated silicate and aluminate phases, new crystals form with smaller volumes and these crystals are a highly porous form of hydrous silica. The decomposition and formation of small crystals cause the deterioration of cement.

Hydrogen sulfide—H_2S is a corrosive material which is produced biologically (by the action of certain microbes) or geochemically (by the reaction of sulfurous minerals). The dissolution of H_2S in brine acidifies the medium and can attack steel and cement.

The electrochemical reaction of steel with H_2S undergoes a cathodic and anodic reaction. The cathode reactions are as follows [36]:

$$H_2S \rightarrow HS^- + H^+ \quad (3.15)$$

$$2HS^- + 2e \rightarrow 2S^{2-} + H_2 \quad (3.16)$$

The anodic reaction is the dissolution of steel and the formation of corrosion product [36]:

$$Fe \rightarrow Fe^{2+} + 2e \quad (3.17)$$

$$Fe + H_2S \rightarrow FeS_{1-x} + xHS^- + (2 - x)H^+ + 2e \quad (3.18)$$

This is an electrochemical reaction of H_2S corrosion which is known as rust. Corrosion causes enormous damage to tubulars and therefore, casing should be protected by a sealing material in order to be accepted as a permanent well barrier element.

H_2S deteriorates cement through two major mechanisms: leaching and sulfidation [37]. As presented in Eq. (3.15), the acidic medium created by H_2S attacks cement and triggers the leaching of Ca^{2+} ions as follows:

$$Ca(OH)_2 + 2H^+ \rightarrow 2H_2O + Ca^{2+} \quad (3.19)$$

$$3CaO \cdot 2SiO_2 \cdot 3H_2O + 6H^+ \rightarrow 3Ca^{2+} + 6H_2O + 2SiO_2 \quad (3.20)$$

It has been proven that H_2S does not drastically decompose the neat cement sheaths but it does interact with the iron containing products of cement hydration such as ferrites to form sulfides; aluminates, and unhydrated di-calcium silicate components [37, 38].

Zhang et al. [39] studied rate of H_2S and CO_2 attack on pozzolan amended class H oilwell cement. Their results shows that that aqueous environment is more favorable for H_2S to attack cement than H_2S in gas phase. In addition, they have shown that in aqueous phase, H_2S penetrates into the cement matrix more rapidly compared to CO_2.

3.2.4.4 Microstructure Analysis

In the petroleum industry, quantification of mechanical properties of materials have been focused on and defined in different standards and guidelines. However, utilization of advanced technologies for quantification and analysis of material microstructure has not been recommended as much as it perhaps should be. Of these technologies one could address light microscopy, X-ray powder diffraction, Scanning Electron Microscopy (SEM), and Transmission Electron Microscopy (TEM) which are vital for analyzing and quantifying microstructure of materials. It is necessary to study the microstructure of materials suggested as WBE over time and their modification at different conditions.

3.2.4.5 Volume Changes

A potential material to be used as a WBE should possess volume stability over time. Any degree of expansion may fracture the adjacent formation and any degree of shrinkage may cause microannuli and or debonding. The volume changes of a WBE may diminish the formation radial stresses and if it falls below the formation pore pressure, the risk of uncontrolled fluid flow is increased [31, 40]. Therefore, measuring the volume changes of well barrier elements as one of the durability parameters is vital.

3.2.4.6 Weight Changes

Degradation of WBE may lead to weight loss or weight gain. The weight change should not compromise the integrity of the WBE and other elements present in the well barrier envelope.

3.2.4.7 Permeability Changes

Any permanent plugging material should possess a very low permeability in the range of cap rocks and maintain the low permeability with an eternal perspective. Considering cement as a permanent plugging material, the permeability changes due to chemical attacks are caused by transformation of C-S-H phases to other phases. The newly formed phases are susceptible to occupying less volume or more volume. Therefore, permeability changes may occur [41].

3.2.4.8 Role of Material Degradation and Tectonic Loads on Durability of Interfaces

After a well is permanently plugged and abandoned, material degradation or induced tectonic stresses may create a small gap at the casing-cement or cement-formation interface, which is referred to as a *microannulus*. One of the processes which creates microannuli is called *debonding*, but it can also be created by residual drilling fluid. The microannulus creates pathways for fluids to escape. The risk of leakage through them is much higher than the risk of leakage through the bulk cement or corroded steel.

Studies show that alteration of the cement-formation interface is a complex problem which depends on rock type in addition to chemicals present in the medium. Theoretical and experimental investigations show that when cement is placed across stable shales in the presence of acidic brine (acidified by CO_2), Portland cement quickly adsorbs pore water present in the shale during the hydration process and after setting. This reaction changes pH of the acidic brine and makes it more acidic. The first and fast reaction which occurs is the dissolution of calcite whereas calcite present at the cement-formation interface has almost disappeared and a microannulus is created [42]. It should be noted that at downhole conditions, the dissolution process of calcite may be slower as the amount of acidified water is less compared to the laboratory case at which this study was performed.

The casing-cement interface is susceptible to degradation and hence, creation of microannuli in acidic environments. Studies show that when CO_2 finds a pathway across cement to steel it starts to degrade the casing-cement interface. Leakage of CO_2 accelerates the degradation of the interface [43].

When tectonic stresses are exerted on formation-cement or casing-cement, debonding may occur and consequently microannuli are created. This is due to differences in elasticity of the materials.

There are models to simulate casing-cement and cement-formation interface debonding. Most of these models are based on assumptions such as linear elasticity of casing, cement, and formation. These models also assume no cement defects at the initial condition [44]. Generally, these models are based on fluid-driven fracture propagation [45] and Coulomb friction [46]. All of these models are developed for well integrity analysis and modeling of cement interface debonding during the production life of wells, and not for permanently abandoned wells. It is therefore necessary to develop models addressing long-term integrity of casing-cement or cement-formation interfaces.

3.2.5 Reparability

The permanent P&A operation is performed to seal the potential fluid flow zones permanently with no intention of well re-entry. As, during the third phase of permanent P&A operations (discussed in Chap. 2), wellhead and conductor are removed, there will be no access to the well for repairing any well integrity issue. Therefore, a candidate plugging material should withstand downhole conditions without compromising its sealing capability. There are some suggestions regarding self-healing or self-repairing materials whereby the material starts to heal itself if some defects are introduced over time. Self-healing cement products are one of these examples.

3.3 Qualification of New Plugging Materials

Any new plugging material designed for permanent P&A needs be qualified prior to being applied in the field. The qualification process may be based on a systematic approach including experimental work and theoretical analysis. The qualification process includes preparation of the material and its placement, verification of its intended functionality when it is in place, and its durability at downhole conditions. The qualification process needs to be quantitative and documented. All the possible failure modes need to be identified and analyzed based on the risks associated with the failure of its functionality over time. When the failure modes and their associated risks are considered, the failure modes are ranked based on the associated risk. Whenever laboratory testing is possible to be carried out, it needs to be performed. Confidentiality of the technology should not limit the availability of data required for qualification.

References

1. NORSOK Standard D-010. 2013. *Well integrity in drilling and well operations*. Standards Norway.
2. Oil and Gas UK. 2015. *Guidelines on qualification of materials for the abandonment of wells*, in *Bond strength*, 48–50. The UK Oil and gas industry association limited: Great Britain.
3. Warren, J.K. 2006. *Evaporites: sediments, resources and hydrocarbons*. Heidelberg: Springer. 978-3-540-26011-0.
4. Nelson, E.B., and Guillot, D. 2006. *Well cementing*. 2nd edn. Sugar Land, Texas: Schlumberger. ISBN-13: 978-097885300-6.
5. Ahmed, T. 2001. *Reservoir engineering handbook*. 2nd edn. Texas: Gulf Publishing Company. 0-88415-770-9.
6. Purcell, W.R. 1949. Capillary pressures—their measurement using mercury and the calculation of permeability therefrom. Journal of Petroleum Technology 01(02). https://doi.org/10.2118/949039-G.
7. Christoffersen, K.R., and C.H. Whitson. 1995. Gas/Oil capillary pressure of chalk at elevated pressures. *SPE Formation Evaluation* 10 (03): 153–159. https://doi.org/10.2118/26673-PA.
8. Hassler, G.L., and E. Brunner. 1945. Measurement of capillary pressures in small core samples. *Transactions of the AIME* 160 (01): 114–123. https://doi.org/10.2118/945114-G.
9. Ward, J.S., and N.R. Morrow. 1987. Capillary pressures and gas relative permeabilities of low-permeability sandstone. *SPE Formation Evaluation* 02 (03): 345–356. https://doi.org/10.2118/13882-PA.

10. Cotthem, A.V., Charlier, R., Thimus, J.F., Tshibangu, J.P. 2006. Multiphysics coupling and long term behaviour in rock mechanics. In *Proceedings of the international symposium of the international society for rock mechanics*. London: Taylor and Francis Group. ISBN 0415410010.
11. Baumgarte, C., M. Thiercelin, and D. Klaus. 1999. Case studies of expanding cement to prevent microannular formation. In *SPE annual technical conference and exhibition*. SPE-56535-MS, Houston, Texas: Society of Petroleum Engineers. https://doi.org/10.2118/56535-MS.
12. De Andrade, J., S. Sangesland, R. Skorpa, et al. 2016. Experimental laboratory setup for visualization and quantification of cement-sheath integrity. *SPE Drilling and Completion* 31 (04): 317–326. https://doi.org/10.2118/173871-PA.
13. Schreppers, G. 2015. A framework for wellbore cement integrity Analysis. in *49th U.S. rock mechanics/geomechanics symposium*. ARMA-2015–349, San Francisco, California: American Rock Mechanics Association.
14. Khalifeh, M., H. Hodne, A. Saasen, et al. 2018. Bond strength between different casing materials and cement. In *SPE Norway one day seminar*. Bergen, Norway: Society of Petroleum Engineers. https://doi.org/10.2118/191322-ms.
15. Evans, G.W., and Carter, L.G. 1962. Bounding studies of cementing compositions to pipe and formations. In *The spring meeting of the Southwestern District, API divisioii of production*. New York, USA: American Petroleum Institute.
16. Becker, H., and Peterson, G. 1963. Bond of cement compositions for cementing wells. In *6th world petroleum congress*. Frankfurt am Main, Germany: World Petroleum Congress.
17. Carter, L.G., and G.W. Evans. 1964. A study of cement-pipe bonding. *Journal of Petroleum Technology* 16 (02): 157–161. https://doi.org/10.2118/764-PA.
18. Khalifeh, M., A. Saasen, H. Hodne, et al. 2017. Geopolymers as an alternative for oil well cementing applications: A review of advantages and concerns. In *International conference on ocean, offshore and arctic engineering*. Trondheim, Norway: ASME.
19. Tettero, F., Barclay, I., and Staal, T. 2004. Optimizing integrated rigless plug and abandonment—A 60 well case study. In *SPE/ICoTA coiled tubing conference and exhibition*. SPE-89636-MS, Houston, Texas, USA: Society of Petroleum Engineers. https://doi.org/10.2118/89636-MS.
20. Soter, K., Medine, F., and Wojtanowicz, A.K. 2003. Improved techniques to alleviate sustained casing pressure in a mature Gulf of Mexico field. In *SPE annual technical conference and exhibition*. SPE-84556-MS, Denver, Colorado: Society of Petroleum Engineers. https://doi.org/10.2118/84556-MS.
21. Liu, X., Nair, S.D., Cowan, M., and van Oort, E. 2015. A novel method to evaluate cement-shale bond strength. In *SPE international symposium on oilfield chemistry*. The Woodlands, Texas, USA: Society of Petroleum Engineers. https://doi.org/10.2118/173802-MS.
22. Scott, J.B. and R.L. Brace. 1966. Coated casing-a technique for improved cement bonding. In *The spring nleetlng of the Mid-Continent District. API division of production*. New York, USA: American Petroleum Institute.
23. Opedal, N., Todorovic, J., Torsaeter, M., et al. 2014. Experimental study on the cement-formation bonding. In *SPE international symposium and exhibition on formation damage control*. Lafayette, Louisiana, USA: Society of Petroleum Engineers. https://doi.org/10.2118/168138-MS.
24. Ladva, H.K.J., B. Craster, T.G.J. Jones, et al. 2005. The cement-to-formation interface in zonal isolation. *SPE Drilling and Completion* 20 (03): 186–197. https://doi.org/10.2118/88016-PA.
25. Subramanian, R. and J.J. Azar. 2000. Experimental study on friction pressure drop for non-newtonian drilling fluids in pipe and annular flow. In *International oil and gas conference and exhibition in china*. SPE-64647-MS, Beijing, China: Society of Petroleum Engineers. https://doi.org/10.2118/64647-MS.
26. Zain, Z.M., A. Suri, and M.M. Sharma. 2000. Mechanisms of mud cake removal during flowback. In *SPE international symposium on formation damage control*. SPE-58797-MS, Lafayette, Louisiana: Society of Petroleum Engineers. https://doi.org/10.2118/58797-MS.

27. McClure, J., Khalfallah, I., Taoutaou, S., et al. 2014. New cement spacer chemistry enhances removal of nonaqueous drilling fluid. Journal of Petroleum Technology 66(10). https://doi.org/10.2118/1014-0032-JPT.
28. Lecolier, E., A. Rivereau, N. Ferrer, et al. 2006. Durability of oilwell cement formulations aged in H_2S-containing fluids. In *IADC/SPE drilling conference*. SPE-99105-MS, Miami, Florida, USA: Society of Petroleum Engineers. https://doi.org/10.2118/99105-MS.
29. Lécolier, E., A. Rivereau, N. Ferrer, et al. 2010. Durability of oilwell cement formulations aged in H2S-containing fluids. SPE Drilling and Completion 25(01). https://doi.org/10.2118/99105-PA.
30. Khalifeh, M., J. Todorovic, T. Vrålstad, et al. 2017. Long-term durability of rock-based geopolymers aged at downhole conditions for oil well cementing operations. *Journal of Sustainable Cement-Based Materials* 6 (4): 217–230. https://doi.org/10.1080/21650373.2016.1196466.
31. Vrålstad, T., J. Todorovic, A. Saasen, et al. 2016. Long-term integrity of well cements at downhole conditions. In *SPE Bergen one day seminar*. SPE-180058-MS, Grieghallen, Bergen, Norway: Society of Petroleum Engineers. https://doi.org/10.2118/180058-MS.
32. ASTM-D1141-98. 2003. Standard practice for the preparation of substitute ocean water. ASTM International: West Conshohocken, PA. https://www.astm.org/DATABASE.CART/HISTORICAL/D1141-98R03.htm.
33. Dugstad, A. 2006. Fundamental aspects of CO_2 metal loss corrosion—Part 1: mechanism. In *CORROSION 2006*. NACE-06111, San Diego, California: NACE International. https://www.onepetro.org/conference-paper/NACE-06111?sort=&start=0&q=NACE+06111&from_year=&peer_reviewed=&published_between=&fromSearchResults=true&to_year=&rows=10#.
34. Taylor, H.F.W. 1992. Cement chemsitry. 1st edn. A.P. Limited: Academic. 0-12-683900-X.
35. Dieguez, E.S., A. Bottiglieri, M. Vorderbruggen, et al. 2016. Self-sealing isn't just for cracks: recent advance in sour well protection. in *SPE annual technical conference and exhibition*. SPE-181619-MS, Dubai, UAE: Society of Petroleum Engineers. https://doi.org/10.2118/181619-MS.
36. Yu, C., X. Gao, and H. Wang. 2017. Corrosion characteristics of low alloy steel under H_2S/CO_2 environment: experimental analysis and theoretical research. *Materials Letters*. https://doi.org/10.1016/j.matlet.2017.08.031.
37. Omosebi, O.A., M. Sharma, R.M. Ahmed, et al. 2017. *Cement Degradation in CO_2-H_2S Environment under High Pressure-High Temperature Conditions*. in *SPE Bergen one day seminar*. SPE-185932-MS, Bergen, Norway: Society of Petroleum Engineers. https://doi.org/10.2118/185932-MS.
38. Fakhreldin, Y.E. 2012. Durability of portland cement with and without metal oxide weighting material in a CO2/ H2S environment. In *North Africa technical conference and exhibition*. Cairo, Egypt: Society of Petroleum Engineers. https://doi.org/10.2118/149364-MS.
39. Zhang, L., D.A. Dzombak, D.V. Nakles, et al. 2014. Rate of H_2S and CO_2 attack on pozzolan-amended Class H well cement under geologic sequestration conditions. *International Journal of Greenhouse Gas Control* 27: 299–308. https://doi.org/10.1016/j.ijggc.2014.02.013.
40. Oyarhossein, M. and M.B. Dusseault. 2015. Wellbore stress changes and microannulus development because of cement shrinkage. In *49th U.S. rock mechanics/geomechanics symposium*. ARMA-2015-118 San Francisco, California: American Rock Mechanics Association.
41. Noik, C. and A. Rivereau. 1999. Oilwell cement durability. In *SPE annual technical conference and exhibition*. SPE-56538-MS, Houston, Texas: Society of Petroleum Engineers. https://doi.org/10.2118/56538-MS.
42. Labus, M., and F. Wertz. 2017. Identifying geochemical reactions on wellbore cement/caprock interface under sequestration conditions. *Environmental Earth Sciences* 76 (12): 443. https://doi.org/10.1007/s12665-017-6771-x.
43. Han, J., J.W. Carey, and J. Zhang. 2012. Degradation of cement-steel composite at bonded steel-cement interfaces in supercritical CO2 saturated brines simulating wellbore systems. In *NACE international*. NACE-2012-1075, Salt Lake City, Utah: NACE International.
44. Feng, Y., E. Podnos, and K.E. Gray. 2016. Well integrity analysis: 3D numerical modeling of cement interface debonding. In *50th U.S. rock mechanics/geomechanics symposium*. ARMA-2016-246, Houston, Texas: American Rock Mechanics Association.

45. Wang, W., and A.D. Taleghani. 2014. Three-dimensional analysis of cement sheath integrity around Wellbores. *Journal of Petroleum Science and Engineering* 121: 38–51. https://doi.org/10.1016/j.petrol.2014.05.024.
46. Bosma, M., K. Ravi, W., and van Driel et al. 1999. Design approach to sealant selection for the life of the well. In *SPE annual technical conference and exhibition*. SPE-56536-MS, Houston, Texas: Society of Petroleum Engineers. https://doi.org/10.2118/56536-MS.

Chapter 4
Types of Permanent Plugging Materials

Portland cement is the prime material used for zonal isolation and permanent P&A of wells. However, there are some concerns which persuade engineers to search for alternative materials to Portland cement. Therefore, this chapter will focus on different material types which might be used as permanent well barrier elements during permanent P&A, Table 4.1 [1, 2]. Some of these materials have already been used as permanent well barrier elements while some other have not. The criteria and requirements for the selection of material types for the permanent P&A varies for different authorities legislating P&A regulations.

4.1 Setting Materials

Throughout history, setting materials have played an important role and were used widely in the ancient world. The Romans found out that a setting material could be made which sets under water and it was used for the construction of marine structures such as harbors. Throughout time and with the development of science, different types of cementitious materials have been developed such as geopolymers, slag, and hardening ceramics [1]. The most known and studied type of cementitious materials is Portland cement.

4.1.1 Portland Cement

In 1824, Joseph Aspdin took out a patent on a setting material he produced by calcining a mixture of limestone and clay at 2640 °F. The produced material looked like Portland stone, a widely-used building stone in England, and therefore he called his invention Portland cement. Since then, different types of Portland cement have been developed for different applications. When limestone (or other materials high in calcium carbonate) and clay or shale are calcined at 2640 °F, partial fusion occurs and clinkers are produced. A few percent of gypsum (Ca_2SO_4) is added to the clinker and

M. Khalifeh and A. Saasen, *Introduction to Permanent Plug and Abandonment of Wells*, Ocean Engineering & Oceanography 12,
https://doi.org/10.1007/978-3-030-39970-2_4

Table 4.1 Different material types which might be used as permanent well barrier element [1–4]

Type	Material	Examples
1	Cements (setting)	Portland cement, pozzolanic cements, blast furnace slag-based cement, phosphate cements, geopolymers, hardening ceramics
2	In situ formation	Shale, salt, claystone
3	Grouts (non-setting)	Unconsolidated sand or clay mixtures, bentonite pellets, barite plugs, calcium carbonate
4	Thermosetting polymers and composites	Resins, epoxy, polyester, vinylesters, including fiber reinforcements, urethane foams, phenol
5	Thermoplastic polymers and composites	Polyethylene, polypropylene, polyamide, Polytetrafluoroethylene (PTFE), Polyether Ether Ketone (PEEK), Polyphenylene Sulfide (PPS), Polyvinylidene Fluoride (PVDF), and polycarbonate, including fiber reinforcements
6	Metals	Steel, other alloys such as bismuth-based materials
7	Modified in situ materials	Barrier materials made from in situ casing and/or formation through thermal or chemical modification
8	Elastomeric polymers and composites	Natural rubber, neoprene, nitrile, Ethylene Propylene Diene Monomer (EPDM), Fluoroelastomer (FKM), Perfluoroelastomer (FFKM), silicone rubber, polyurethane, PUE and swelling rubbers, including fiber reinforcements
9	Gels	Polymer gels, polysaccharides, starches, silicate-based gels, clay-based gels, diesel/clay mixtures
10	Glass	

the blend is finely ground to make the cement. The gypsum controls the setting rate and can be replaced by other forms of calcium sulphate [5]. The major components of clinker are approximately 67% CaO, 22% SiO_2, 5% Al_2O_3, 3% Fe_2O_3, and 3% of other components.[1] The clinker mainly contains four major phases: alite, belite, aluminate phase, and ferrite phase. The alite phase is tricalcium silicate ($3CaO{\cdot}SiO_2$ or "C_3S") and constitutes 50–70% of normal Portland cement clinkers. The belite phase is dicalcium silicate ($2CaO{\cdot}SiO_2$ or "C_2S") and constitutes 15–30% of normal Portland cement clinkers. The aluminate phase is a tricalcium aluminate ($3CaO{\cdot}Al_2O_3$ or "C_3A") and represents 5–10% of the most normal Portland cement clinkers. The ferrite phase is a tetracalcium aluminoferrite ($4CaO{\cdot}Al_2O_3Fe_2O_3$ or "C_4AF") and

[1]Clinker for construction cement.

represents 5–15% of normal Portland cement clinkers. Although there are several other phases such as alkali sulfates and calcium oxides, they represent minor amounts.

API categorizes the identified cements into nine different classes [6]:

- *API Class A*: This an ordinary Portland cement which is intended for use from surface to 6000 (ft) depth. It is not sulfate resistant and may be used when no special properties are required and well conditions allow.
- *API Class B*: It is ordinary Portland cement which is intended for use from surface to 6000 (ft) depth. It is available both as moderate and high sulfate-resistant.
- *API Class C*: It is called high early strength cement and used where early strength cement is required. It is intended for use from surface to 6000 (ft) and available as ordinary, moderate, and high sulfate-resistant types.
- *API Class D*: It is a retarded cement type which is intended for use from 6000 to 10000 (ft) depth and under conditions of moderate to high temperatures and pressures. It is available both as moderate and high sulfate-resistant types.
- *API Class E*: This is a retarded cement which is intended for use from 10000 to 14000 (ft) depth and conditions of extremely high temperatures and pressures. It is available both as moderate and high sulfate-resistant types.
- *API Class F*: It is intended for use from 10000 to 16000 (ft) depth and conditions of ultra-high pressures and temperatures. This class is available both as moderate and high sulfate-resistant types.
- *API Class G*: Intended for use as basic cement from surface to 8000 (ft) depth. It is manufactured in such a way that accelerators or retarders can be used to cover a wider range of well depths and temperatures. This class is available both as moderate and high sulfate-resistant types.
- *API Class H*: It is intended for basic cement use from surface to 8000 (ft) depth and can be used with retarders and accelerators to cover a wide range of well depths and temperatures. It is only available as moderate sulfate-resistant type.
- *API Class J*: It is known as special order only and intended for use from 12000 to 16000 (ft) depth. This class is for ultra-high pressure and temperature conditions and with accelerators or retarders wider ranges of well depths and temperatures can be covered.

Table 4.2 tabulates the properties of common oil well cements identified and classified by API. Cement class D, E, and F are seldom used for oil well cementing. The cement classes G and H are now the most common.

Table 4.3 presents typical physical properties of the various API classes of cement, given in Table 4.2, and cured at different pressures and temperatures.

There are some cementitious materials which have been or are used in oil well cementing effectively but they do not fall into any specific API category. These materials include [7]: (a) pozzolanic-Portland cements, (b) pozzolan-lime cements, (c) resins or plastic cements, (d) gypsum cements, (e) diesel oil cements, (f) expanding cements, (g) calcium aluminate cements, (h) latex cement, and (i) cement for permafrost environments.

Pozzolanic-Portland cements—This is a kind of blended cement which is produced by either intergrinding ordinary Portland cement clinker with gypsum and

Table 4.2 API classification and properties of common oil well cements [7]

Type	Range of usage (ft)	Static temp. (°F)	Water ratio (gal/sk)	Slurry weight (lb/gal)	Volume (ft^3/sk)	Remarks
Class A (Portland cement)	6000	60°–170°	5.2	15.6	1.18	No sulfate resistance. May be used well conditions allow
Class B (Portland cement)	6000	60°–170°	5.2	15.6	1.18	Moderate sulfate resistance
Class C (High early strength)	6000	60°–170°	6.3	14.8	1.32	Available in ordinary, moderate, and high sulfate-resistance types
Class G (Basic cement)	8000	200°	5.0	15.8	1.15	Compatible with accelerators or retarders for usage to cover the classes A through E
Class H (Basic cement)	8000 8000	200° 200°	4.3 5.2	16.4 15.6	1.06 1.18	Higher density, higher and lower water volume

pozzolanic materials or preparing each part separately and then blending them. Pozzolans are either natural or artificial reactive siliceous materials, processed or unprocessed, which start to hydrate in the presence of lime and water and develop cementitious properties. The source of most natural pozzolanic materials are volcanic ashes. The artificial pozzolans are produced by calcination of natural siliceous materials such as for example: clays, shales, rice husk ash, and certain siliceous rocks [8]. Fly ash is one of the artificial pozzolanic materials which is a combustion by-product of coal. In oilwell cementing, fly ash is added to cement to improve its strength and water-tightness.

Pozzolan-lime cement—Silica-lime or pozzolan-lime cements are blends of siliceous materials (e.g. fly ash), hydrated lime, and small quantities of a chemical activator (e.g. calcium chloride), which hydrate with water to produce calcium silicate. Their reaction rate is very slow at low temperatures compared to Portland cements. Therefore, these type of cements are recommended for wells where moderate to high temperatures are encountered. The use of these materials is not recommended in wells where temperature is less than 140 °F [9]. The reaction can be either accelerated or retarded by use of additives to cover a wide range of well conditions. Light weight, strength stability at high temperatures, low slurry cost, and less CO_2 emission are features of Pozzolan-lime cements.

Table 4.3 Typical physical properties of the various API classes of cement [7]

Properties of API classes of cement							
			Class A	Class C	Classes G and H	Classes D and E	
Specific gravity (average)			3.14	3.14	3.15	3.16	
Surface area (range), cm^2/g			1500–1900	2000–2800	1400–1700	1200–1600	
Weight per sack, lbm			94	94	94	94	
Bulk volume, ft^3/sk			1	1	1	1	
Absolute volume, gal/sk			3.6	3.6	3.58	3.57	
Temperature (°F)	Pressure (psi)		Portland	High early strength	API Class G	API Class H	Retarded
			Typical compressive strength (psi) at 24 h				
60	0		615	780	440	325	a
80	0		1470	1870	1185	1065	a
95	800		2085	2015	2540	2110	a
110	1600		2925	2705	2915	2525	a
140	3000		5050	3560	4200	3160	3045
170	3000		5920	3710	4830	4485	4150
200	3000		a	a	5110	4575	4775
			Typical compressive strength (psi) at 72 h				
60	0		2870	2535	–	–	a
80	0		4130	3935	–	–	a
95	800		4670	4105	–	–	a
110	1600		5840	4780	–	–	a
140	3000		6550	4960	–	7125	4000
170	3000		6210	4460	5685	7310	5425
200	3000		a	a	7360	9900	5920
Depth (ft)	Temperature (°F)						
	Static	Circulating	*High-Pressure thickening time (hours: minutes)*				
2000	110	91	4:00+	4:00+	3:00+	3:57	a
4000	140	103	3:26	3:10	2:30	3:20	4:00+
6000	170	113	2:25	2:06	2:10	1:57	4:00+
8000	200	125	1:40[a]	1:37[a]	1:44	1:40	4:00+

[a]Not generally recommended at this temperature

Resins or plastic cements—They are a mixture of either an API class A, B, G, or H cement with a mix of water liquid resins and a catalytic converter. These types of cements are used in small volumes for plugging open holes, and squeezing perforations. The recommended range of temperature is between 60 and 200 °F.

Gypsum cements—These types of cements are a mixture of API class A, G, or H cement with 8–10% of gypsum. The gypsum can be either in the hemihydrate form ($CaSO_4 \cdot ½ H_2O$) or the dihydrate form ($CaSO_4 \cdot 2H_2O$). The gypsum cements set rapidly and they have high early strength with positive expansion properties. The expansion is in the range of 0.3%. Gypsum cements are used in lost circulation zones. The associated challenge with the use of gypsum cements is water solubility of the hardened cement and they can therefore only be used in wells where water does not exist. A solution for minimizing the water solubility of the hardened gypsum cements is to use equal volume of cement and gypsum [7].

Diesel oil cements—The diesel oil cements have been developed to selectively block off unwanted water production during drilling or in producing wells [10]. During the slurry design, an API cement class A, B, G, or H is mixed with diesel oil with a surface-active agent. These types of cements have unlimited pumping time as long as water does not meet the slurry. Water-in-oil emulsion cements are another type of diesel oil cement in which cement is mixed in a liquid phase consisting of oil as external or continuous phase and existing water as droplets. Low-fluid-loss characteristic, less damage to the oil-producing zones, and less damage to water-sensitive formations are some of the advantages of diesel oil cements and water-in-oil emulsion cements [11, 12].

Expanding cements—One of the drawbacks of Portland cement is its shrinkage which can create microannuli. Some degree of expansion can compensate the wellbore stress changes and improve the hydraulic and shear bond strengths [13]. Therefore, expansive cements or additive agents such as sodium sulfate (Na_2SO_4), pozzolans, anhydrous calcium sulfoaluminate ($4CaO \cdot 3Al_2O_3 \cdot SO_3$), calcium sulfate ($CaSO_4$), and lime are introduced to the slurry. When sulfates and calcium aluminate components of cement are present, ettringite crystals, which are a hydrous calcium aluminum sulfate mineral are formed. As a result of the formation of the mineral crystals, pressure is developed and is the main expansion mechanism [14].

Calcium aluminate cements—These types of cements are known as high-alumina cements in which bauxite (aluminum ore) or other aluminous materials and limestone are heated in a furnace to be liquefied. Compared to Portland cement, these types of cements are low in silica content [5]. The calcium aluminate cements have high resistance to corrosive environments, rapid hardening properties, and are stable at high temperatures [15]. Addition of Portland cement to calcium aluminate cements results in flash setting.

Latex cement—Latex is commonly used to control gas migration, fluid loss and enhance the bonding properties of cements [16]. A latex cement is a blend of classes A, G, or H cement with latex either in liquid or powder form. The identified latexes are polyvinyl acetate, polyvinyl chloride, and butadiene styrene emulsions [7]. Butadiene styrene latexes are commonly used in oil well cementing; however, they are sensitive to temperature, mechanical energy, and free ions. As the latex has consisted of charged particles, latex demulsifies and precipitates in the presence of salts such as sodium chloride and calcium chloride [17]. One mitigation is the use of an anionic surfactant as an additive to the cement slurry to stabilize the latex cement in the presence of salts. Some investigators have shown the anti-corrosion ability of latex cements [18].

Cement for permafrost environments—Cementing conductor and surface casing in sub-freezing zones is a challenging task as Portland cement freezes, or never sets,

Table 4.4 Compressive strength of gypsum cement for sub-freezing temperatures [21]

			Curing temperature, 20 °F				
Pumping time (h)	Sodium Chloride, Percent[a]	Water Ratio (Ft^3/sk)	4 h	1 Day	3 Days	7 Days	14 Days
			(psi)				
2	10	048	470	855	615	600	1095
			Curing temperature, 15 °F				
				4 h	6 h	8 h	24 h
				(psi)			
2	10	048		345	530	635	545
3	10	048		38	ND	530	632
3	18	048		195	540	555	690

[a]By weight of mixing water
ND—Not determined

or permafrost melts as results of hydration heat [19]. There are some mitigations such as adding calcium chloride salt, short-chain alcohols such as methyl, propyl, or isopropyl to depress the freezing-point [20]. However, adding calcium chloride accelerates the hydration reaction and causes fast setting. Therefore, when calcium chloride is used to depress the freezing-point, retarders are used to postpone the setting time. There are four different types of blended cements available for permafrost environments; classes A and G API cements with calcium chloride, calcium aluminate cements with fly ash, refractory cements, and gypsum-cement blends [21]. Of these four, the gypsum-cement blends and refractory cements are mainly used in sub-freezing environments. Shryock and Cunningham [21] measured pumpability and compressive strength development of a gypsum-cement blend prepared for permafrost zones and found recipes that are applicable in permafrost areas, Table 4.4.

4.1.1.1 Durability

A plugging material is intended to withstand downhole conditions in an eternal perspective. Therefore, as eternal perspective is impossible, use of long-term durability knowledge of intended plugging materials is important. It is necessary to expose plugging materials to downhole chemicals, at downhole conditions, for different time intervals; sometimes up to a few years [22–25]. Then, the mechanical properties of plugging materials are characterized at different time intervals. Unfortunately, there is no international standard describing testing of plugging materials to qualify them for an eternal perspective. Thus, different researchers have selected different chemicals with different dosages to study the degradation of oilwell cements at downhole conditions. Vralstad et al. [25] aged neat class G cement, at downhole conditions, by exposing the samples to brine, crude oil, and H_2S dissolved in brine for different

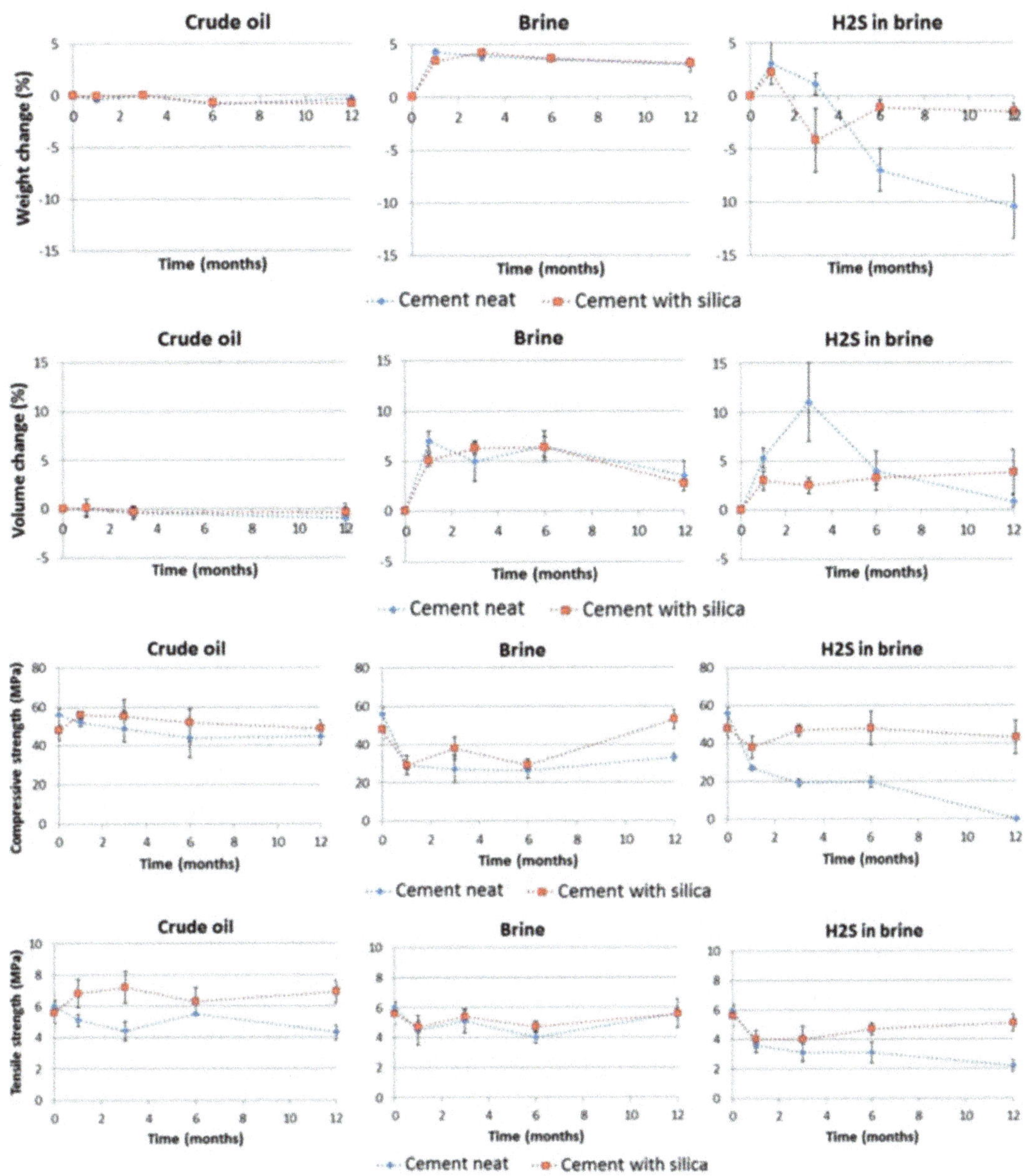

Fig. 4.1 Durability of neat class G cement exposed to downhole chemicals [25]

time intervals, up to 12 months. They also studied the weight and volume changes of the cement. Figure 4.1 shows the obtained results by Vrålstad et al. [25].

4.2 In Situ Formation (Formation as Barrier)

In some hydrocarbon fields, traditional sonic logs and ultrasonic azimuthal bond logs provide information of good bonding above the theoretical top of cement and at depths where there is no cement or where a poor cement job has been reported.

In addition, performed extended leak off (XLOT) tests show qualified seals [26]. The question is how this could be happened? In all of these cases, the used drilling fluids were known to have no setting properties. In addition, no casing collapses were reported. The only main remaining parameter is in situ formations which may have the potential to move toward or expand into the annular space and create a good seal, see Fig. 4.2. If the in situ formation moves and creates a good seal with sufficient strength, then it can be regarded as a most suitable permanent plugging material; as it has been standing in the overburden with intact long-term durability.

In order for the formation to move and create a good seal in the annular space, it should deform. *Deformation* is defined as changes in the shape or position of a rock body in response to stress. Stress is the internal resistance of rocks against the forces applied to deform the rock. Stress can be divided into two different stress types; *confining stress*, and *differential stress*. Whenever stresses acting on a rock are larger than its strength, rock can experience four different phenomena; folding, flowing, fracturing, or faulting. In a confining stress scenario, rock experiences a uniform stress in all direction. As a result of the acting confining stress, rock can expand or contract. Consider an in situ formation adjacent to an uncemented annulus which experiences equal stresses from overburden and the annulus fluid. The formation would not be able to move toward casing to seal the uncemented interval, see Fig. 4.3. So, confining stress is not the interest of this section and therefore, it is not discussed further. In differential stress scenarios, rocks experience unequal stresses in different directions. The resultant force acting on a rock may cause a *compressional stress*, *tensional stress*, or *shear stress*. Compressional stress is applied inward and triggers the rock to be squeezed, see Fig. 4.4a. Tensional stress is an outward stress which acts on a rock, the rock is pulled and subsequently it is elongated (see Fig. 4.4b). Shear stress is applied from one direction and cause movement of one part of the rock to pass the other which is still, (see Fig. 4.4c).

When the differential stresses act on a rock sample over time, deformation occurs. The deformation can be reversible or irreversible. *Reversible deformation* is a temporary shape change that is self-reversing after load is removed. In other words, the rock

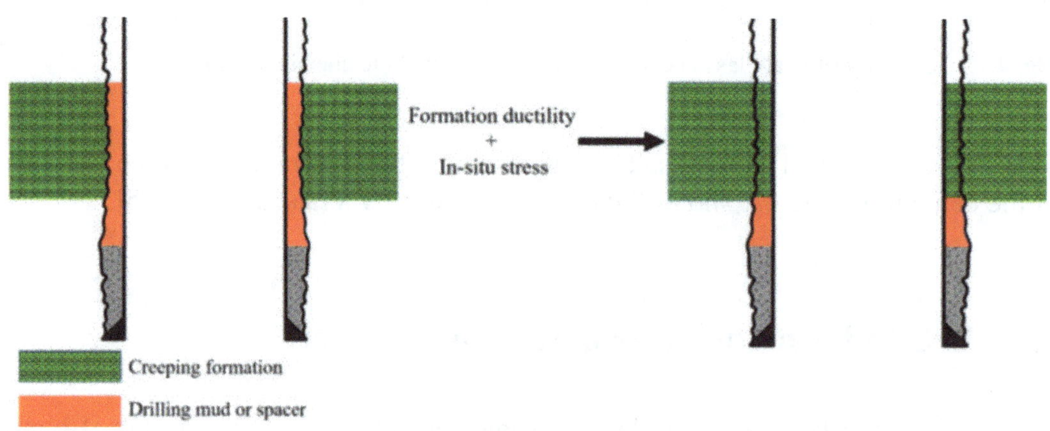

Fig. 4.2 In-situ formation moves toward the casing and create a seal

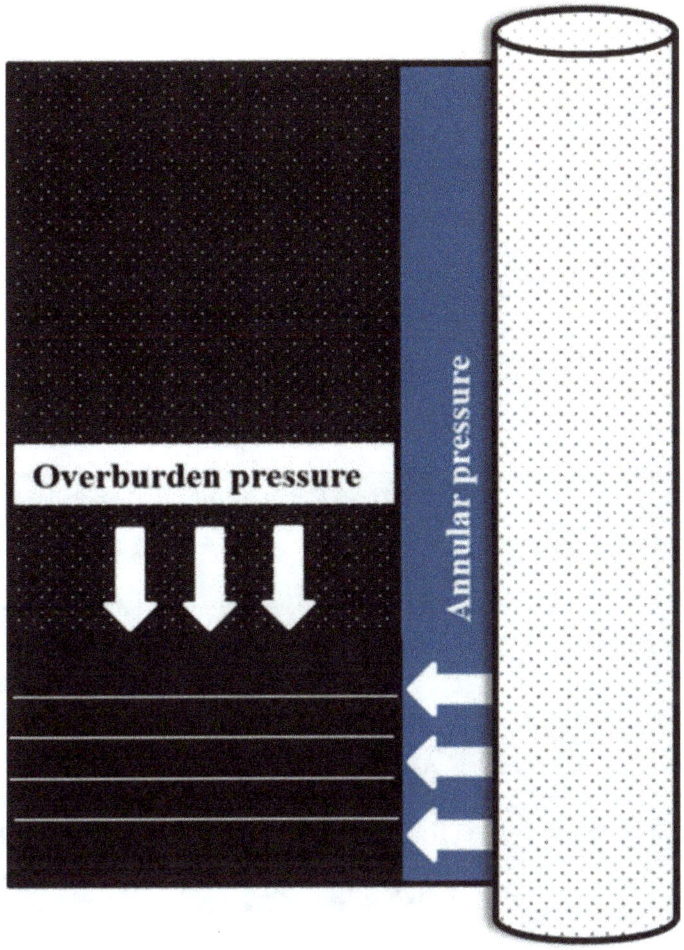

Fig. 4.3 Equal stresses acting on a rock from all directions; overburden and annulus

returns to its original shape when the acting forces are removed. This type of deformation is known as *elastic deformation*. Elastic deformation happens at low levels of stress and it is recoverable after the stress is removed. In elastic deformation, the bonds between individual atoms and lattices are stretched, allowing the material to deform. Irreversible deformation is a permanent shape change that is not reversible when the load is removed. In other words, the rock does not return to its original shape when the acting forces are removed. The irreversible deformation is known as *plastic deformation*. In plastic deformation, the applied stress on a material cause microscopic dislocations such as edge and screw dislocations in the material lattices.

As we discussed earlier, when a differential stress is applied on a rock and the stress is higher than the rock strength, the rock deforms. The change in length of the rock, caused by stress, is called *strain*. In rock mechanics, a Stress-Strain diagram is plotted to measure the mechanical properties of rocks including elastic and plastic deformation limits. On a stress-strain diagram, when a stress is applied on a rock, strain behaves proportionally to the applied stress. The region where the stress-strain plot corresponds to an elastic deformation is linear, see Fig. 4.5. The slope is known as *modulus of elasticity, E*:

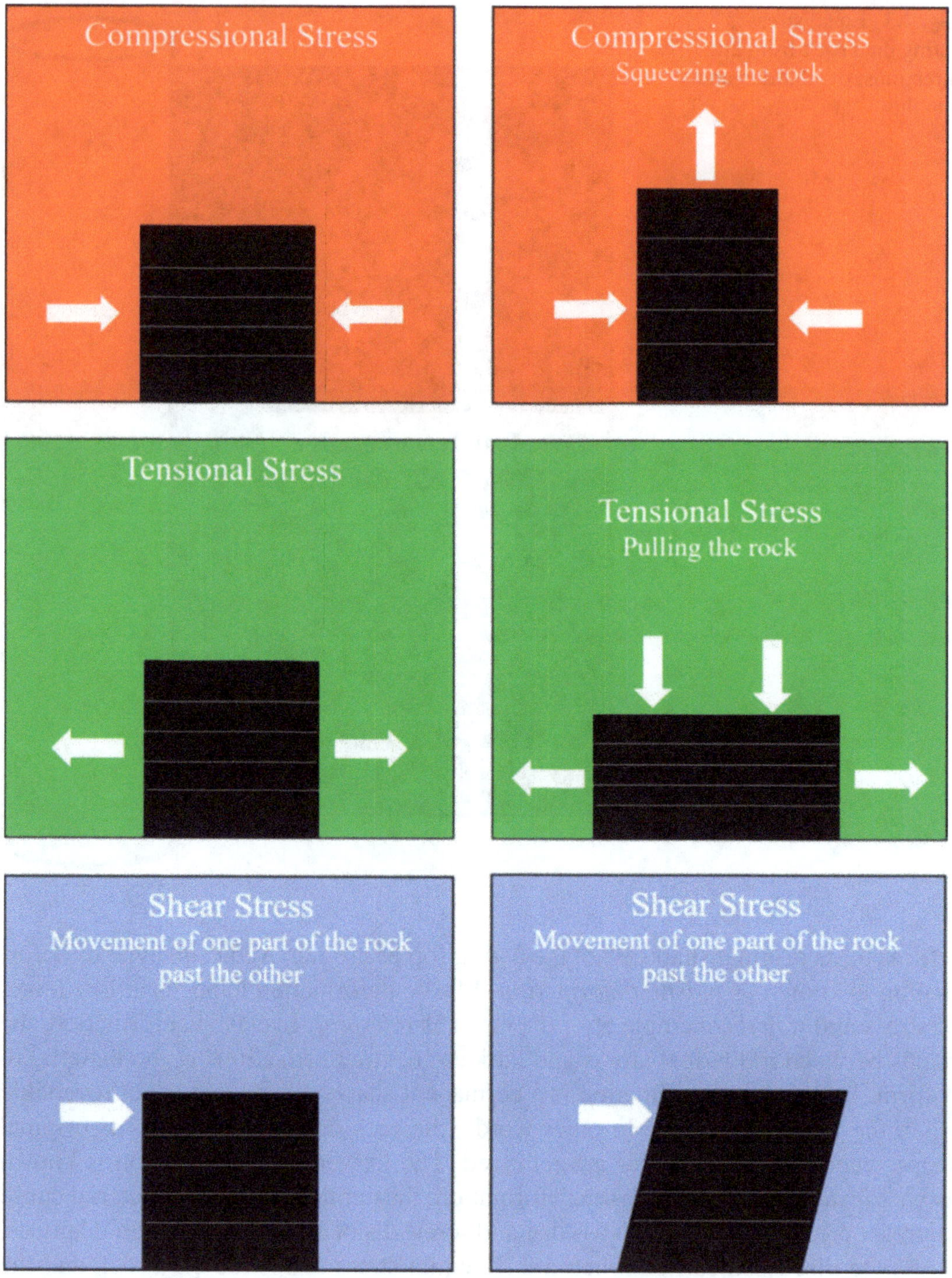

Fig. 4.4 A rock sample which experiences unequal stresses from different directions; **a** compressional stress, **b** tensional stress, and **c** shear stress

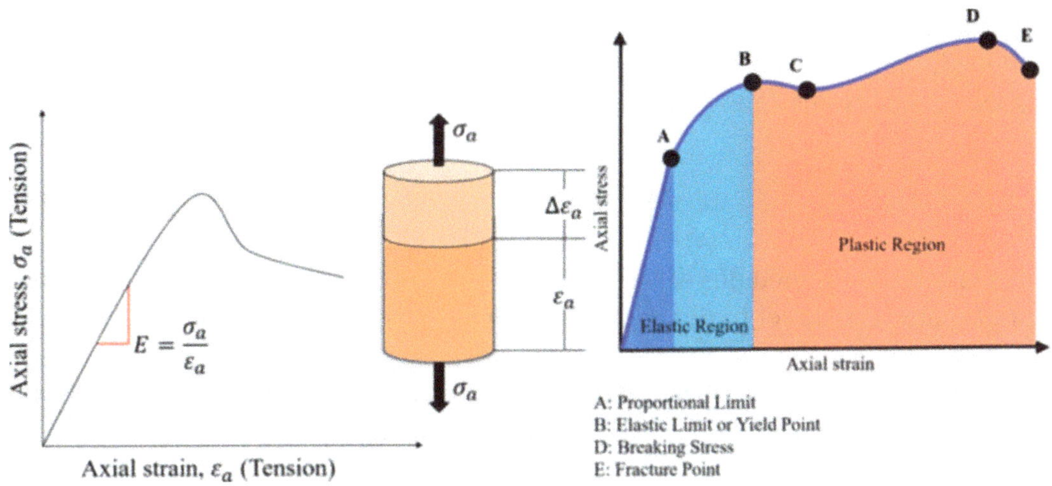

Fig. 4.5 A stress-strain diagram showing the elastic region (red) and plastic region (green)

$$Stress = E \times Strain \tag{4.1}$$

The stress where deformation shifts from elastic to plastic, is called the *yield stress*. The *yield strength* is the stress that is required to cause plastic deformation. In the plastic region, the stress-strain relationship is not linear and the material deforms much more rapidly compared to the elastic region, (see Fig. 4.5).

When a material experiences compressive stress or tensile stress, stress exists through the object, it can be elongated or shortened (see Fig. 4.6). The change in length of the material is estimated by Young's modulus.

$$\Delta l = \frac{1}{E} \times \frac{F}{A} \times l_0 \tag{4.2}$$

where E is the Young's modulus of the material, F is the applied force, A is the cross section where the force is applied, and l_0 is the initial length of the material.

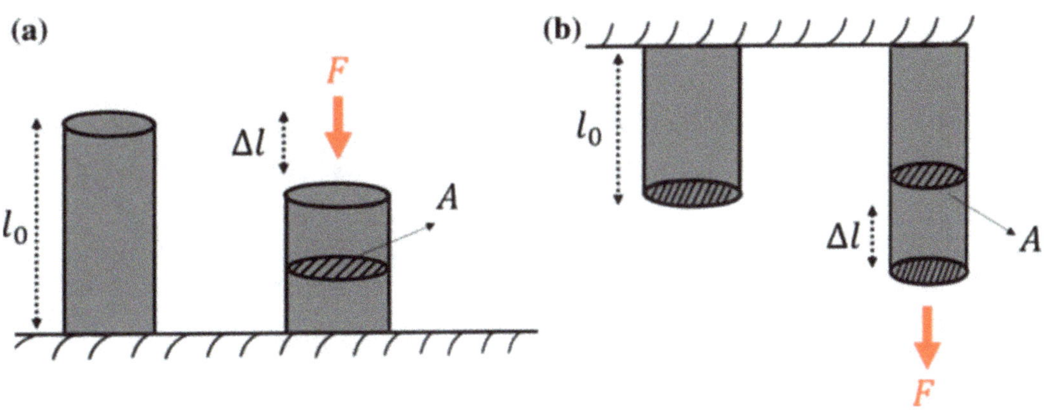

Fig. 4.6 Changes in length of a sample; **a** experiencing a compressive stress, **b** tensile stress

Shear modulus—When a material experiences a shear stress, its length can be shortened, Fig. 4.7. The length changes caused by a shear stress is characterized by its shear modulus, also known as modulus of rigidity:

$$\Delta l = \frac{1}{G} \times \frac{F}{A} \times l_0 \tag{4.3}$$

where G is the shear modulus of material, F is the applied shear force, A is the surface area of material parallel to direction of the applied shear force, and l_0 is the initial length of sample.

Bulk modulus—When a material experiences confining forces, from all directions, its volume can be reduced (Fig. 4.8). The volume change of the material is given by its bulk modulus:

$$\Delta V = -\frac{1}{B} \times \frac{F}{A} \times V_0 \tag{4.4}$$

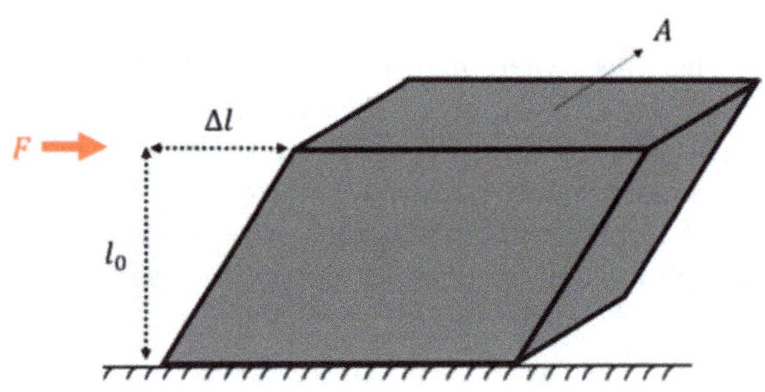

Fig. 4.7 Shear stress acting on a sample

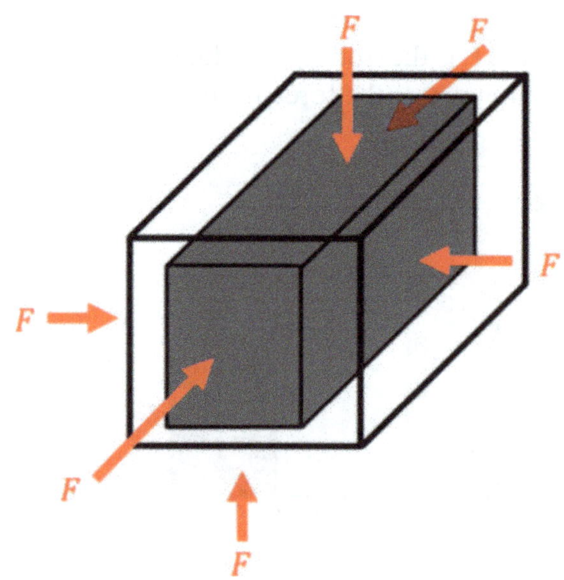

Fig. 4.8 Compressional forces acting on a material, equally, from all directions

where B is the bulk modulus of material, F is the confining pressure, A is the surface which force is applied on, and V_0 is the initial volume of the material. The volume change can also be expressed in form of applied pressure or pressure change:

$$\Delta V = -\frac{\Delta P}{B} \times V_0 \tag{4.5}$$

Poisson's ratio—When a material experiences longitudinal stress from one direction, it will experience lateral strain. Therefore, the material will contract in one direction while elongating in a perpendicular direction, Fig. 4.9. The ratio of transverse contraction strain to longitudinal extension strain, in the direction of stretching force, is known as Poisson's ratio, ϑ, given by:

$$\vartheta = -\frac{d\varepsilon_{tran}}{d\varepsilon_{axial}} \tag{4.6}$$

where $d\varepsilon_{tran}$ is transverse strain (lateral strain), and $d\varepsilon_{axial}$ is axial strain (longitudinal strain). Poisson's ration of materials is $0 \le \vartheta \le 0.5$.

Failure of a material can occur in elastic region or plastic region of the axial stress-axial strain curve. If the failure occurs in the elastic region, it is known as brittle failure and if it occurs in the plastic region, it is known as a ductile failure, Fig. 4.10.

In order to utilize in situ formations, which flow toward the annulus behind the casing, as a permanent well barrier element, the in situ formation should deform plastically and create a good seal. The plastic deformation which is time, stress and temperature dependent is known as *creep*. The creep, in rocks, is a slow deformation which can normally take long time. So it could be said that when in situ formation creeps toward the casing, it can create a seal and the seal can be used as a permanent plugging material.

If in situ formations experience a counter force from the annular pressure which is higher than the overburden stress, creep will not occur. Therefore, a formation barrier

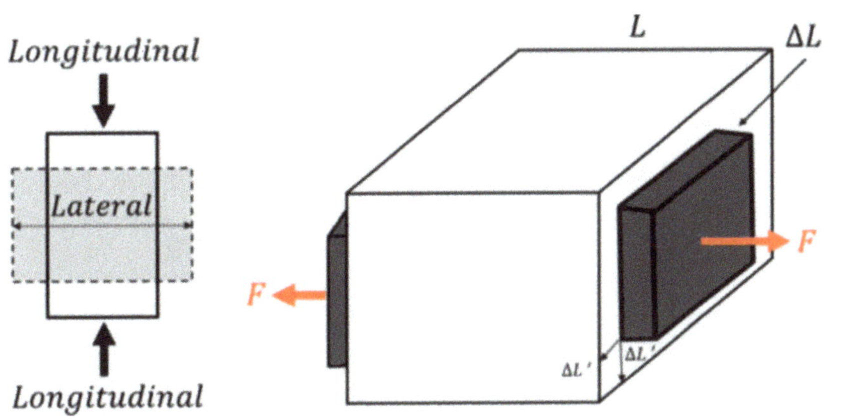

Fig. 4.9 Lateral strain and longitude strain acting on a sample

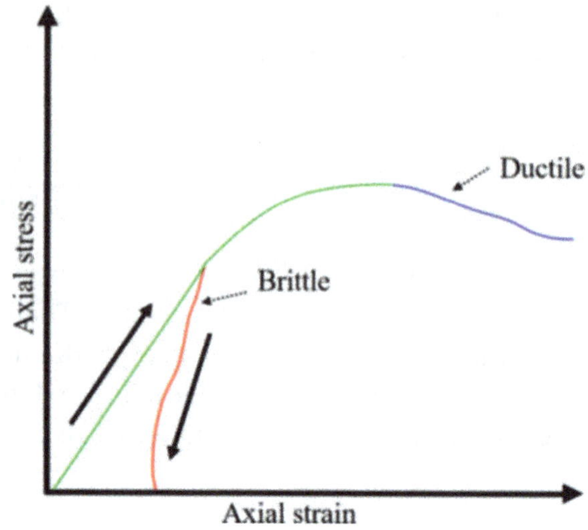

Fig. 4.10 Brittle failure versus ductile failure

diagram is defined, Fig. 4.11. If the resultant stress of annular pressure and in situ stress falls in the "closed gap" region, a seal is created. However, if the resultant force falls in the "opened gap" region, then the annulus remains open and no seal is created.

It is important to differentiate between two phenomena; creep and swelling. Swelling is caused by the hydration of shales; heterogeneous porous media which has a matrix mainly of clays. Different theories have been presented as the driving mechanisms for swelling which include capillary pressure, hydraulic pore pressure imbalance, osmosis pressures, and the polar attraction of water molecule by the charged clay surfaces within the shale matrix [28–30]. According to the last theory, when water molecules move into a saturated shale body which is under constant contraction stress, the total volume of the body increases. Therefore, swelling strains develop at boundaries of clay layers. The swelling phenomenon is a reversible process upon dehydration. In other words, swelled shale is contracted as water molecules are drained. So, if swelling shale creates a seal in the annular space behind casing, it may not be a suitable material to be selected as a permanent plugging material.

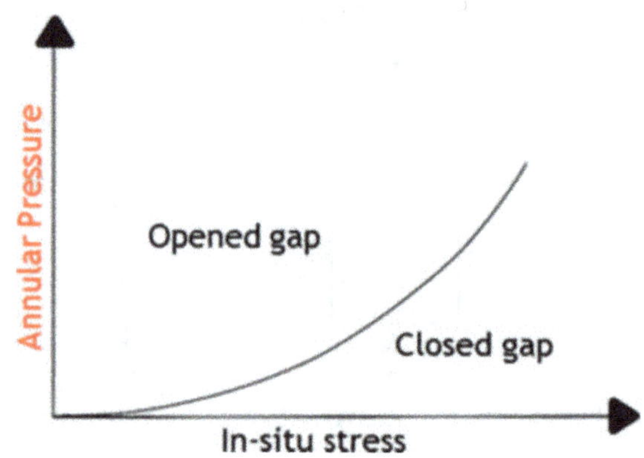

Fig. 4.11 Formation barrier map [27]

It is worthwhile to mention that not every formation creeps naturally but some do. As an example, in the Norwegian sector of the NCS a creeping formation has been reported in Statfjord A and Grane fields; some degrees of formation creep has been observed in all the fields North Sea and Norwegian Sea. However, no naturally creeping formation has been reported in the central/eastern Barents Sea, see Fig. 4.12.

Six different displacement mechanisms are suggested as driving mechanisms of formation movements [26, 31]:

- Shear or tensile failure,
- Compaction failure,
- Liquefaction,
- Thermal effect or expansion,
- Chemical effects,
- Creep.

Shear or tensile failure—Whenever exerted pressure by annular fluids is lower than the in situ overburden pressure, the formation falls into an unstable condition.

Fig. 4.12 An overview of creeping formation and non-creeping formation on the Norwegian sector of the NCS [32]

If the differential pressure is high enough, then the formation experiences shear or tensile failure. Annular pressure drop is a common phenomenon in post-construction periods as the mud density of the annular fluid is reduced over time due to segregation of heavier components in the mud. Bond logs from some wells have shown that there are some lithological layers, between advancing formations, which have not moved. However, upper and lower formations behind have moved in the same well [26]. Thus, this driving mechanism alone is not the driver for formation movement.

Compaction failure—When a porous rock experiences a high hydrostatic pressure or pore pressure change, the grains may loosen or break. The movement of grains into the open spaces, which can be regarded as reorientation, results in a closer packing. This process is known as compaction failure [31]. Compaction failure is common in highly porous rocks such as sandstones. This mechanism is believed to be the subsequent response to movement and not the triggering process.

Liquefaction—Generally, any process that causes a non-liquid phase to behave in accordance with fluid dynamics is termed *liquefaction*. In rock mechanics, when a highly porous rock, which is loose (uncompacted), is fully or partially saturated and substantially loses its strength and starts to flow in response to any applied stresses, the process is called liquefaction. Recorded bond logs have shown good bonding which means that a solid material fills the annular space. Therefore, liquefaction is not the driving mechanism.

Thermal effect or expansion—Generally, increasing the temperature eases rock movement and causes some degree of expansion. However, during the production life of wells, the temperature change is small and besides, formation movement has been recorded in shallow depths where temperatures are not high. So, thermal effect cannot be the triggering mechanism.

Chemical effects—The movement of formation toward casing has been recognized in different wells which have been drilled with oil-based muds and water-based muds. So, the chemical effect cannot be the major contributor or be regarded as the main driving mechanism.

Creep—It is a time dependent parameter and happens over a long time. In some fields, the process of rock movement has been observed to be slow, however, in some wells is has been relatively quick. It can be concluded that creep can be regarded as the main mechanism in some fields while in others a combination of creep and shear failure can simultaneously be the driving mechanisms.

So far, in the North Sea, the lithology of identified creeping formations ranges in age from Oligocene (upper Tertiary) to Upper Jurassic. Another example of the use of creeping formation as well barriers is in the Gulf of Mexico where salt formations are used commonly as an exterior barrier [32].

As utilization of creeping formations is cost effective and a safe method, researchers have pursued an understanding of the mechanisms and conditions at which formations may start to creep. Studies show that the following parameters may activate or accelerate the formation creep: thermal treatment, chemical activation, changing annular fluid, and sudden pressure drop. Of these, sudden pressure drop has shown the potential to be a swift activation mechanism [27]. Changing the annular fluid may create a swelling effect and therefore, it might be of less interest.

Table 4.5 Advantages and possible limitations for use of creeping formation as annular barrier

Advantages	Possible limitations
• No section milling is required • Cost effective P&A method • Reliable and durable plugging material • No HSE issue • Durable	• Not every formation creeps • Activation mechanisms of non-creeping formations are not well understood • Required length of plug defined by authorities • Qualification method is not clear • Artificially activation of formation may compromise its mechanical properties and long-term durability

4.2.1 Durability

Formations have shown their long-term durability as cap rock. Therefore, no research work is performed to analyze the durability of creeping formations. However it is important to note that the artificial activation of a formation to creep may change its properties. Therefore, it is recommended to investigate the mechanical properties and the long-term integrity of activated formations. There are several advantages associated with utilization of formations as annular barriers, but also some limitations which are listed in Table 4.5.

4.3 Non-setting (Grouts)

Permanent plugging materials can be subjected to different stress scenarios during abandonment and post-abandonment such as tectonic stress, reservoir compaction, and temperature changes. Portland cement is the prime material used for permanent P&A. However, the associated concerns regarding Portland cement include but are not limited to brittleness, shrinkage, gas migration through the bulk material, long-term degradation by exposure to high temperatures and chemical substances, persuades researchers to seek for alternative materials. As solidified materials have some degree of brittleness, engineered unconsolidated materials have been suggested as an alternative plugging material [33]. These types of materials are also known with other names such as *grouts*, and *non-setting* materials [1]. Of these materials, unconsolidated sand or clay mixtures [33], bentonite pellets [34], calcium carbonate [35], and barite plugs [36] are the most well-known.

4.3.1 Unconsolidated Sand Slurries

Unconsolidated sand slurries consist of two phases; solid phase and liquid phase. The solid phase is sand with engineered Particle Size Distribution (PSD). The liquid phase, known as the conducting fluid, is an inert fluid consisting of water, a small amount of dispersant and viscosifier which provides pumpability of the mixture. The mixture usually consists of approximately 75% solid and 25% of carrier fluid by volume [33], with a density of 17.9 (ppg). The slurry is a Bingham-plastic material which behaves as a rigid body at low stresses but flows at high stresses, Fig. 4.13.

The particles are kept packed by the electrostatic forces (Zeta potential[2]) between the solid particles and the conducting fluid [37]. Figure 4.14 shows a commercialized unconsolidated sand slurry which has been used for both temporary abandonment and permanent abandonment operations.

Figure 4.15 presents Scanning Electron Microscopy (SEM) image of a dehydrated unconsolidated sand slurry; different particle sizes are obviously visible.

An unconsolidated sand slurry does not set after placement and subsequently it does not shrink. As the material does not solidify, any introduced stress cannot fracture the material. When the downhole shear forces exceed the material limit, the material starts to flow and shear forces are reduced below the yield strength. This eventuates reshaping the material and the whole process is mechanical. As the unconsolidated sand slurry is made of quartz, it is thermodynamically stable and in the absence of carrier fluid, the plug remains homogenous. Table 4.6 presents advantages and possible limitations associated with unconsolidated sand slurries with regards to utilization for permanent P&A.

As the slurry does not set, the material could be used for zonal isolation and also during well construction provided a solid foundation exists. In addition, permeability of the plug cannot be measured directly, therefore, the Blake-Kozeny model is used [38]:

$$k = \frac{\varepsilon^3}{(1-\varepsilon)^2} \frac{d_p^2}{150} \tag{4.7}$$

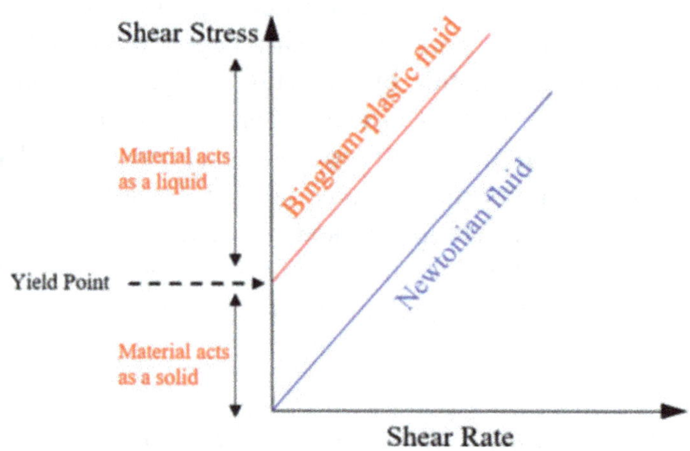

Fig. 4.13 Flow behavior of a Bingham-plastic material

[2]The potential difference between the surface of solid particles immersed in a conducting liquid.

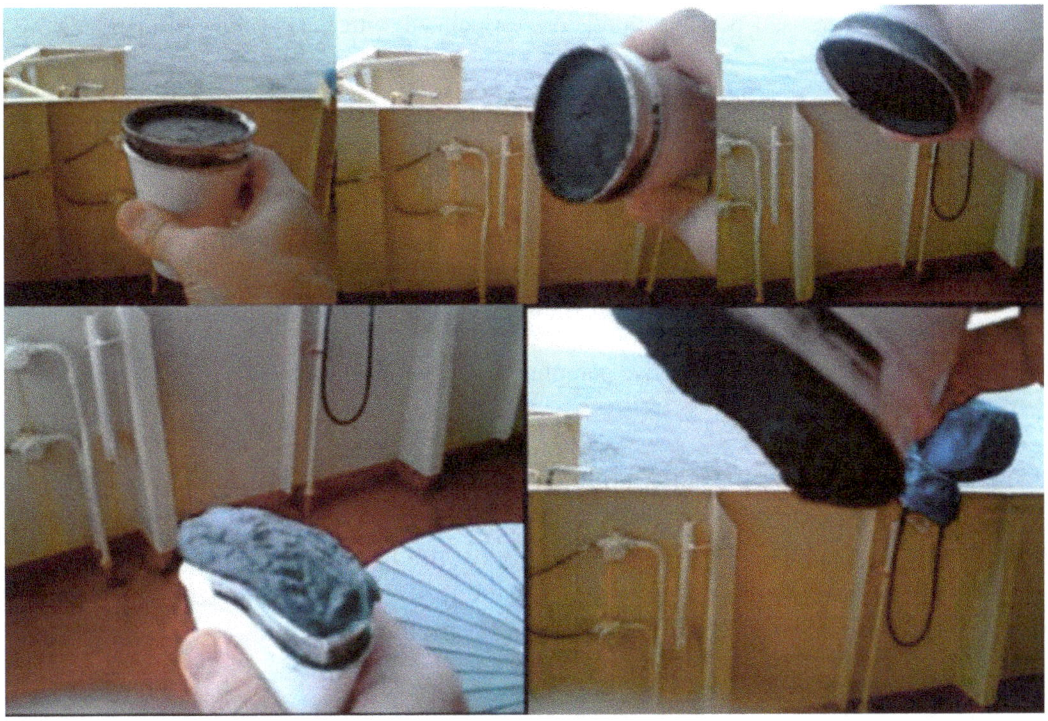

Fig. 4.14 Unconsolidated sand slurry as an alternative plugging material. (Courtesy of FloPetrol Well Barrier)

Fig. 4.15 SEM images of dehydrated unconsolidated sand slurry. (Courtesy of FloPetrol well barrier)

Table 4.6 Advantages and possible limitations associated with unconsolidated materials with regards to permanent P&A

Advantages	Possible limitations
• Flexible • Non-degradable • Non-shrinking • Non-toxic • Self-healing • Gas tight • No waiting on setting	• High Yield Stress may cause difficulties with regards to pumpability • It needs a permanent foundation to be used for permanent P&A • No chemical bond strength to formation nor casing • No casing or formation seal if material is not confined • Conventional verification methods may not be applicable to find top of plug • Ineffective pumping may cause pack-off

where d_p is the effective particle diameter, ε is porosity of the medium, and 150 is an empirical factor which includes the geometrical terms. The Blake-Kozeny shows that the maximum permeability will be defined with micron-sized particles. By application of Darcy's law and substituting the permeability term from Eq. (4.7), the flow velocity of an incompressible fluid through the medium is:

$$v = \frac{d_p^2}{150\mu}\frac{\varepsilon^3}{(1-\varepsilon)^2}\frac{\Delta P}{\Delta L} \tag{4.8}$$

Example 4.1 Consider an unconsolidated sand slurry with a porosity of 0.25 and an effective particle diameter of 1 and 0.1 micron. Estimate the permeability of the slurry.

Solution By using the Blake–Kozeny equation with a porosity of 0.25 and effective particle diameter of 1 micron, the permeability will be:

$$k = \frac{0.25^3}{(1-0.25)^2}\frac{1^2}{150} = 1.85 \times 10^{-4} Darcy$$

By selecting the effective particle diameter as 0.1 micron, the estimated permeability will be:

$$k = \frac{0.25^3}{(1-0.25)^2}\frac{0.1^2}{150} = 1.85 \times 10^{-6} Darcy$$

4.3.1.1 Pumpability

Placeability of any plugging material at downhole conditions is a primary requirement. Research activities have shown that the pumpability of unconsolidated sand slurries can be adjusted by adjusting the PSD design. The increase of conducting fluid volume should be considered carefully as inappropriate liquid content may increase the distance between particles and consequently change the porosity and permeability. As unconsolidated sand slurries possess a very high yield stress, imposed pump friction may be challenging. Experiments have shown that the yield stress can be controlled by adjusting the PSD of sand slurries without increasing the volume of the conducting fluid [33, 39].

4.3.1.2 Durability

As permanent plugging materials are intended to withstand downhole conditions with an eternal perspective, their long-term durability should be examined. The durability should normally be tested in accordance with recognized standards, however, there is no such standard for alternative materials to cement [37]. As unconsolidated sand slurries consist of quartz sand, silica fume, and crushed rocks, their interactions with downhole chemicals seems to be less probable.

4.4 Thermosetting Polymers

Thermosetting materials, also known as thermoset polymers, are organic compounds which are characterized by their three-dimensional structures and low molecular weight (<10000 g/mol). Figure 4.16 shows the chemical relationship between thermosetting polymers in chemistry. The term polymer is used to describe a macromolecule made of many monomers, repeating units, Fig. 4.17a. The rheological and mechanical properties of polymers depend on several factors including the monomer unit, the linkages between each monomer, intermolecular, and intramolecular forces which exists between polymers. Resins, synthetic polymers, are divided into two categories: thermosetting resins and thermoplastic resins.

Thermosetting polymers are cross-linked to one another, Fig. 4.17b, and due to the cross-links, these materials develop strength. The cross-links can break by heating or chemical interaction however, to break these bonds the conditions need to be severe.

Thermosetting resins set in the presence of catalysts, by application of heat and pressure or combination of these. The setting process is irreversible. It means that

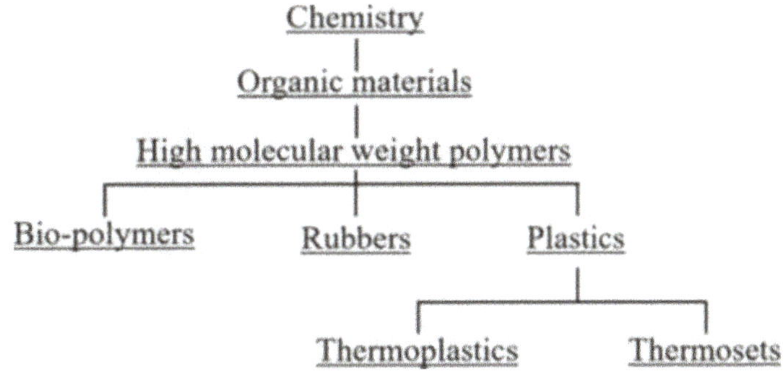

Fig. 4.16 Relationship of thermosetting materials in chemistry science [40]

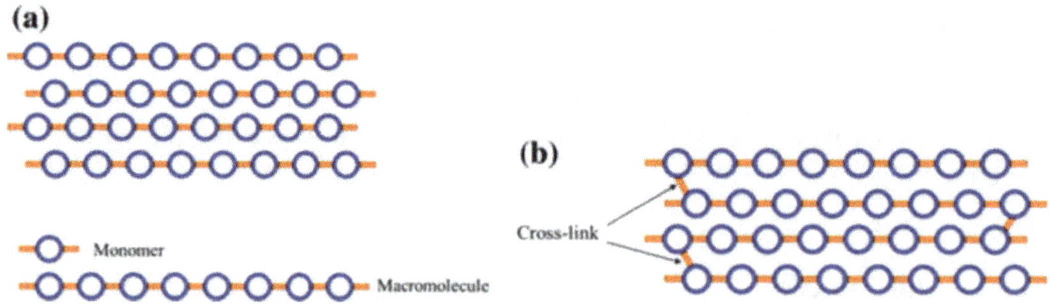

Fig. 4.17 Structure of polymers made of macromolecules and monomers

the resin cannot be reheated and remolded after setting. Thermoset plastic polymers are usually in liquid phase prior to curing, Fig. 4.18. The chemical reaction occurs during heating and results in the formation of strong covalent bonds, cross-links. Table 4.1 lists some examples of thermosetting plastic polymers. Thermoset plastic polymers usually undergo permanent or plastic deformation under load.

The stability of thermosetting materials depends on the density of cross-links and aromatic content of the polymer but generally speaking, there is a concern related to brittleness of the product with increase of cross-link density [40].

The history of thermosetting polymers for use in petroleum industry goes back to the 1960s when resins were suggested for remediating sand production [42]. Since then, they have been proposed and used for mud-loss control, remedial operations particularly for sealing tight channels, casing leaks, perforations and shoe and liner top squeezes, sand control, and production of resin-based cement [4, 41, 43–45]. As each wellbore is unique with respect to depth, downhole pressure and temperature, inclination, wellbore geometry, and formation strength, a range of additive materials are necessary to obtain the appropriate rheological and mechanical properties for

Fig. 4.18 Physical and color appearance of a thermoset resin [41]. (Courtesy of WellCem AS)

thermosetting polymers. These materials include catalysts, accelerators, inhibitors, weight fillers, expansion agents and viscosifiers [46]. Catalyst or hardener is liquid which promotes the chemical reaction without itself being consumed. By increasing the concentration of catalysts, the rate of reaction of thermosets increases. The catalysts used for thermosetting polymers used in well cementing are usually acid-based. Accelerators are liquids which increase the rate of reaction at low temperatures. As elevated temperatures increase the rate of reaction, inhibitors or retarders are used to postpone the setting time of thermosetting polymers. The inhibitors are usually basic solutions. Thermosetting polymers have a wide range of densities so to modify their density, hollow spheres can be added to lower their density or heavy particles can be added to increase their density. As thermosetting polymers have relatively low, constant viscosity, addition of heavy particles may cause particle segregation. Therefore, viscosifiers are introduced to the resin mix to increase the viscosity and consequently the lifting capacity of the resin. Table 4.7 presents properties of thermosetting resins used for zonal isolation.

In order to wash the batch mixer, pumps, lines and all equipment, cleaners are used. Cleaners are chemical solutions, usually xylene or alcohol solutions, which cannot be disposed to sea or surroundings. Therefore, cleaners need to be handled and properly managed during and after use.

Table 4.8 presents a comparison between neat G Portland cement and a thermosetting resin developed for zonal isolation of hydrocarbon wells. However, shrinkage factor, shear bond strength, and hydraulic bond strength data for thermosetting resins are not publicly available.

Table 4.7 Properties of a commercialized thermosetting synthetic polymer for cementing applications [46]

Property	Range
Density	6.2–20.8 ppg
Viscosity	10–2000 cp
Right angle set	Yes
Target temperature	68–300 °F
Pumpable through pipe	Yes
Miscible with water or well fluids	No
Decomposition temperature	900 °F
Setting time	Depends on curing temperature

Table 4.8 Properties of hardened thermosetting resins designed for zonal isolation versus Neat G-cement [47]

	Compressive strength (kpsi)	Flexural strength (kpsi)	E-modulus (kpsi)	Failure flexural strain (%)	Permeability (mD)	
					Water	Oil
Thermosetting resin	11.2	6.5	325	1.9	<0.5 × 10^{-6}	<0.5 × 10^{-6}
Portland cement	8.4	1.4	537	0.32	1.6 × 10^{-3}	[a]

[a]Not available

4.4.1 *Main Degradation Mechanisms*

Synthetic polymers degrade by different mechanisms depending on the structure and exposure conditions. Generally, organic polymers have five main degradation mechanisms: (a) physical, (b) chemical, (c) thermal, (d) hydrothermal, and (e) biodegradation [48, 49]. Physical degradation occurs as disruption of polymer morphology due to mechanical stresses, temperature, and time. It results in physical property changes which are reversible. Chemical degradation mainly occurs due to exposure to elevated temperatures, pollutants and micro- and macro-organisms. This type of degradation drastically changes the physical properties of organic polymers. This type of degradation is irreversible and occurs at the molecular level. Thermal degradation occurs above glass transition temperature and is irreversible. Hydrothermal degradation occurs at high temperatures in the presence of moisture and results in permanent physical property changes. In this condition, water molecules penetrate into the polymeric matrix and, at elevated temperatures, degenerates the interaction between polymer chains. One consequence could be swelling of the matrix and plasticization of the polymeric matrix. Biodegradation mechanism includes microorganisms to breakdown of thermosetting organic materials. With regards to utilization of thermosetting organic polymers, in petroleum industry, for zonal isolation and permanent P&A, a combination of these degradation mechanisms may be synergistic or antagonistic. Significant research is needed to understand the durability of thermosetting synthetic polymers at downhole conditions.

4.4.2 *Long-Term Integrity of Thermosetting Resins*

4.4.2.1 Exposure to Downhole Chemicals

Long-term durability of any suggested plugging material needs to be investigated comprehensively as re-entry and repairing the barrier may be risky, time consuming and even impossible. Interactions of thermosetting resins with downhole chemicals including brines, crude oil, H_2S, CO_2, and thermogenic gas at downhole conditions must be documented.

Preliminary experiments have been performed and published on durability of a commercially available thermosetting resin designed for oil well applications, Table 4.9. Crude oil, CO_2, H_2S, and methane gas were used as representatives of wellbore chemicals. Aging test results of thermosetting resin systems when exposed to brine are not available. However, hydrothermal degradation is one of the main degradation mechanisms.

Table 4.9 shows that crude oil degraded the thermosetting resin system at 212 and 266 °F but methane gas was inert and did not interact with the resin system. CO_2 did not affect the compressive strengths at both 212 and 266 °F but the flexural strength was affected at 266 °F. H_2S reduced both the compressive and the flexural strengths at 212 and 266 °F.

There is a need to study weight and volume changes of thermosetting resin systems, during aging tests, when they are exposed to downhole chemicals.

4.4.2.2 Thermal degradation

Usually in crystalline polymers, upon heat, the material transits from a hard and solid material to a liquid phase (see Fig. 4.19a). However, in amorphous polymers a reversible transition from a hard and relatively brittle state into a viscous and rubbery state occurs above specific temperatures (see Fig. 4.19b). This temperature is known as the glass transition temperatures (T_g). Degradation kinetics of thermosetting resins above their glass transition temperature needs also to be evaluated.

For evaluation of thermal stability and degradation of thermosetting resins, two main parameters are studied: the activation energy (E_a) and the pre-exponential constant (A). The Arrhenius equation is used to quantify E_a and A above the glass transition temperature:

$$k = Ae^{-\frac{E_a}{RT}} \tag{4.9}$$

where k is the rate constant (s^{-1}) for a particular reaction, T is the absolute temperature (Kelvins), and R is the universal gas constant. The natural logarithm of Eq. (4.9) yields:

$$\ln(k) = \ln(A) - \frac{E_a}{R}\frac{1}{T} \tag{4.10}$$

Equation (4.10) is the plot of a straight line with a slope of $-\frac{E_a}{R}$ and an intercept at *ln(A)*, see Fig. 4.19. So, k is the required parameter to find E_a and A. The thermal degradation of a resin system is modeled as a first order reaction as follows:

Table 4.9 Exposure of thermosetting resin to different downhole chemicals [47]

Temperature (°F)	Property	Initial value	1-month	3-month	6-month	12-month
Chemical: Crude oil 38° API (Curing pressure of 7250 psi)						
212	Permeability (nD)	<0.5	ND	ND	ND	<4
212	Uniaxial compressive strength (psi)	11170 ± 725	7106 ± 1305	6091 ± 435	5800 ± 145	6526 ± 725
212	Flexural strength (psi)	6236 ± 435	4786 ± 725	3770 ± 145	3625 ± 145	3916 ± 145
266	Permeability (nD)	<0.5	ND	ND	ND	<20
266	Uniaxial compressive strength (psi)	11170 ± 725	5511 ± 145	5221 ± 435	4931 ± 145	5366 ± 145
266	Flexural strength (psi)	6236 ± 435	3625 ± 145	3625 ± 145	3190 ± 145	2900 ± 290
Chemical: Methane 100% gas (Curing pressure of 7250 psi)						
212	Permeability (nD)	<0.5	ND	ND	ND	<128
212	Uniaxial compressive strength (psi)	11170 ± 725	11458 ± 725	11312 ± 1015	11167 ± 725	13488 ± 580
212	Flexural strength (psi)	6236 ± 435	7687 ± 1015	6381 ± 2900	3335 ± 1595	7977 ± 1015
266	Permeability (nD)	<0.5	ND	ND	ND	Not possible[1]
266	Uniaxial compressive strength (psi)	11170 ± 725	11022 ± 1595	11748 ± 1885	12473 ± 1160	12183 ± 1305
266	Flexural strength (psi)	6236 ± 435	6816 ± 2900	8557 ± 3480	7832 ± 1450	8702 ± 435
Chemical: $CO_2$5% in N_2gas (Curing pressure of 7250 psi)						
212	Permeability (nD)	<0.5	ND	ND	ND	<79
212	Uniaxial compressive strength (psi)	11170 ± 725	10732 ± 870	11748 ± 1160	10587 ± 580	13343 ± 435
212	Flexural strength (psi)	6236 ± 435	4351 ± 1160	3770 ± 1450	3335 ± 1450	8122 ± 870
266	Permeability (nD)	<0.5	ND	ND	ND	Not possible[1]
266	Uniaxial compressive strength (psi)	11170 ± 725	7977 ± 435	Not possible[a]	10732 ± 435	11022 ± 435
266	Flexural strength (psi)	6236 ± 435	2755 ± 1160	Not possible[1]	5076 ± 145	5511 ± 290

(continued)

Table 4.9 (continued)

Temperature (°F)	Property	Initial value	1-month	3-month	6-month	12-month
Chemical: H_2S 5000 ppm (Curing pressure of 145 psi)						
212	Permeability (nD)	<0.5	ND	ND	ND	Not possible[a]
212	Uniaxial compressive strength (psi)	11170 ± 725	5511 ± 145	7251 ± 290	7396 ± 290	7832 ± 435
212	Flexural strength (psi)	6236 ± 435	1885 ± 1015	2320 ± 290	2755 ± 1450	4496 ± 1160
266	Permeability (nD)	<0.5	ND	ND	ND	666
266	Uniaxial compressive strength (psi)	11170 ± 725	6961 ± 290	7832 ± 290	9282 ± 580	7251 ± 145
266	Flexural strength (psi)	6236 ± 435	2755 ± 1015	3190 ± 2755	2900 ± 1740	4206 ± 870

ND: Not determined
[a]Due to difficulties

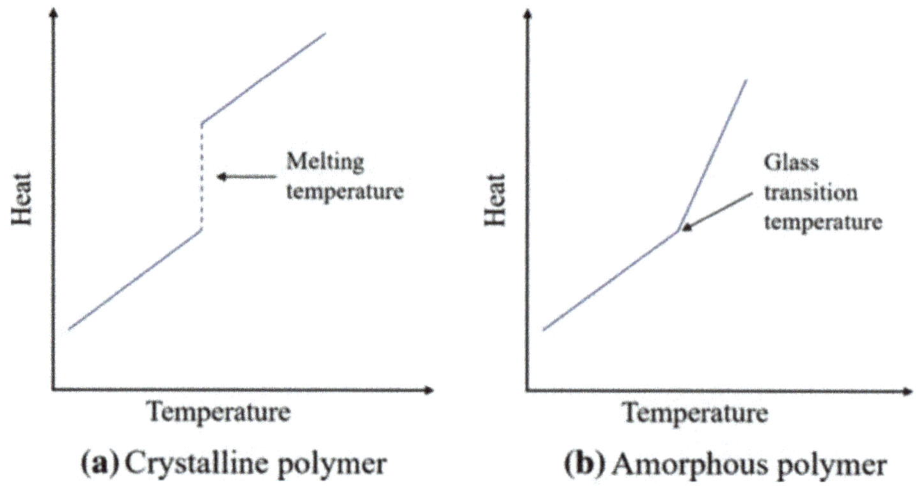

Fig. 4.19 Heat-Temperature plot for crystalline and amorphous polymers

$$-\frac{d[w]}{dt} = k[w] \tag{4.11}$$

where $[w]$ is the relative weight to the initial weight of resin at time zero $[w]_0$, t is the time. An integration of Eq. (4.11) yields:

$$\ln[w] = -kt + \ln[w]_0 \tag{4.12}$$

So, a plot of $\ln[w]$ versus time yields the $-k$. Once the E_a, and A are known, the lifetime of a resin system at any temperature can be estimated. Research studies conducted on thermal characterization of different resin systems show that the lifetime of a resin system might be a concern for permanent P&A applications where downhole temperature is above the glass transition temperature (T_g) [49]. If downhole temperature is above the glass transition temperature, the polymer has a higher free volume and higher permeability. Jones et al. [49] studied experimentally the thermal degradation of three resin systems at temperatures above their T_g. They estimated that for these systems, it will take between 60 and 150 years for the resins to lose 10% of their weight. However, the effect of high pressure on thermal degradation was not studied. They also concluded that above T_g, the resin degradation is intensified by increasing temperature. The T_g needs to be considered during the design of resin systems particularly for application in permanent P&A of wells. Advantages and possible limitations associated with the use of thermosetting resins for oil well cementing are listed in Table 4.10.

Table 4.10 Advantages and possible limitations associated with thermosetting resins with regards to permanent P&A [47, 49, 50]

Advantages	Possible limitations and disadvantages
• Gas tight (very low permeability) • Strong bond to formation and steel • Good mechanical properties • Chemically inert to wellbore fluids and rocks • Outdoor storage has no detrimental effects • No special equipment is required to prepare the mix • Low waiting on setting time • High tensile strength • The resins themselves are solid free (no grains)	• Usually brittle in solid state. Their brittleness is a function of polymer and curing pressure and temperature • Partly unknown long-term durability behavior • Interaction with brine and depolymerization • HSE issue regarding toxicity • Limited data available on pumpability • Unknown verification method when used for cementing casing • Possible interaction with workover fluid or mud • High-pH medium can deteriorate thermosetting polymers • Chemical shrinkage • No data available on hydraulic bond strength to formation and steel

4.5 Metals

As discussed in Chap. 3, casing steel is not qualified as a permanent plugging material unless it is protected internally and externally by cement or another suitable material. There are some other types of metals with low melting point which have been suggested as permanent plugging materials including metal Bismuth, Gallium, Antimony, or low-melt-point eutectic alloys (Cerro alloys) [51, 52]. A eutectic alloy is a formulation of metal elements which melts and solidifies at a single temperature, which is lower than the melting points of the separate elements or of any combination. Eutectic alloys have no solidus or liquidus transition phases, completely solid or completely liquid. The most known eutectic alloy in the petroleum industry is bismuth based alloy. There are some concepts which suggest the use of eutectic alloys such as bismuth as a permanent barrier. Therefore, bismuth alloys are considered in detail in the following.

Bismuth is a metallic element with symbol Bi and atomic number of 83. Bismuth is brittle and a very weak radioactive material. Bismuth alloys have been laboratory tested and tested in a few field trials for use as a permanent plugging material, remediating sustained casing pressure, and shutting off water producing zones [52–54]. Bismuth based alloys are developed to create a metal to metal seal and their use in the petroleum industry goes back to the Schlumberger brothers in the 1930s. Bismuth based alloys are metals with very low melting point compared to other metals. A pure bismuth element has a melting point of 520 °F at ambient pressure and it expands upon solidification by 3%. However its alloys are reported to have much lower melting points, down to 174 °F, and with lower, albeit distinct, expansion factors.

As the metal alloys have very high densities, the liquid metal requires a foundation as a base. The expanding alloy is designed in a way that it has a melting temperature which is higher than the maximum anticipated well temperature. There have been two different techniques for placement of bismuth based alloys; the molten alloy is lowered to the desired depth within a container or the solid alloy is lowered to the desired depth and heated downhole. In the first technique, the molten alloy is carried with a container which can provide temperatures above the alloy melting point. When the alloy is at the desired depth, the container hatch is opened and the liquid alloy exits the container. The second technique is the most common and carried out in different ways including: heating once downhole using electric resistive or electromagnetic induction, in situ exothermic chemical reaction, or heated steam injection [51, 55].

One of the challenges concerning bismuth based alloys is the control of vertical heat propagation during installation of the plug when an in situ exothermic reaction is applied. A recent development employs a wireline operation as a bismuth alloy plug placement technique (see Fig. 4.20). The plug assembly consists of four main parts: ignition system, alloy jacket, inner tube and skirt. The inner tube, filled with thermite, passes through the bismuth alloy jacket. On ignition, the thermite reaction generates heat and once heated, the bismuth alloy jacket is melted. As the melted bismuth alloy has a high density and its positioning is not maintained, the skirt provides a mechanical support until the bismuth alloy plug cools down and solidifies. Using this method, the radial and vertical heat control is achieved more effectively.

Table 4.11 lists a wide selection of the expandable bismuth alloys with different ranges of melting temperatures.

There are advantages and some possible limitations associated with the utilization of alloy based plugging materials in permanent P&A operations, Table 4.12. As bismuth alloys create no physical bonding with casing, it relies on expansion to take mechanical and hydraulic loads. In addition, if the exerted force on the casing, due to expansion, is high then potential deformation of casing cement may occur which may put the integrity of the cement at risk.

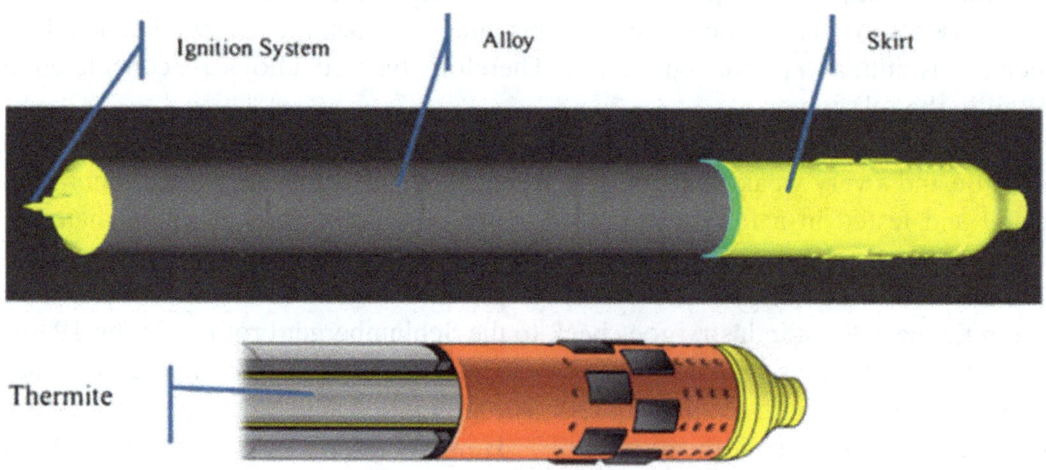

Fig. 4.20 The bismuth alloy plug placement assembly [55]

Table 4.11 Some typical melting point temperature of bismuth based alloys [56–58]

Alloys	X: chemical formula	Range of melting point (°F)
$Bi_{100-x}Sn_x$	0–5	464–520
$Bi_{100-x}Cu_x$	0–45	520–1562
$Bi_{100-x}Hg_x$	0–45	530–520
$Bi_{100-x}Sn_x$	5–42	280–520
$Bi_{100-x}Pb_x$	0–44.5	255–622
$Bi_{100-x}Cd_x$	0–40	284–610

Table 4.12 Advantages and possible limitations of bismuth based alloys for use in permanent P&A

Advantages	Possible limitations and disadvantages
• Very low permeability or impermeable • Rigless operation • Non-explosive • No shrinkage	• No data available on sealing capability • No data available on durability • No chemical bonding to formation or casing • Uncertainty regarding downhole fluid displacement • Controlling vertical heat propagation during installation • No data available on hydraulic bond strength to formation and steel • Relatively brittle for a metal • Barrier verification method is not clear • Limited maximum length of barrier • Toxic if mercury or lead is used in the alloy

4.6 Modified In Situ Materials

Recently a concept has been developed to do permanent P&A operations rigless and efficiently. In this concept, a target interval, in the wellbore, is selected and all the in situ elements are melted. Upon cooling, a solidified barrier is created from the in situ materials, Fig. 4.21.

To melt the in situ materials, thermite is used as the source of energy to generate the required amount of heat. The term "thermit" was first introduced by Goldschmidt in 1903 [60]. *Thermite* is a metal powder which produces an effect by heat. The reaction is an exothermic reduction-oxidation reaction and it is ignited by heat. The reaction is written in a general form as [60]:

$$M + AO \rightarrow MO + A + \Delta H \tag{4.13}$$

where M is a metal or an alloy, A is either a metal or a non-metal, MO and AO are their corresponding oxides, and ΔH is the generated heat during the reaction.

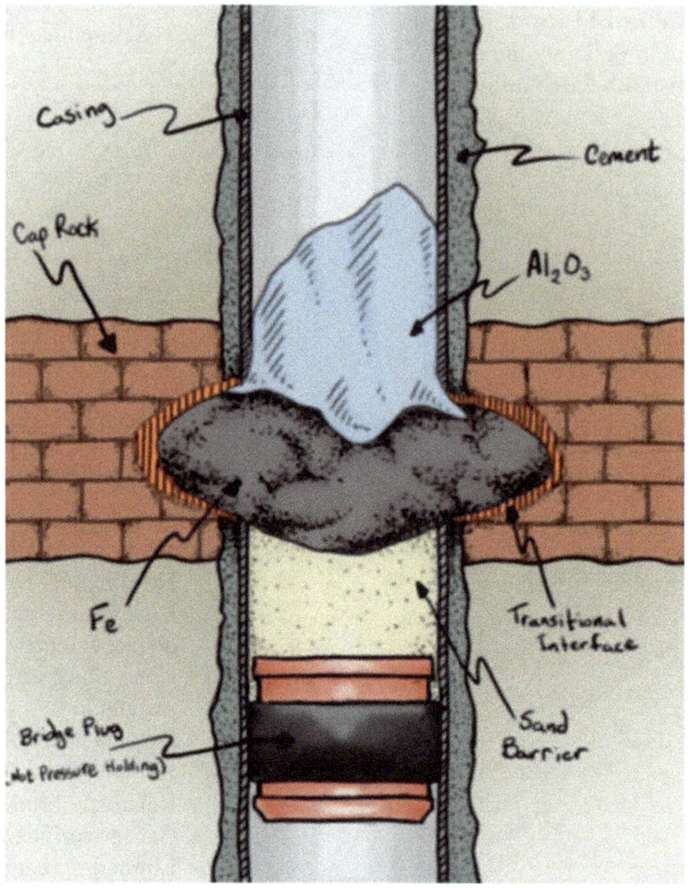

Fig. 4.21 Conceptual illustration of modified in situ materials as permanent barrier [59]

For example, the following is a well-known reaction:

$$2Al + Fe_2O_3 \rightarrow 2Fe + Al_2O_3 + \Delta H \quad (4.14)$$

The reaction is a single-replacement reaction in which aluminum reacts with iron oxide and takes the place of iron in the compound. Aluminium is more reactive compared to iron, it donates its electrons easier, and as a result of replacement, a large amount of energy is generated. The above-mentioned reaction creates temperatures in excess of 5432 °F. The reaction consumes small amounts of oxygen and it is self-sustained, Eq. (4.14). Generally speaking, a thermite reaction requires fuel metals and an oxidizer. The fuel metals include, but are not limited to, aluminium, titanium, magnesium, boron, zinc, and silicon. Of these, aluminium is of most interest as it has low cost and high boiling point. The oxidizers include, but are not limited to, bismuth (III) oxide, boron (III) oxide, silicon (IV) oxide, iron (II) oxide, and copper (II) oxide. Table 4.13 presents adiabatic temperatures for some thermite reactions, T_{ad}, and melting point of their products.

Table 4.13 Adiabatic combustion temperatures and melting points of the product metals after Wang et al. [60]

Reaction	T_{ad} (K)[a]	T_{mp} of metal (K)[b]
I. Formation of common structural metals		
$Al + \frac{1}{2}Fe_2O_3 \rightarrow Fe + \frac{1}{2}Al_2O_3$	3622	1809
$Al + \frac{3}{2}NiO \rightarrow \frac{3}{2}Ni + \frac{1}{2}Al_2O_3$	3524	1726
$Al + \frac{3}{4}TiO_2 \rightarrow \frac{3}{4}Ti + \frac{1}{2}Al_2O_3$	1799	1943
$Al + \frac{3}{8}Co_3O_4 \rightarrow \frac{9}{8}Co + \frac{1}{2}Al_2O_3$	4181	1495
II. Formation of refractory metals		
$Al + \frac{1}{2}Cr_2O_3 \rightarrow Cr + \frac{1}{2}Al_2O_3$	2381	2130
$Al + \frac{3}{10}V_2O_5 \rightarrow \frac{6}{10}V + \frac{1}{2}Al_2O_3$	3785	2175
$Al + \frac{3}{10}Ta_2O_5 \rightarrow \frac{6}{10}Ta + \frac{1}{2}Al_2O_3$	2470	3287
$Al + \frac{1}{2}MoO_3 \rightarrow \frac{1}{2}Mo + \frac{1}{2}Al_2O_3$	4281	2890
$Al + \frac{1}{2}WO_3 \rightarrow \frac{1}{2}w + \frac{1}{2}Al_2O_3$	4280	3680
$Al + \frac{3}{10}Nb_2O_5 \rightarrow \frac{6}{10}Nb + \frac{1}{2}Al_2O_3$	2756	2740
III. Formation of other metals and non-metals		
$Al + \frac{1}{2}B_2O_3 \rightarrow B + \frac{1}{2}Al_2O_3$	2315	2360
$Al + \frac{3}{4}PbO_2 \rightarrow \frac{3}{4}Pb + \frac{1}{2}Al_2O_3$	>4000	600
$Al + \frac{3}{4}MnO_2 \rightarrow \frac{3}{4}Mn + \frac{1}{2}Al_2O_3$	4178	1517
$Al + \frac{3}{4}SiO_2 \rightarrow \frac{3}{4}Si + \frac{1}{2}Al_2O_3$	1760	1685
IV. Formation of nuclear metals		
$Al + \frac{3}{16}U_3O_8 \rightarrow \frac{9}{16}U + \frac{1}{2}Al_2O_3$	2135	1405
$Al + \frac{3}{4}PuO_2 \rightarrow \frac{3}{4}Pu + \frac{1}{2}Al_2O_3$	796	913

[a]Adiabatic temperature
[b]Melting point temperature

Different combinations of the fuel metals and oxidizers generate different energy levels in a controlled manner. This method is known as "*dilution*" to control the amount of generated heat. When sufficient energy is produced, downhole equipment such as casing, cement, control lines and a portion of the in situ formation is melted. However, achieving high enough energy to melt the downhole equipment requires a certain amount of thermite, depending on the type of fuel and oxidizer.

The product of the reaction is usually a heaver metallic phase and a lighter oxide phase. Due to gravity, the lighter phase migrates upward and the heavier phase moves downward. For oil well applications, the freezing point is critical whereas solidification of the product prior to migration of the lighter oxide phase may result in a discontinuous barrier.

4.6.1 Barrier Establishment

A mechanical plug is required to be installed as a foundation for the molten materials prior to cooling. Because the high temperature may damage a mechanical plug, sand is placed on top of it. The wireline tool is equipped with a jacket which contains thermite and an igniter. This method employs a wireline operation to establish the barrier. The tool consists of four main parts: heavy load, separator, thermite pool, and igniter, (see Fig. 4.22a). The wireline provides the electrical power required to ignite the thermite mixture. When the thermite reaction is initiated, the thermite available in the thermite pool gets consumed in the reaction, and the reaction heats and melts the adjacent equipment. As the generated energy is high, rapid expansion of downhole fluids, at shallow depths, may be a concern. Therefore, a heavy load is applied to the thermite pool to compress the barrier while cooling (see Fig. 4.22b). As the high energy may melt the heavy load, a separator protects it. When the barrier is established, the heavy load and separator are retrieved (see Fig. 4.22c).

Controlled heat propagation, both in radial and vertical direction, is one of the challenges associated with this method. When a barrier is established, the transition area between the modified and non-modified materials needs to be qualified. Currently, there is no established methodology. In addition, limited data availability is another potential limitation besides, concerns associated with durability of the barrier, Table 4.14.

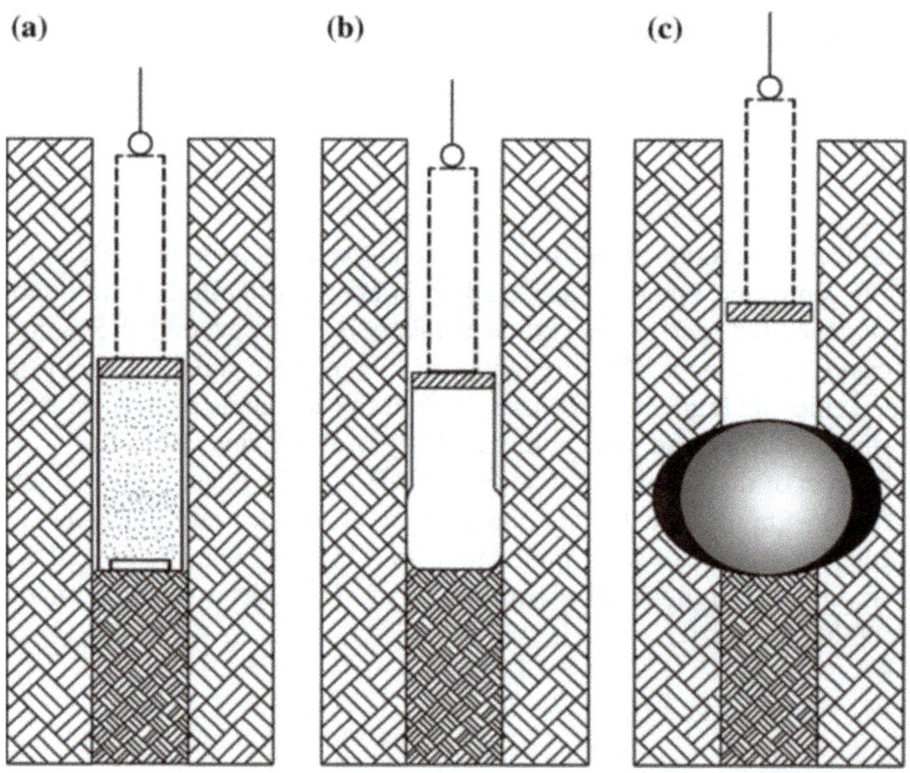

Fig. 4.22 Permanent barrier establishment by modified in situ materials method

Table 4.14 Advantages and possible limitations concerned with modified in situ materials [59–62]

Advantages	Possible limitations and disadvantages
• Rigless based concept • Safe to handle and low equipment intensity	• Barrier mainly consists of iron which associates to a long term durability concern • Presence of downhole fluids may compromise the sealability of barrier while establishing barrier • Maximum length of established barrier • Minimum hole diameter to run the tool in • Gravity force, in deviated sections, can cause segregation of the plug when it is in liquid phase • Contraction upon cooling and solidification may introduce micro cracks • Limited data availability • Not commercialized

References

1. Oil & Gas UK. 2015. Guidelines on qualification of materials for the abandonment of wells. In *Bond strength*. The UK Oil and gas industry association limited: Great Britain p. 48–50.
2. Vrålstad, T., A. Saasen, E. Fjær, et al. 2018. Plug & abandonment of offshore wells: Ensuring long-term well integrity and cost-efficiency. *Journal of Petroleum Science and Engineering*. https://doi.org/10.1016/j.petrol.2018.10.049.
3. Khalifeh, M., A. Saasen, H. Hodne, et al. 2019. Laboratory evaluation of rock-based geopolymers for zonal isolation and permanent P&A applications. *Journal of Petroleum Science and Engineering* 175: 352–362. https://doi.org/10.1016/j.petrol.2018.12.065.
4. Wasnik, A.S., S.V. Mete, and B. Ghosh. 2005. Application of resin system for sand consolidation, mud loss control & channel repairing. In *SPE international thermal operations and heavy oil symposium*. SPE-97771-MS, Calgary, Alberta, Canada: Society of Petroleum Engineers. https://doi.org/10.2118/97771-MS.
5. Taylor, H.F.W. 1992. *Cement Chemsitry*. 1st A.P. Limited. Academic Press. 0-12-683900-X.
6. American Petroleum Institute. Production Dept. 1988. *API specification 10A—Specification for materials and testing for well cements, in Section 2*. American Petroleum Institute.
7. Smith, D.K. 1990. *Cementing*. 2nd ed. Monograph, ed. Henry L. Doherty Series. vol. 4. New York City: Society of Petroleum Engineers. 1-55563-006-5.
8. Østnor, T. 2007. *"Alternative pozzolans" as supplementary cementitious materials in concrete—State of the art*. SINTEF: Trondheim—Norway. 25. https://www.sintef.no/globalassets/sintef-byggforsk/coin/sintef-reports/sbf-bk-a07032_alternative-pozzolans_-as-supplementary-cementitious-materials-in-concrete.pdf.
9. Smith, D.K. 1956. *A new material for deep well cementing*. Society of Petroleum Engineers: American Institute of Mining, Metallurgical, and Petroleum Engineers, Inc.
10. Smith, D.K. 1974. Cements and cementing, In *Cements and Cementing (1974 DPM Chapter 16)*. Society of Petroleum Engineers. 0-87814-057-3.
11. Haarmon, J.A. and H.D. Woodard. 1955. *Use of oil-cement slurries for decreasing water production*. American Petroleum Institute.
12. Patchen, F.D., R.F. Burdyn, and I.R. Dunlap. 1959. *Water-In-Oil emulsion cements*. Society of Petroleum Engineers.
13. Oyarhossein, M. and M.B. Dusseault. 2015. Wellbore stress changes and micro annulus development because of cement shrinkage. In *49th U.S. rock mechanics/geomechanics symposium*. ARMA-2015-118 San Francisco, California: American Rock Mechanics Association.

14. Carter, L.G., H.F. Waggoner, and C. George. 1966. Expanding cements for primary cementing. *Journal of Petroleum Technology* 18 (05): 551–558. https://doi.org/10.2118/1235-PA.
15. Benge, G. 2005. *Cement* designs for high-rate acid gas injection wells. In *International petroleum technology conference*. IPTC-10608-MS, Doha, Qatar: International Petroleum Technology Conference. https://doi.org/10.2523/IPTC-10608-MS.
16. Pike, W.J. 1997. Cementing multilateral wells with latex cement. Journal of Petroleum Technology, 49(08): p. 849 - 849. https://doi.org/10.2118/0897-0849-JPT.
17. Zeng, J., F. Sun, P. Li, et al. 2012. Study of Salt-tolerable Latex Cement Slurries. In *The twenty-second international offshore and polar engineering conference*. Rhodes, Greece: International Society of Offshore and Polar Engineers.
18. Sun, F., G. Lv, and J. Jin. 2006. Application and research of latex tenacity cement slurry system. In *International oil & gas conference and exhibition in China*. SPE-104434-MS, Beijing, China: Society of Petroleum Engineers. https://doi.org/10.2118/104434-MS.
19. Bykov, V.V., S.A. Paleev, and Y.V. Medvedev. 2016. Improving the quality of cementing lines and conductors in the conditions of permafrost in the fields in eastern Siberia. In *SPE Russian Petroleum technology conference and exhibition*. SPE-181937-MS, Moscow, Russia: Society of Petroleum Engineers. https://doi.org/10.2118/181937-MS.
20. Morris, E.F. 1970. Evaluation of cement systems for permafrost. In *Annual Meeting of the American Institute of Mining, Metallurgical, and Petroleum Engineers*. SPE-2824-MS, Denver, Colorado: Society of Petroleum Engineers. https://doi.org/10.2118/2824-MS.
21. Shryock, S.H. and W.C. Cunningham. 1969. Low-temperature (Permafrost) cement composition. In *Drilling and Production Practice*. API-69-048, Washington, D.C.: American Petroleum Institute.
22. Duguid, A. 2009. An estimate of the time to degrade the cement sheath in a well exposed to carbonated brine. *Energy Procedia* 1 (1): 3181–3188. https://doi.org/10.1016/j.egypro.2009.02.101.
23. Lécolier, E., A. Rivereau, G. Le Saoût, et al. 2007. Durability of hardened portland cement paste used for oilwell cementing. *Oil & Gas Science and Technology*, 62 (3). https://doi.org/10.2516/ogst:2007028.
24. Noik, C. and A. Rivereau. 1999. Oilwell cement durability. In *SPE annual technical conference and exhibition*. SPE-56538-MS, Houston, Texas: Society of Petroleum Engineers. https://doi.org/10.2118/56538-MS.
25. Vrålstad, T., J. Todorovic, A. Saasen, et al. 2016. Long-term integrity of well cements at downhole conditions. In *SPE bergen one day seminar*. SPE-180058-MS, Grieghallen, Bergen, Norway: Society of Petroleum Engineers. https://doi.org/10.2118/180058-MS.
26. Williams, S.M., T. Carlsen, K.C. Constable, et al. 2009. Identification and qualification of shale annular barriers using wireline logs during plug and abandonment operations. In *SPE/IADC drilling conference and exhibition*. SPE-119321-MS, Amsterdam, The Netherlands: Society of Petroleum Engineers. https://doi.org/10.2118/119321-MS.
27. Fjær, E. and T.G. Kristiansen. 2017. *Activated creeping shale to remove the open annulus*. https://www.norskoljeoggass.no/Global/PAF%20seminar%202017/10%20Activated%20creeping%20shale%20to%20remove%20the%20open%20annulus%20-%20Erling%20Fj%C3%A6r,%20Sintef%20-%20Tron%20Kristiansen,%20AkerBP-20171022192329.pdf?epslanguage=no. Cited 2018 January 14.
28. Al-Bazali, T.M., J. Zhang, M.E. Chenevert, et al. 2005. Measurement of the sealing capacity of shale caprocks. In *SPE annual technical conference and exhibition*. SPE-96100-MS, Dallas, Texas: Society of Petroleum Engineers. https://doi.org/10.2118/96100-MS.
29. Chenevert, M.E. and S.O. Osisanya. 1992. Shale swelling at elevated temperature and pressure. In *The 33th U.S. Symposium on Rock Mechanics (USRMS)*. ARMA-92-0869, Santa Fe, New Mexico: American Rock Mechanics Association.
30. Santos, H., A. Diek, J.C. Roegiers, et al. 1996. Can shale swelling be (easily) controlled? In *ISRM international symposium—EUROCK 96*. ISRM-EUROCK-1996-014, Turin, Italy: International Society for Rock Mechanics.

31. Fjar, E., R.M. Holt, A.M. Raaen, et al. 2008. *Petroleum related rock mechanics*. Developments in Petroleum Science. vol. 53. The Netherlands: Elsevier Science. 9780444502605.
32. Fredagsvik, K. 2017. *Formation as barrier for plug and abandonment of wells*. Department of Petroleum Engineering University of Stavanger: Stavanger, Norway.
33. Saasen, A., S. Wold, B.T. Ribesen, et al. 2011. Permanent abandonment of a north sea well using unconsolidated well-plugging material. *SPE Drilling & Completion* 26 (03): 371–375. https://doi.org/10.2118/133446-PA.
34. Englehardt, J., M.J. Wilson, and F. Woody. 2001. New abandonment technology new materials and placement techniques. In *SPE/EPA/DOE exploration and production environmental conference*. SPE-66496-MS, San Antonio, Texas: Society of Petroleum Engineers. https://doi.org/10.2118/66496-MS.
35. Ogata, S., S. Kawasaki, N. Hiroyoshi, et al. 2009. Temperature dependence of calcium carbonate precipitation for biogrout. In *ISRM regional symposium—EUROCK*. ISRM-EUROCK-2009-052, Cavtat, Croatia: International Society for Rock Mechanics.
36. Messenger, J.U. 1969. Barite plugs effectively seal active gas zones. In *Drilling and production practice*. API-69-160, Washington, D.C.: American Petroleum Institute.
37. Vignes, B. 2011. Qualification of well barrier elements—long-term integrity test, test medium and temperatures. In *SPE European health, safety and environmental conference in oil and gas exploration and production*. SPE-138465-MS, Vienna, Austria: Society of Petroleum Engineers. https://doi.org/10.2118/138465-MS.
38. Pacella, H.E., H.J. Eash, B.J. Frankowski, et al. 2011. Darcy permeability of hollow fiber bundles used in blood oxygenation devices. *Journal of Membrane Science* 382 (1): 238–242. https://doi.org/10.1016/j.memsci.2011.08.012.
39. Godøy, R., Svindland, A., Saasen, A., et al. 2004. Experimental analysis of yield stress in high solids concentration sand slurries used in temporary well abandonment operations. In *Annual transactions of the nordic rheology society* 12: Nordic Rheology Society. https://nordicrheologysociety.org/Content/Transactions/2004/Experimental%20analysis%20of%20yield.pdf.
40. Dodiuk, H. and S.H. Goodman. 2014. *Handbook of thermoset plastics*, 3rd ed. United States of America: Elsevier. 978-1-4557-3107-7.
41. Knudsen, K., G.A. Leon, A.E. Sanabria, et al. 2015. First application of thermal activated resin as unconventional LCM in the middle east. In *Abu Dhabi international petroleum exhibition and conference*. SPE-177430-MS, Abu Dhabi, UAE: Society of Petroleum Engineers. https://doi.org/10.2118/177430-MS.
42. Strohm, P.J., M.A. Mantooth, and C.L. Depriester. 1967. Controlled injection of sand consolidation plastic. *Journal of Petroleum Technology* 19 (04): 487–494. https://doi.org/10.2118/1485-PA.
43. Degouy, D. and M. Martin. 1993. Characterization of the evolution of cementing materials after aging under severe bottom hole conditions. *SPE Drilling & Completion*, 08 (01). https://doi.org/10.2118/20904-PA.
44. Morris, K., J.P. Deville, and P. Jones. 2012. Resin-based cement alternatives for deepwater well construction. In *SPE deepwater drilling and completions conference*. SPE-155613-MS, Galveston, Texas, USA: Society of Petroleum Engineers. https://doi.org/10.2118/155613-MS.
45. Sinclair, A.R. and J.W. Graham. 1978. An effective method of sand control. In *SPE Symposium on Formation Damage Control*. SPE-7004-MS, Lafayette, Louisiana: Society of Petroleum Engineers. https://doi.org/10.2118/7004-MS.
46. Sanabria, A.E., K. Knudsen, and G.A. Leon. 2016. Thermal activated resin to repair casing leaks in the middle east. In *Abu Dhabi international petroleum exhibition & conference*. SPE-182978-MS, Abu Dhabi, UAE: Society of Petroleum Engineers. https://doi.org/10.2118/182978-MS.
47. Beharie, C., S. Francis, and K.H. Øvestad. 2015. Resin: An alternative barrier solution material. In *SPE Bergen One Day Seminar*. SPE-173852-MS, Bergen, Norway: Society of Petroleum Engineers. https://doi.org/10.2118/173852-MS.
48. Bakir, M., C.N. Henderson, J.L. Meyer, et al. 2018. Effects of environmental aging on physical properties of aromatic thermosetting copolyester matrix neat and nanocomposite foams.

Polymer Degradation and Stability, 147 (Supplement C): 49–56. https://doi.org/10.1016/j.polymdegradstab.2017.11.009.

49. Jones, P., C. Boontheung, and G. Hundt. 2017. Employing an arrhenius rate law to predict the lifetime of oilfield resins. In *SPE international conference on oilfield chemistry*. SPE-184557-MS, Montgomery, Texas, USA: Society of Petroleum Engineers. https://doi.org/10.2118/184557-MS.
50. Khoun, L., and P. Hubert. 2010. Characterizing the cure shrinkage of an epoxy resin in situ. *Plastic Research Online*. https://doi.org/10.2417/spepro.002583.
51. Bosma, M., E.K. Cornelissen, K. Dimitriadis, et al. 2010. *Creating a well abandonment plug*. U.S. Patent, Editor. https://www.google.com/patents/US7640965.
52. Carpenter, R.B., M.E. Gonzalez, V. Granberry, et al. 2004. Remediating sustained casing pressure by forming a downhole annular seal with low-melt-point eutectic metal. In *IADC/SPE drilling conference*. SPE-87198-MS, Dallas, Texas: Society of Petroleum Engineers. https://doi.org/10.2118/87198-MS.
53. Carpenter, R.B., M. Gonzales, and J.E. Griffith. 2001. Large-scale evaluation of alloy-metal annular plugs for effective remediation of casing annular gas flow. In *SPE Annual Technical Conference and Exhibition*. SPE-71371-MS, New Orleans, Louisiana: Society of Petroleum Engineers. https://doi.org/10.2118/71371-MS.
54. M2M, W.-L. *Alaska pilot test*. 2017. https://www.norskoljeoggass.no/Global/PAF%20seminar%202017/09%20Field%20experience%20of%20alternative%20sealing%20materials%20and%20barriers%20-%20Paul%20Carragher,%20BiSN-20171022192332.pdf?epslanguage=no. Cited 13 Dec 2017.
55. Abdelal, G.F., A. Robotham, and P. Carragher. 2015. Numerical simulation of a patent technology for sealing of deep-sea oil wells using nonlinear finite element method. *Journal of Petroleum Science and Engineering*, 133 (Supplement C): 192–200. https://doi.org/10.1016/j.petrol.2015.05.010.
56. Braga, M.H., J. Vizdal, A. Kroupa, et al. 2007. The experimental study of the Bi–Sn, Bi–Zn and Bi–Sn–Zn systems. *Calphad* 31 (4): 468–478. https://doi.org/10.1016/j.calphad.2007.04.004.
57. Chakrabarti, D.J. and D. Laughlin, E.. 1984. The Bi-Cu (Bismuth-Copper) system. *Bulletin of Alloy Phase Diagrams* 5 (2): 148–155. https://link.springer.com/content/pdf/10.1007/BF02868951.pdf.
58. Senapati, M.R. 2006. *Advanced engineering chemistry*. New Delhi, India: Laxmi Publications (P) LTD. 8170088895.
59. Mortensen, F.M. 2016. *A New P&A technology for setting the permanent barriers*. Department of Petroleum Engineering. University of Stavanger: Norway. p. 98.
60. Wang, L.L., Z.A. Munir, and Y.M. Maximov. 1993. Thermite reactions: their utilization in the synthesis and processing of materials. *Journal of Materials Science* 28 (14): 3693–3708. https://doi.org/10.1007/BF00353167.
61. Stein, A. 2016. Meeting the demand for barrier plug integrity assurance & verification of well abandonment barriers. In *SPE Asia Pacific Oil & Gas Conference and Exhibition*. SPE-182468-MS, Perth, Australia: Society of Petroleum Engineers. https://doi.org/10.2118/182468-MS.
62. Wittberg, S.A. 2017. *Expanding the well intervention scope for an effective P&A Operation, in department of industrial economy*, 118. Norway: University of Stavanger.

Chapter 5
Different Categories of Working Units

When considering permanent plug and abandonment of hydrocarbon wells, location of the well plays a critical role; location can be either onshore or offshore. For onshore wells, the well depth, downhole pressure, and complexity of operation dictate the type of working unit. For offshore wells, type of facility, water depth, downhole pressure, and working unit serviceability are the governing factors for selection of the working unit. The facility can be either platform-based or subsea-based. This chapter will familiarize the reader with different types of working units, for permanent P&A purposes, based on the well location and type of facility (see Fig. 5.1). In addition, to drilling rigs, vessels are also reviewed as a new generation of working unit but as they are not counted as rigs, they are not included in Fig. 5.1.

5.1 Onshore Units

Land wells are the most common drilled hydrocarbon wells. History of the first known land hydrocarbon well, goes back to China where the earliest well was drilled in 347 CE [1]. Accordingly, many oil wells were drilled until 1859, when Edwin L. Drake drilled the first commercially successful oil well. Since then, with the increase of need for energy, drilling activities for hunting hydrocarbons have speeded up and countless wells have been drilled. Subsequently, depth of penetration has been increased and thus, different types of land rigs have been developed. Land rigs are designed based on portability and maximum operating depth and are divided into two main categories: *conventional rigs* and *mobile rigs*.

M. Khalifeh and A. Saasen, *Introduction to Permanent Plug and Abandonment of Wells*, Ocean Engineering & Oceanography 12,
https://doi.org/10.1007/978-3-030-39970-2_5

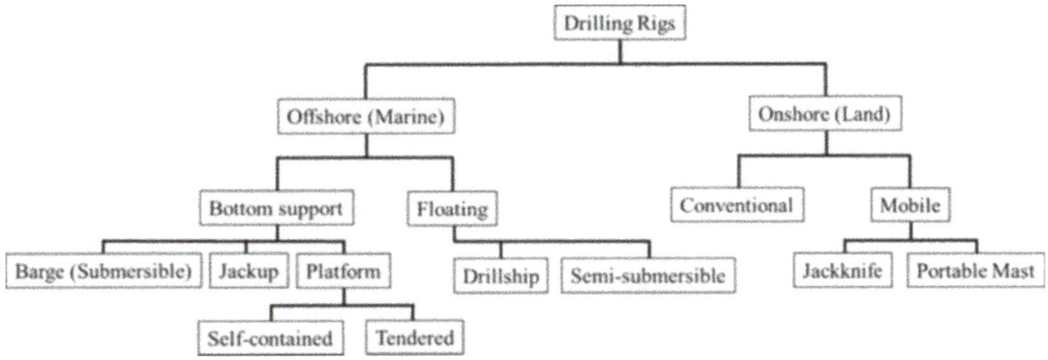

Fig. 5.1 Different working units based on the well location

5.1.1 Conventional Land Rigs

Conventional land rigs are built on location and left on site after the well is completed. The rig can be used for workover activities during the life-cycle of the well. However, due to the high cost of rig construction, mobile rigs were introduced where the derrick can be moved and reused. Figure 5.2 illustrates an onshore rotary drilling rig and its main components.

5.1.2 Mobile Land Rigs

Mobile land rigs are categorized as *jackknife* and *portable mast*. The jackknife, also known as *cantilever derrick*, is assembled on location, on ground, and then raised to vertical position by utilization of the rig-hoisting equipment or the drawworks. The portable mast is usually mounted on wheeled-trucks as a single unit and transported to the location, and raised in the vertical position by using hydraulic pistons on the carrier unit. Different types of land rigs are designed and available, depending on well location, depth of operation, and horsepower requirement. Fit-for-purpose rigs are a class of land rigs specially designed for remote areas where few people wish to venture. These remote areas such as deserts, and arctic areas may have few or no highways.

The main components of a rotary rig are: a power system, a hoisting system, a circulating system, a rotary system, and a well control system [2]. All of these components are necessary for drilling and permanent P&A operations and are therefore, comprehensively discussed in this chapter.

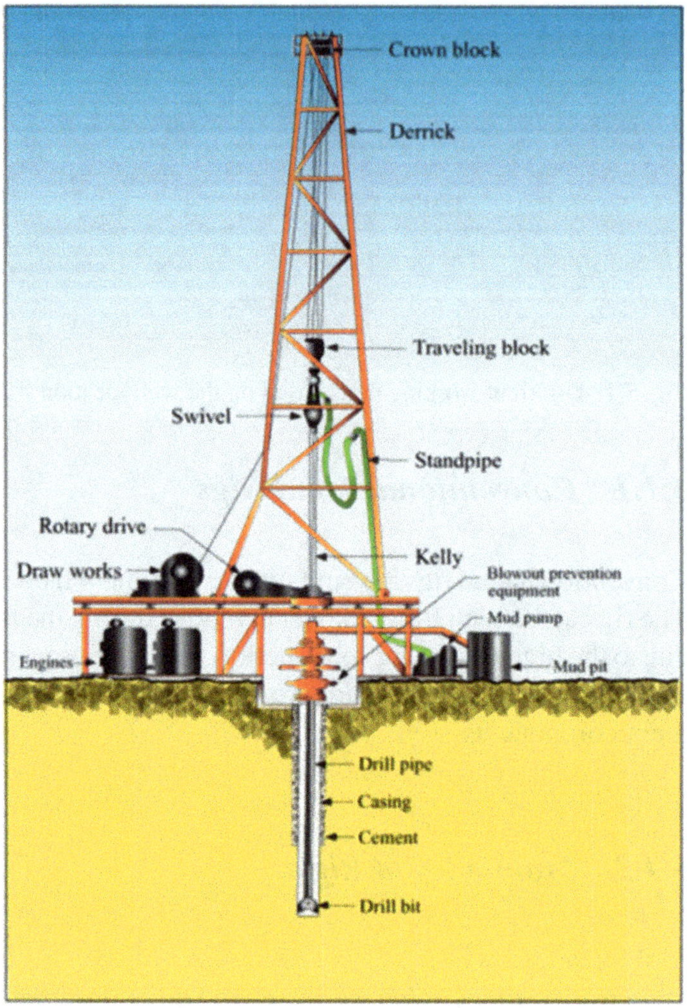

Fig. 5.2 Onshore rotary drilling rig (Taken from Weebly)

5.2 Offshore Units

With the increase in world energy needs for fossil fuels, exploration and production of hydrocarbons have been extended to remote areas such as offshore locations. Although the main intended purpose of a drilling rig and its main systems may not be influenced by well location, the water depth requires modification of land rigs. Consequently, mobile offshore drilling units (MODUs or rigs) or marine rigs were developed and introduced. The main design features for offshore rigs are portability and maximum water depth of operation. Offshore rigs are classified broadly as *floating* or *bottom support*. The floating rigs are categorized as semisubmersible, and drillship. Bottom supported rigs are categorized as barge, jackup, and platform rigs [3].

5.2.1 Submersible/Barge Rigs

These types of rigs are used for drilling at shallow water depths. The operational water depth of these submersible barge rigs is less than 40 (ft) and where there is no severe wave action. The rig is installed on a barge, large pontoon-like structure, and towed to the location. When on location, the pontoons are filled with water, the platform sinks partly or fully, and rests on its anchors. When the drilling operation is completed, water is pumped out and the platform is ready to move to a new location. If the barge rests on the seafloor, then it is counted as a bottom supported drilling rig.

5.2.2 Semisubmersible Rigs

Semisubmersible (see Fig. 5.3) rigs are capable of performing drilling operations while resting on the seafloor as well as being in a floating position. In other words, the drilling rig is on a barge similar to submersible rigs. Compared to submersible rigs (known also as bottle-type semisubmersible rig), the semisubmersible rigs (known as column-stabilized semisubmersible rigs) are designed with good stability and seakeeping characteristics. These types of rigs are usually used at larger water depths where a rig cannot rest on the seafloor. When the semisubmersible rig cannot

Fig. 5.3 A semisubmersible drilling rig towed to location. (Courtesy of Seadrill)

rest on the seafloor, the unit is either *anchored* onto the position or kept on location with dynamic positioning systems. The construction and operational cost of semisubmersible rigs are higher than for submersible rigs.

5.2.3 Drillship

A drillship is a type of floating vessel where the drilling rig is mounted on a merchant ship (see Fig. 5.4). The drillship is usually used for offshore exploration and equipped with advanced dynamic positioning systems. As drillships benefit from the dynamic positioning systems, they are usually much more costly compared to semisubmersible rigs. In recent years, drillships have been used for operation in deepwater and ultra-deepwater areas. There are some generations of drillships, which are equipped with only mooring systems or general dynamic positioning systems that have lower cost compared to semisubmersible rigs. Another challenge for using a drillship is its susceptibility to severe waves, wind and currents. A benefit of using drillships is their efficient mobilization and high speed between drilling locations.

Recently, riserless well intervention vessels have been used for small activities such as coring [4]. These types of vessels are small sized drillships which have the capability to be equipped with well intervention equipment such as coiled tubing units. The cost of these vessels is much lower than cost of other types of rigs; however,

Fig. 5.4 Drillship on location (Courtesy of Seadrill)

Fig. 5.5 A jackup rig on location. (Courtesy of Seadrill)

time spent waiting on weather is higher compared to other types of drilling rigs. The vessels will be reviewed later in this chapter.

5.2.4 *Jackup Rig*

Jackups are the most common bottom-supported rigs. The rig consists of a barge-type hull (triangular barge form) and three legs, Fig. 5.5. When the rig is in place, legs are lowered to adjust to a given clearance. Jackups are self-contained rigs that can be mobilized and demobilized easily. Depending on their size, they can operate in water depths up to 500 (ft) [5].

5.2.5 *Platform Rigs*

Platform rigs are usually employed during development phase where an economically viable offshore field is exploited. Many directional wells can be drilled from a platform. Large platforms are capable of accommodating drilling rigs or modular

Fig. 5.6 A platform rig in operation. (AkerBP)

rigs and therefore are known as *self-contained* (see Fig. 5.6). Rig-up time of platform rigs are usually less compared to most of the MODUs as no mooring system nor dynamic positioning system is required. But there are some circumstances when the rig-up time can increase due to waiting on weather.

5.2.6 Tendered Rigs

There are circumstances where the platform is small and not capable of accommodating all the components of a drilling rig or storage facilities. In this situation, a floating vessel is anchored next to the platform (see Fig. 5.7). The floating vessel is known as the *rig tender*. The rig tender can contain storage facilities, many of the rig components and the living quarters.

5.2.7 Vessels

Vessels are small sized merchant ships which offer some basic operations such as well intervention activities and anchor handling. Compared to drillships, the day rate of vessels are much lower. These types of vessels are categorized as light well intervention vessel and anchor handing vessels.

Fig. 5.7 A tender rig in operation while the anchored vessel is in service. (Courtesy of Seadrill)

5.2.7.1 Light Well Intervention Vessels

Light Well Intervention Vessels (LWIVs) have been used for over 25 years in the North Sea. LWIVs are typically monohull, flexible and extremely cost efficient and can be used for a single or multi-well (a campaign) of subsea wells. They can accommodate a wireline unit and coiled tubing unit, Fig. 5.8.

Well integrity and suspension operations including mechanical plug setting, mechanical repair or well maintenance, perforating and setting cement plugs, wellhead cutting and removal, logging, Remotely Operating Vehicle (ROV) services, and pumping operations are typical activities which are conducted by use of LWIVs [6]. The future approach for the use of LWIVs is to perform the complete permanent P&A operations. However, there are some limitations to be solved before reaching to the goal, see Table 5.1.

5.2.7.2 Anchor Handler Vessels (AHVs)

Anchor handling operations may contribute 10–20% of the total well costs of offshore exploration drilling [8]. In a conventional anchor handling operation, the rig's winches are used to tension the anchors. AHV transports and deploys the anchors, connects the required chains, wires and polyester ropes. AHV can pre-lay the anchors before the rig arrives, and more time can be dedicated to drilling or P&A operations.

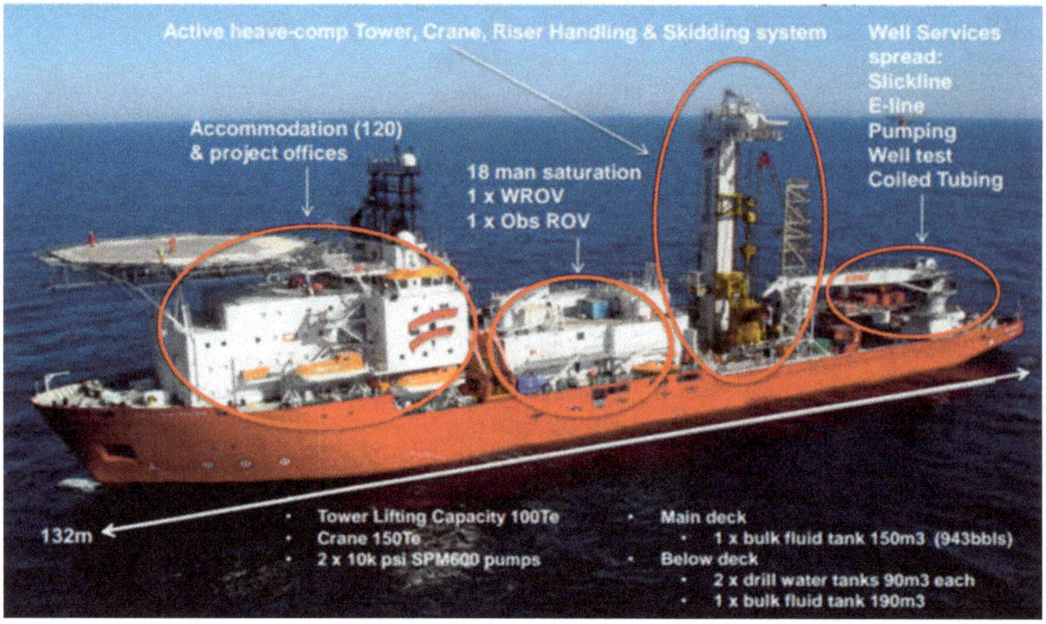

Fig. 5.8 Light well intervention vessel. (Courtesy of Helix Energy Solutions Group)

Table 5.1 Advantages and possible limitations of LWIVs for use in permanent P&A operations [7]

Advantages	Possible limitations
• Equipped with well control package • Wireline operations • Coiled tubing operations • Wellhead cut and removal • Activities for establishment of temporary abandonment • Pipe handling • Cementing adaptor tool • Flexible and cost efficient	• Limited pulling capacity • Waiting on weather is high due to the small size • Unable to work full bore 7-in. • Limited deck space • High motions add more risk

5.3 Types of Offshore Wells

Depending on the field development planning, offshore wells can be completed as either subsea wells or platform wells. Depending on well type, subsea or platform, the plug and abandonment operation will be different. Therefore, it is important to review the major differences between subsea and platform wells.

5.3.1 Subsea Wells

In a *subsea well*, the wellhead, XMT, and production-control equipment are located on the seabed. Subsea wells may be drilled and completed individually, in clusters, or on a template.

Individual subsea wells—An individual subsea well is a well which is drilled and completed as a single well. Every time a well is completed, the drilling unit is demobilized and mobilized to the next well and consequently, associated costs are increased.

Clustered subsea satellite wells—The concept of *clustered subsea satellite wells* is that individual wells are drilled but they are connected to a manifold, Fig. 5.9, and then the manifold is connected to a production unit. In this case, some costs associated with field development are saved because of flow line and control umbilical savings.

Multiwell template subsea wells—The multi-well template is another subsea field development concept where wells are drilled from one location by utilization of a drilling template. In this concept, the drilling unit stays in place while drilling several wells through the template. Therefore, costs associated with demobilization and mobilization will be minimized.

Subsea wells are equipped with *templates*, a large supportive structure which is made of steel. The template is used as a *temporary guide base* and serves as the anchor for guidelines for a permanent guide base. The template has opening(s) which the bit passes through and drilling can be performed, Fig. 5.10. A subsea permanent

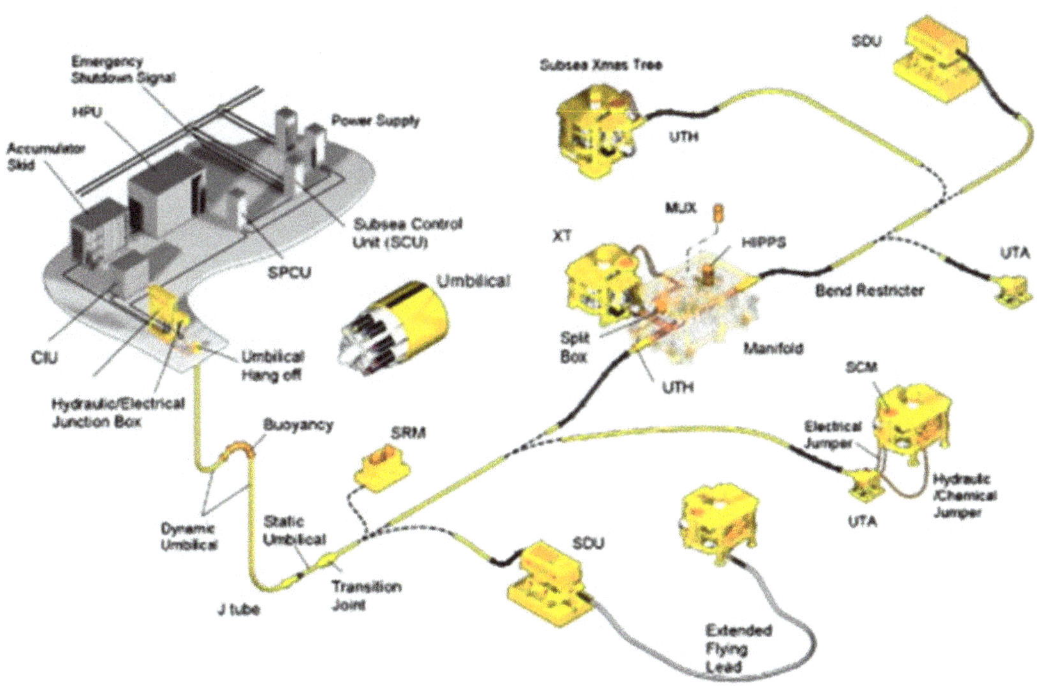

Fig. 5.9 Clustered subsea satellite wells connected to a manifold. (Courtesy of TechnipFMC)

Fig. 5.10 Subsea multi-well template below a moon pool. (Courtesy of Claxton)

guide base is initially used for drilling, hanging off and supporting conductor, wellhead, and subsea tree. In addition, templates provide a base for protective structures. The *permanent guide base* is a steel structure which seats in and is attached to the temporary guide base, Fig. 5.11.

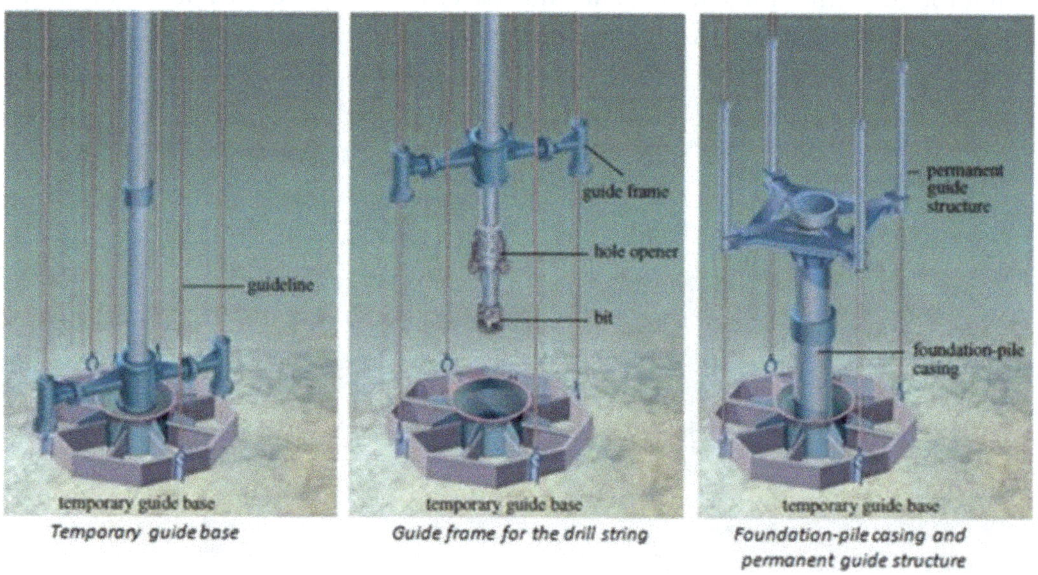

Fig. 5.11 Single-well temporary and permanent guide bases. (Taken from Encyclopedia of Hydrocarbons Eni)

5.3.2 Platform Wells

For a *platform well*, the wellhead, Christmas tree, and production-control equipment are located on the production platform. Platform size depends on number of wells, water depth, and facilities to be installed on top side such as the drilling rig, living quarters, Helipad, etc.

5.4 Types of Offshore Production Units

Offshore production units can be divided into two main categories: bottom supported and vertically moored structures, and floating production systems. Figure 5.12 shows different categories of offshore platforms.

5.4.1 Bottom Supported and Vertically Moored Structures

This category of offshore platforms can be divided into four major types (see Fig. 5.13):

- Fixed platform
- Compliant tower
- Tension leg platform
- Mini-Tension leg platform

Fixed platform—These platform types are, built on concrete or steel legs, or both, and directly anchored to the seabed. They are designed and built for long-term use in moderate water depths up to 400 m. Steel jacket, concrete caisson, floating steel, and floating concrete are various types of fixed platforms. Steel jackets are vertical sections made of tubular steel members, which provides a protective layer around pipes, and are usually piled into the seabed. Fixed platforms typically have a main deck, a cellar deck, and a Helideck which comprise the deck structure. The deck

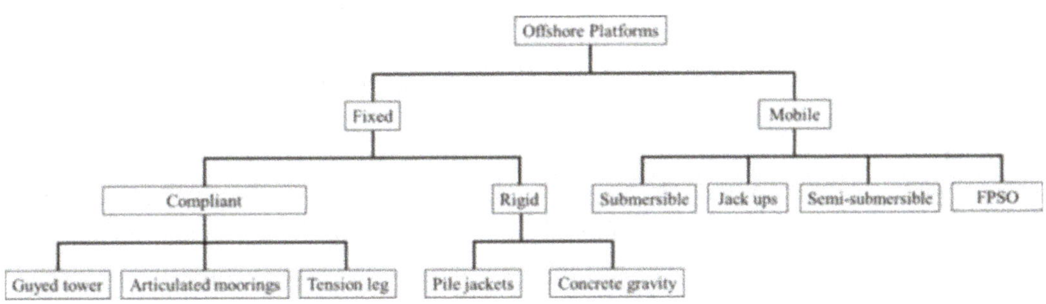

Fig. 5.12 Different categories of offshore platforms

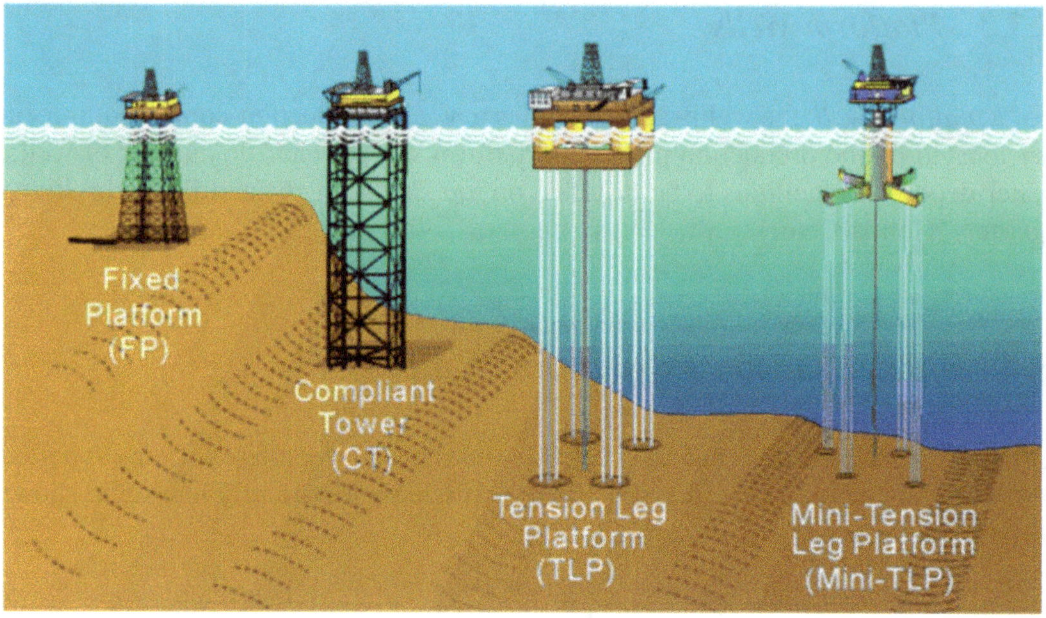

Fig. 5.13 Various types of bottom supported and vertically moored offshore platforms. (Courtesy of BOEM)

structure is standing on deck legs which are connected to the top of the piles. The piles, which are located inside the legs of a jacket, penetrate into soil and are extended above the mean sea level.

Compliant towers—This type of platform is capable of moving along with the external forces acting on the structure. Therefore, a flexibility is given to the structure and it responds to the applied external forces [9]. A compliant tower platform consists of a narrow and flexible (compliant) tower which is supported by piled foundations. The pilled foundations (connected to the sea floor and allowing the structure to move freely with current, waves, and wind) support the deck which accommodates the drilling rig and production facility. However, they are not usually designed for drilling operations but exceptions may exist. The compliant tower platforms are designed and built for deep water depths ranging from 1400 to 3000 (ft). Guyed towers (either piled or spud can foundation), articulated towers, and tension leg platforms are different types of compliant tower platforms.

Tension leg platforms—These type of platforms, known as TLPs, may also be noted as a subcategory of compliant towers as they can move horizontally (see Fig. 5.14). A TLP is a 4-column design whereas each column is moored permanently to the seabed by tethers or tendons. A tether is a vertical steel tube. A group of tethers is called a tension leg, and are designed in such a way that vertical movement of the platform is eliminated. In other words, all the tethers are in pre-tension. This feature allows the wellhead to be placed on deck and connected to the subsea well by use of a rigid riser. As the legs are in tension, the platform is sensitive to topside load variations.

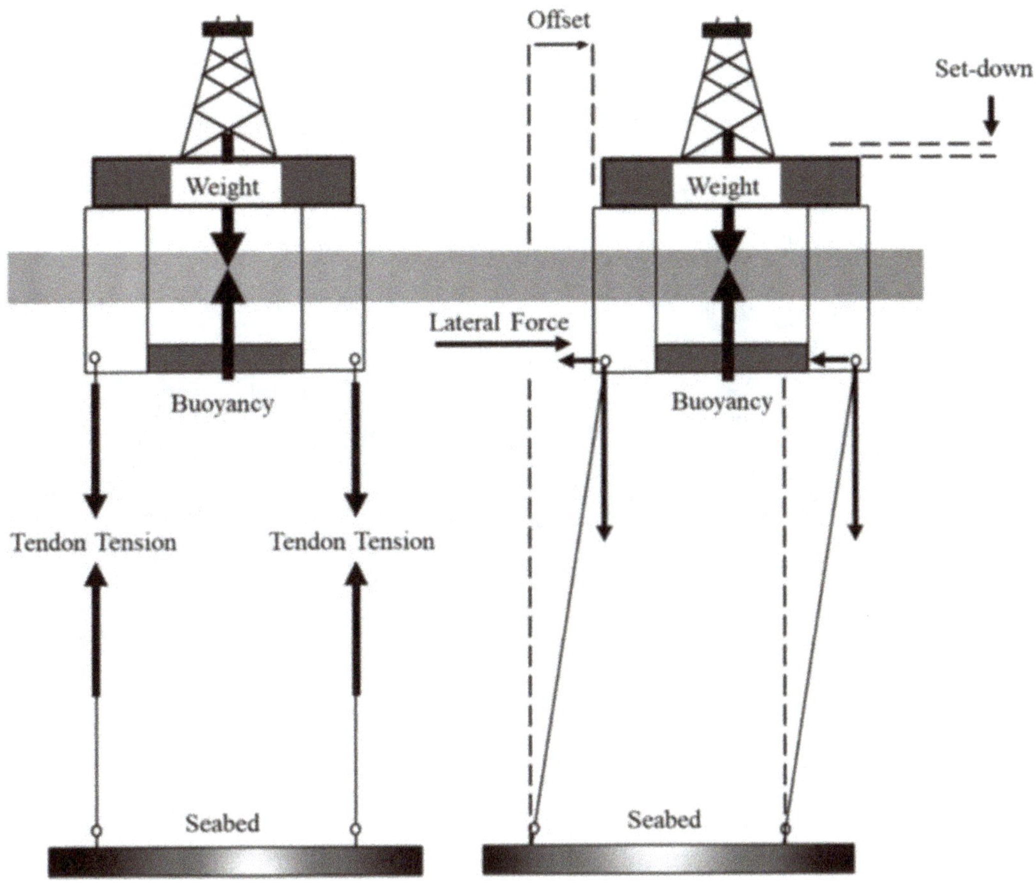

Fig. 5.14 A tension leg platform which is restrained in vertical direction but highly flexible in horizontal plane

Mini-Tension leg platforms—These type of platforms combine the simplicity of a SPAR platform (a floating platform) and favorable features of a TLP [10]. The platform consists of decks, tower, hull, horizontal pontoons, and tethers. Typically, a mini TLP has a low water plane and subsequently experiences less environmental loads and has good response characteristics.

5.4.2 Floating Production Systems

This category of offshore platforms can be divided into three major types (see Fig. 5.15):

- Spar platforms
- Floating production systems
- Floating, production, storage and offloading (FPSO) vessels

Spar platforms—A Spar platform is a type of floating production facility made of a large-diameter, single vertical cylinder (hard tank), which supports a deck on

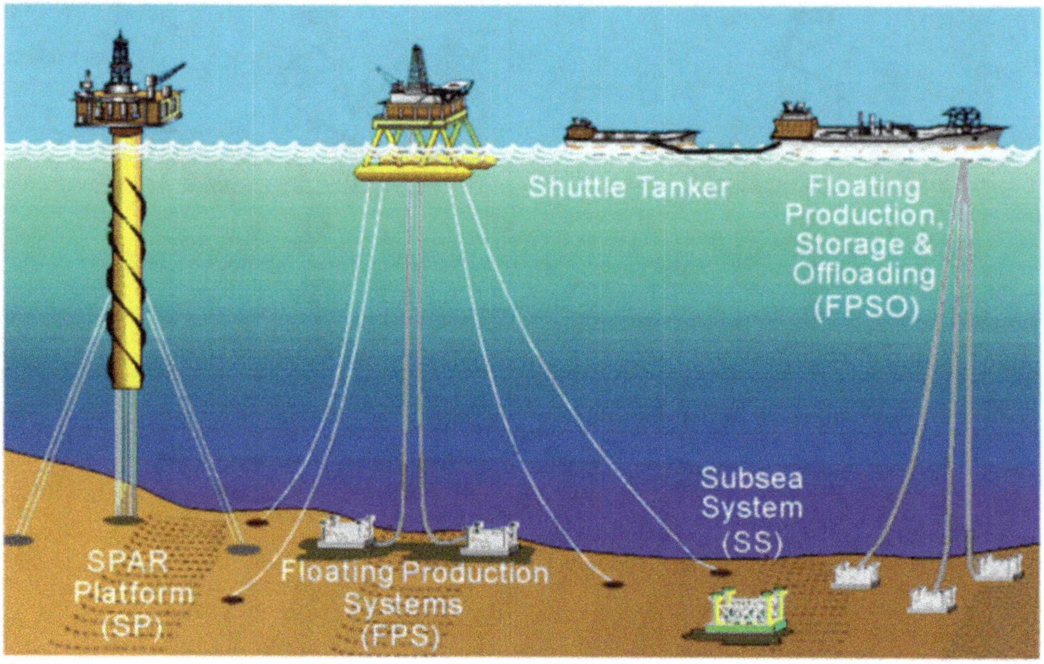

Fig. 5.15 Overview of floating production systems (Courtesy of BOEM)

top. Spar platforms are permanently anchored to the seabed, vertically, by a spread moored system. There are four different types of Spar platforms: classic Spar, truss Spar, cell Spar, and mini-DOC Spar (see Fig. 5.16). One of the major differences of these types is related to size and design of the hard tank. Among these, types, the truss Spar platforms are the most common.

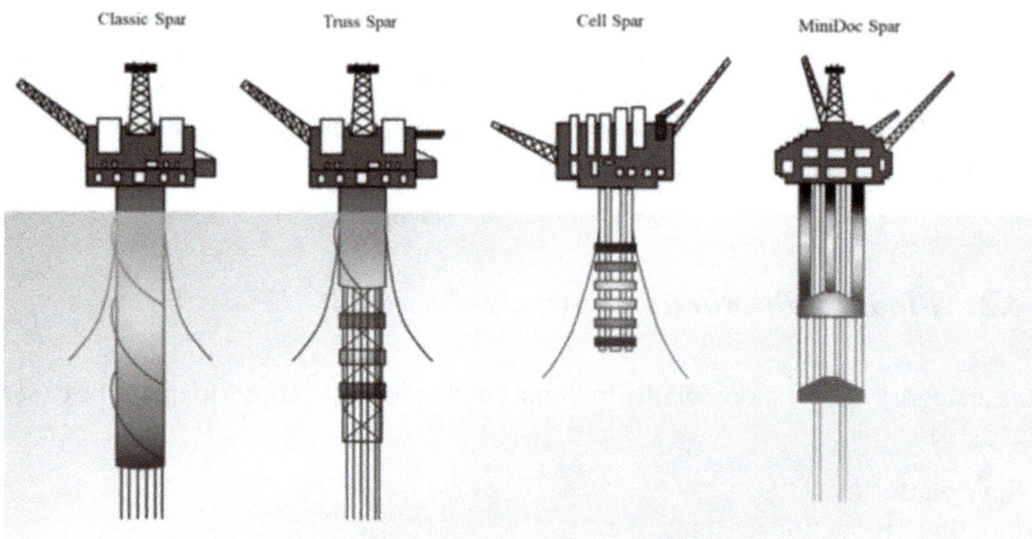

Fig. 5.16 Four different types of Spar platforms

It provides accommodation, crane facilities and usually a drilling rig. The Christmas tree can be either on the seabed (wet tree) or on the platform (dry tree).

Floating production systems (FPSs)—FPSs consist of monohull structures and are equipped with processing facilities. FPSs are moored and can be mobilized and reused after the abandonment of wells. The FPSs are usually used for subsea wells. There are different types of systems and floating, production, storage and offloading systems are a variant.

Floating, production, storage and offloading vessels (FPSO)—FPSOs are a generation of the FPSs. These vessels are ship shaped floaters and do not provide rig or intervention units [11]. The FPSOs are used for subsea wells.

5.5 Manned and Unmanned Platforms

Fixed platforms can be categorized in two types: manned platforms and normally unmanned platforms.

5.5.1 Manned Platforms

All the offshore construction facilities, which accommodate at least one person routinely for more than 12 h for 24 h periods, are known as manned platforms. Such facilities provide an area for a well intervention unit or supporting drilling rig.

5.5.2 Unmanned Platforms

Unmanned platforms are a type of automated offshore platform which primarily operate remotely and without the continuous presence of personnel. Such platforms are operated remotely from onshore bases. They can be categorized into five different types by considering the number of available wells, helideck availability, fire water system, and crane availability (see Table 5.2).

These types of platforms are small in size and may provide a helipad on top but they do not possess accommodation, except for shelters to address personnel emergencies. If a crane is available, they are usually light-weight and not rated for lifting heavy units such as coiled tubing units. As such platforms are small, when the platform is manned to carry out routine activities such as maintenance and well intervention activities, a supply vessel or jackup unit stands by the platform. The standby unit provides enough deck space and accommodation for personnel on board (POB). When an unmanned platform is manned for activities which require more POB, additional safety measures are necessary as the platform may not provide enough rescue boats or fixed fire water systems.

Table 5.2 Five different types of unmanned platform [12]

Type			Specifications
Type 0:	Complex platform with helideck	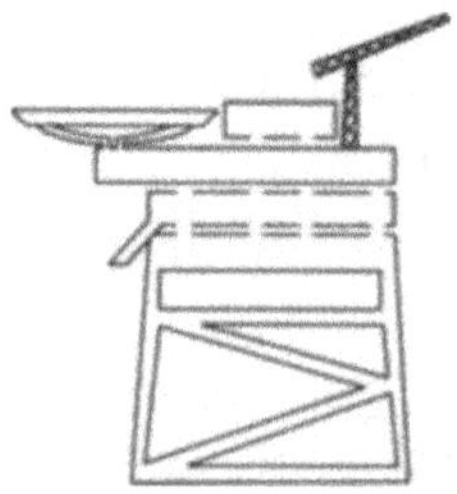	• Equipped with fixed fire water system • Equipped with various process equipment including crane (lifting capacity of 50–60 tonnes) • Automated • Allows remote operation for typically 1–5 weeks • Designed for both coiled tubing and wireline operations
Type 1:	Simple platform with helideck		• Supports typically 2–12 wells • Crane is available (lifting capacity of 10–50 tonnes) • No fire water system • Equipped with test separator or multiphase metering • Allows remote operation for typically 2–3 weeks • May be designed for coiled tubing and wireline operations or only wireline operations
Type 2:	Simple platform without helideck	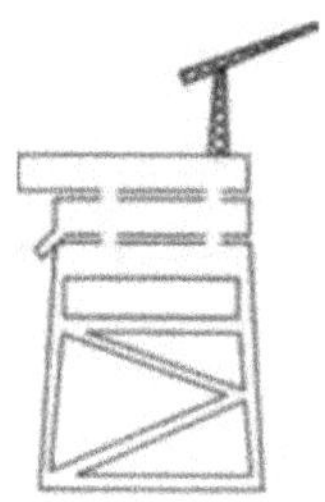	• Supports typically 2–10 wells • Small crane is available (lifting capacity of 1–2 tonnes) • No fire water system • No process facility • Allows remote operation for typically 3–5 weeks
Type 3:	Minimalistic platform		• Supports typically 2–12 wells • No crane • No fire water system • No process facility • Allows remote operation for typically 6 months up to 2 years • All well intervention operations require an offshore support rig
Type 4:	Super minimalistic platform	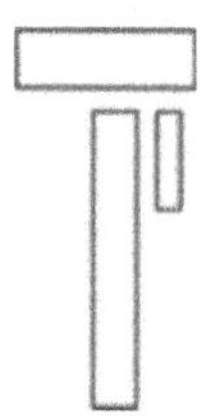	• Supports typically 1 well • One small deck • Well is connected directly connected to pipeline • All well intervention operations require an offshore support rig

It is a common practice to design and construct simplified offshore installations to keep the initial costs low. Consequently, when considering operational activities for unmanned platforms, some major factors should be considered including: safety-critical systems, deck space, POB, and weather. Due to the design of unmanned platforms, they are not equipped with all of the safety-critical systems such as fixed firewater pumps and larger capacity life boats. The deck space is also very limited due to the compact design of the platforms. Consequently, during operations, a minimum of personnel are permitted to work on unmanned platforms due to safety and emergency response, unless the operation is carried out from a standby working unit. Weather is another major factor to be considered for executing operations on unmanned platforms. Due to constraints including deck space, lifting capacity, etc., an offshore support rig (working unit) is employed to perform the operations. If the employed working unit is not a bottom supported unit, bad weather could cause disastrous consequences such as a collision between the platform and working unit or compromising pressure control procedures. When employing floating working units for intervention and P&A activities, a weather downtime of up to 50% is reported for unmanned platforms in North Sea. However, this depends on the season.

Challenges associated with unmanned platforms can be listed as personnel accommodation, equipment limitations (such as number, size, and weight), and fast crew transfer. Most of these challenges can be overcome by proper selection of a supplementary working unit. Normal anchor handler tug (AHT) vessels, supply vessels, and dynamic positioning vessels are some options besides offshore drilling units [12, 13].

5.6 Mooring Systems for Floating Units

When considering offshore activities, unit motion becomes a critical subject which increases the operation cost and risk. For floating platforms and floating working units, motion means weather downtime and subsequently, weather downtime means increased operation cost. In other words, the primary task of mooring is to reduce the motion of platform or working unit. Fixed platforms and fixed working units do not require mooring system. Studies show that mooring operations can contribute up to 25% of drilling cost. Therefore, an efficient mooring system needs to be considered during P&A of subsea wells or platform wells that may require a complementary floating unit. A mooring system consists of: mooring chain (chain cable) and fiber ropes, windlass, anchors, and mooring winches.

Mooring systems can be either temporary or permanent. A temporary mooring system provides service for relatively short periods of time. The periods can be weeks or months at a time. Most mobile units employed to carry out P&A operations benefit from a temporary mooring system. However, permanent mooring systems provide station-keeping for several years. Typically, permanent mooring systems are utilized to tether floating production facilities. The differences between permanent and temporary mooring systems can be referred to as criteria considered in the design

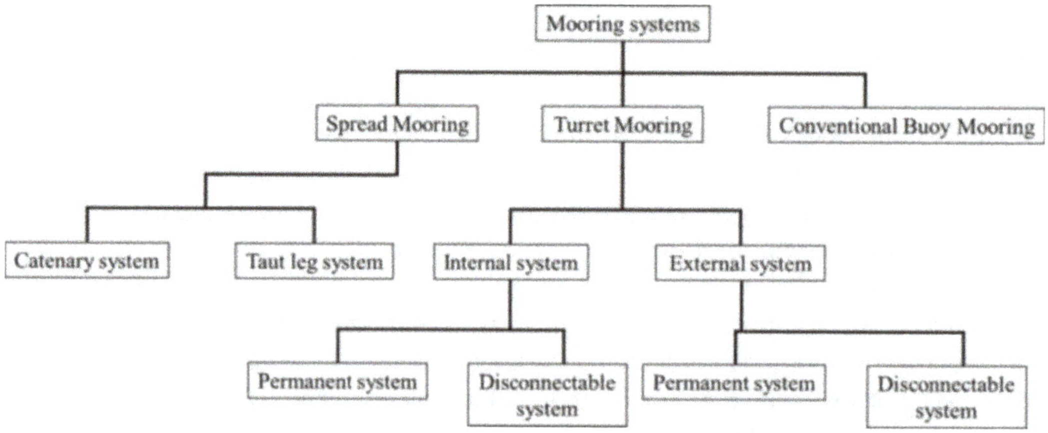

Fig. 5.17 Three main categories of mooring systems

process of the system, type and size of mooring components, type of system analysis, installation methods, and inspection and maintenance philosophy. Generally, mooring systems are categorized into three main categories (see Fig. 5.17): spread mooring systems (Fig. 5.18a), turret mooring systems (Fig. 5.18b), and conventional buoy mooring system (Fig. 5.18c).

5.6.1 Spread Mooring Systems

In this system, mooring lines are spread over multiple points and the system maintains the working unit or platform on location with a fixed heading. In spread mooring systems, two different configurations of mooring line are distinguishable (see Fig. 5.19): catenary system, and taut leg mooring system.

In a catenary system, a parabolic geometry of cables are anchored to the seabed (see Fig. 5.19a). In this configuration, lines are laid down on the seabed and then leave the seabed to the connectors on the unit. Usually, lines are steel chains which subsequently occupy a large space and their transportation is a challenging task. Corrosion of chains is another issue to be considered when utilizing steel lines.

In taut system, lines are stretched between two points; one point on the seabed and another point to the connectors of the unit (see Fig. 5.19b). The lines are polyester ropes which have several advantageous over steel chains. Advantageous include: polyester ropes are lighter and less challenging with regards to accommodation and transportation, they give a softer mooring system, better vortex induced motion response to loop currents, lower product cost, no concerns associated with corrosion, and reduction in mooring pre-tension [14].

Fig. 5.18 **a** Spread mooring system, **b** Internal turret mooring system, and **c** Buoy mooring system. (Courtesy of Offshore Magazine)

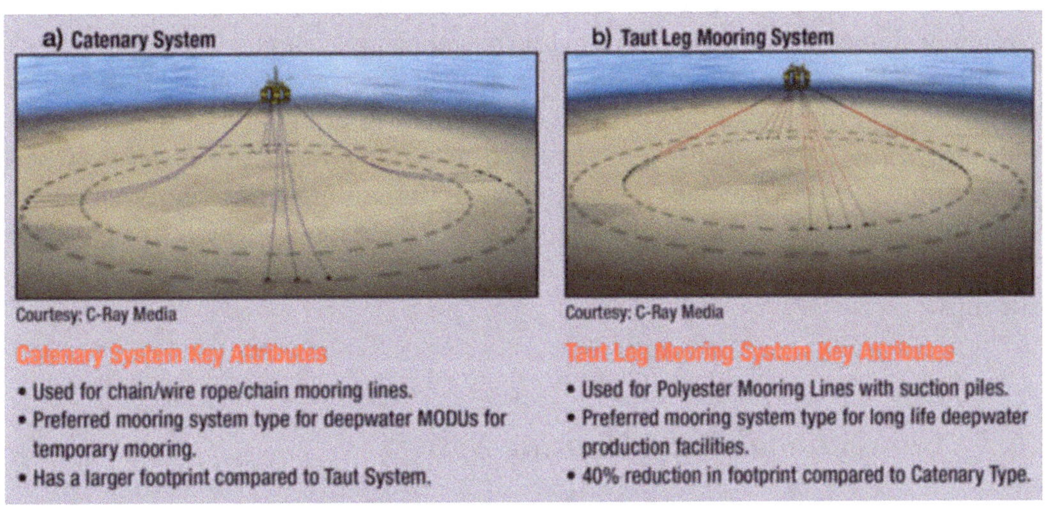

Fig. 5.19 Mooring line configuration used in spread mooring systems; **a** Catenary system, **b** Taut leg mooring system. (Courtesy of Offshore Magazine)

Fig. 5.20 External turret with dry mooring table. (Courtesy of Offshore Magazine)

5.6.2 Turret Mooring Systems

Turret mooring systems are divided into two main categories: internal turret and external turret (Fig. 5.20). Internal turret mooring systems are the most common for extreme design conditions. The internal turret mooring system is positioned inside the hull (see Fig. 5.18b) and it is either permanent or disconnectable [15–17]. Permanent internal turret mooring systems are located in a moonpool. Internal turret mooring systems are designed for moderate to deep water depths and locations where a large number of flexible risers are required. External turret mooring systems are also categorized as permanent or disconnectable. Turret mooring system are usually used for Floating, Production, Storage, and Offloading (FPSO) floating systems and drillships.

5.6.3 Conventional Buoy Mooring System

A conventional buoy mooring (CBM) system (see Fig. 5.18c) typically consists of buoys, mooring legs, and anchor points. A typical CBM consists of 3 to 4 buoys which are moored to the seabed by chain legs, high holding power anchors, or piles.

5.6.4 Offshore Mooring Patterns

There are different types of offshore mooring patterns for temporary and permanent mooring systems, Fig. 5.21. Depending on the intended use of the floating offshore unit, type of operation, and location, different configurations are available.

A mooring line is made of different components (see Fig. 5.22). Manufacturing and selection of the components depends on duration of tethering, size of floating offshore unit, location, water depth, etc. Weight and space allocation for mooring lines is important during the designing process which may influence the mooring configuration.

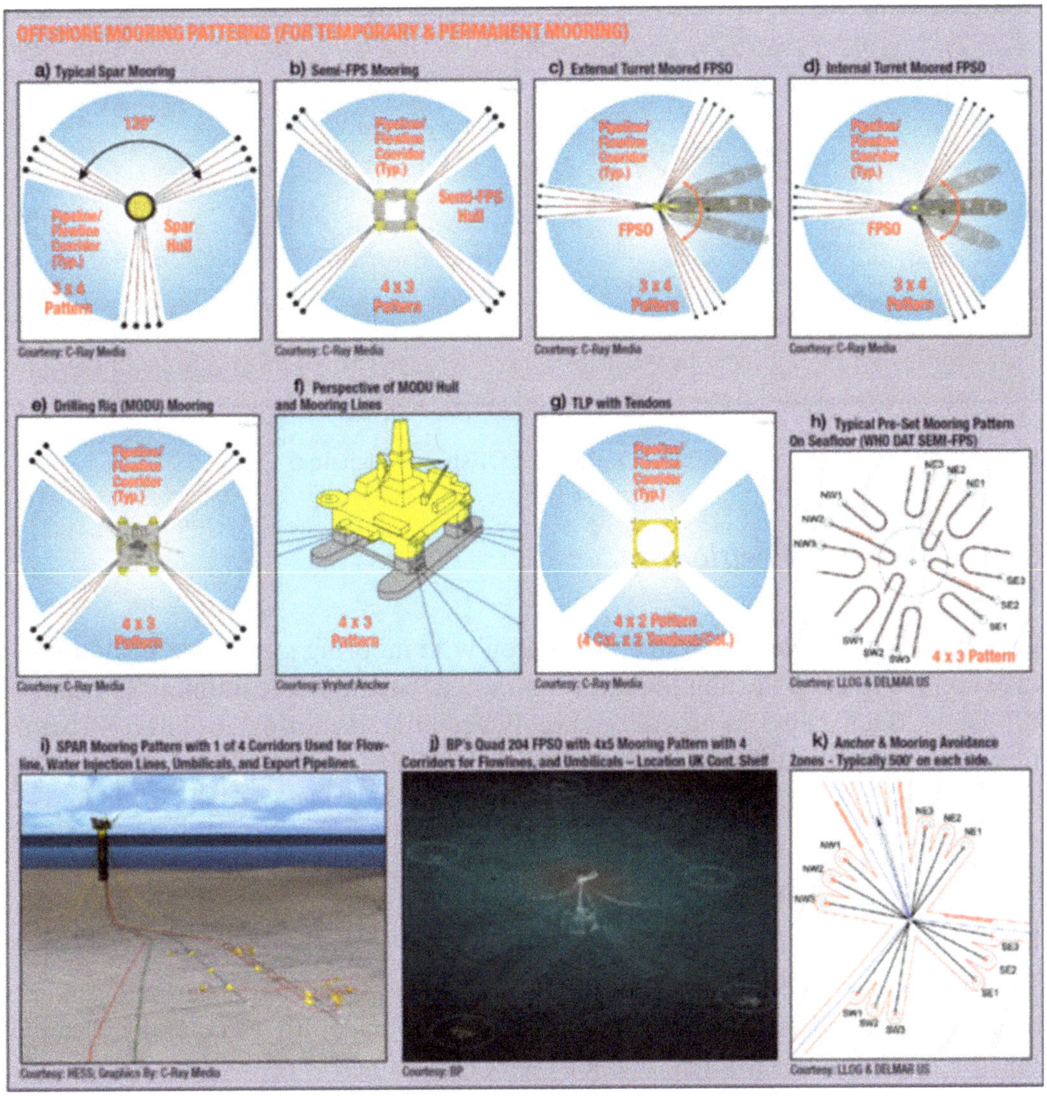

Fig. 5.21 Temporary and permanent offshore mooring configurations. (Courtesy of Offshore Magazine)

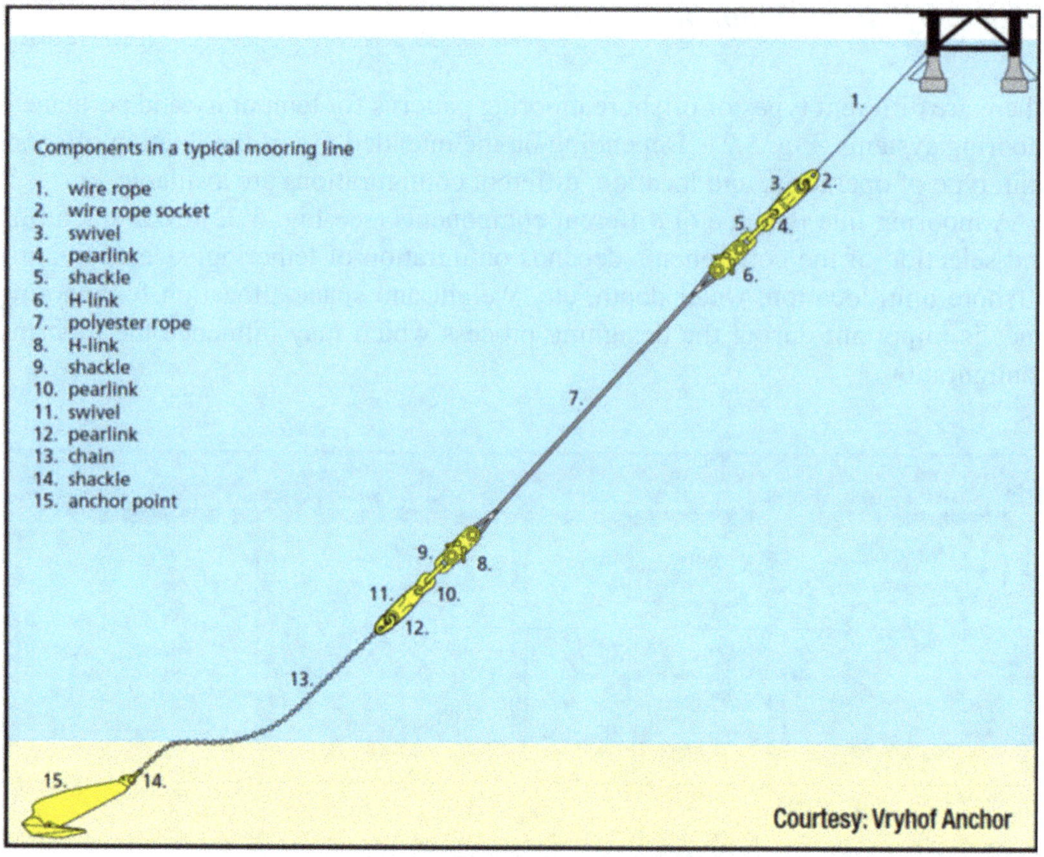

Fig. 5.22 Mooring line components. (Courtesy of Offshore Magazine)

5.6.5 Dynamic Positioning

When thrusters, standalone or in combination with mooring systems, are simultaneously applied to keep the unit in place, it is called "Dynamic Positioning" system or DP system. Such a system provides a highly versatile anchoring system for floating units at deep and ultra-deep locations [18].

5.7 Anchoring Types

Mooring systems need to be anchored to the seabed. The marine ground-anchors are designed based on their capacity for withstanding uplift force and horizontal drag force. There are different anchor types including clump weight, driven pile, drag anchor, suction pile, torpedo pile (drop anchor), and vertical load anchor (see Fig. 5.23).

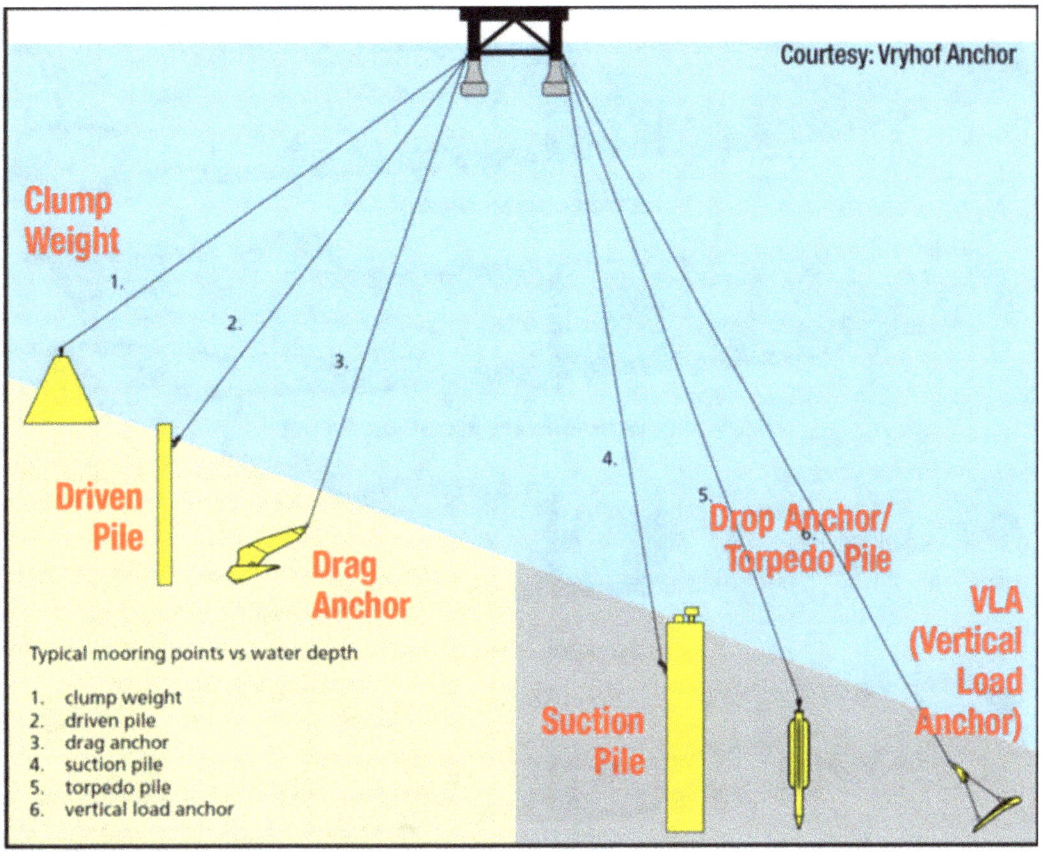

Fig. 5.23 Different anchor types. (Courtesy of Offshore Magazine)

Pre-laid mooring system
In *traditional mooring systems*, the mooring system is established while the working unit is in place. However, due to the day-rate hire cost of the rig, this is not of interest. *Pre-laid mooring system* is a cost efficient alternative scenario. In this approach, prior to mobilizing the working unit, a vessel spreads and establishes the mooring system.

5.8 Moonpool

The moonpool is an open space located in the hull of a vessel or a drillship, which provides access to water entry. The moonpool can have different configurations varying from rectangular to an inverted funnel-like shape, Fig. 5.24. Size, configuration and number of available moonpool can impact the efficiency of a P&A operation as the number of operations which can run simultaneously and crew numbers depend on such factors. When a working unit is in operating mode, at zero speed, the moonpool is opened and a large volume of water known as *entrained water* enters into it. The entrained water has motion which appears as two modes, oscillation and sloshing.

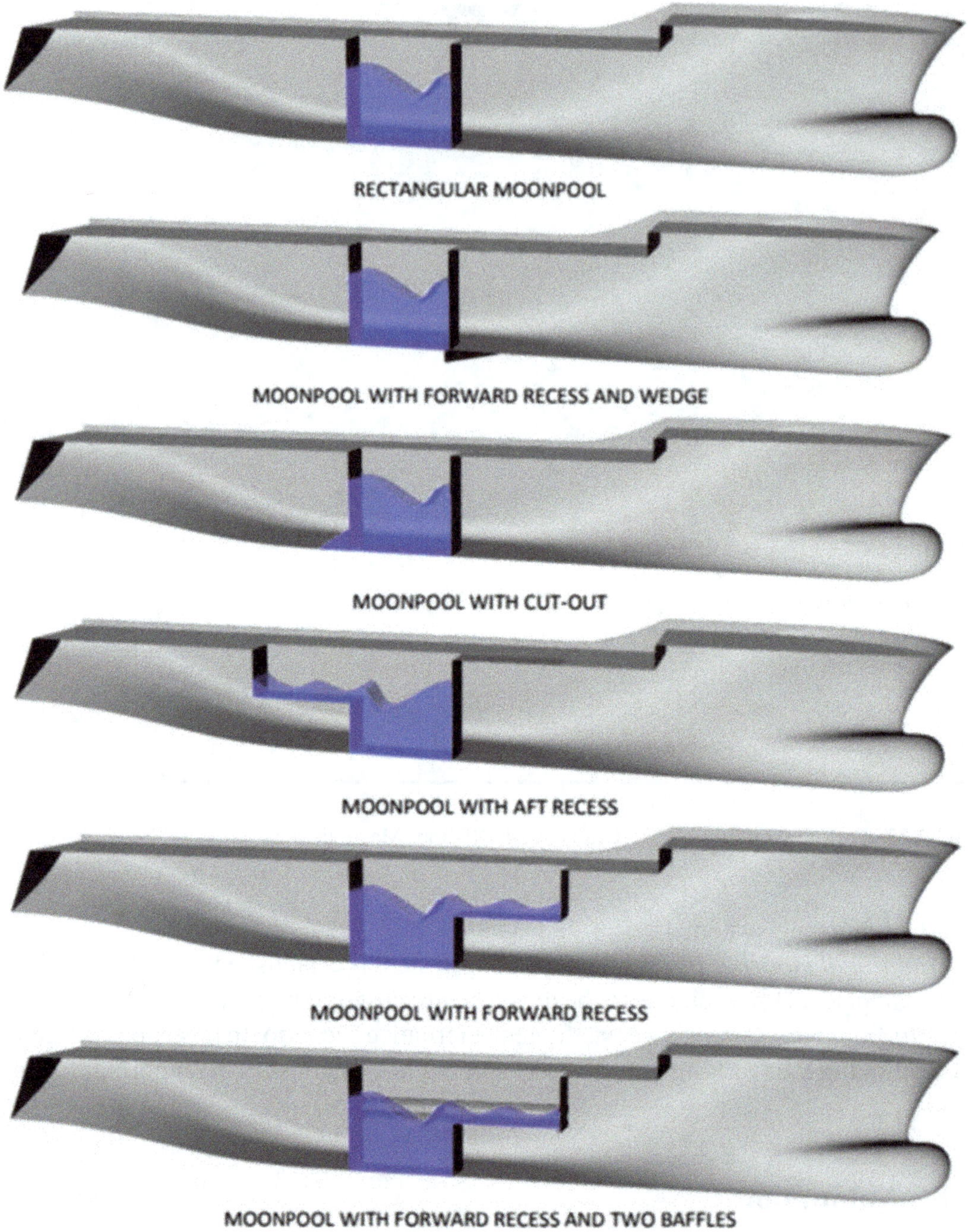

Fig. 5.24 Some types of moonpool configurations. (After Hammargren and Tornblom) [19]

The oscillation mode is when there is a vertical motion of water column. The sloshing mode is when water moves in a longitudinal direction. There are situations where the water motion inside the moonpool can be so strong that the water level reaches the deck and can cause harm to personnel.

References

1. Totten, G.E. 2004. *A timeline of highlights from the histories of ASTM committee D02 and the petroleum industry.* https://www.astm.org/COMMIT/D02/timeline.pdf. Cited 18 Dec 2017.
2. A.T. Bourgoyne Jr., K.K. Millheim, M.E. Chenevert, et al. 1991. Applied drilling engineering. *Society of Petroleum Engineers*. 978-1-55563-001-0.
3. Kaiser, M.J. and B. Snyder. 2013. The five offshore drilling rig markets. *Marine Policy*, 39 (Supplement C), 201–214. https://doi.org/10.1016/j.marpol.2012.10.019.
4. Wilson, A. 2016. Core drilling using coiled tubing from a riserless light-well-intervention vessel. *Society of Petroleum Engineers*, 68(06). https://doi-org.ezproxy.uis.no/10.2118/0616-0051-JPT.
5. Englehardt, J., M.J. Wilson, and F. Woody. 2001. New abandonment technology new materials and placement techniques. In *SPE/EPA/DOE exploration and production environmental conference*. SPE-66496-MS, San Antonio, Texas: Society of Petroleum Engineers. https://doi.org/10.2118/66496-MS.
6. Bosworth, P. and O. Willis. 2013. Rigless intervention: Case studies, UK and Africa. In *Offshore technology conference*. Houston, Texas, USA: Offshore Technology Conference. https://doi.org/10.4043/24065-MS.
7. Franklin, R., I. Collie, and C. Kochenower. 2000. *Subsea Intervention From a NonDrilling-Rig-Type Vessel*. in *Offshore Technology Conference*. OTC-12126-MS, Houston, Texas, USA: Offshore Technology Conference. https://doi.org/10.4043/12126-MS.
8. Saasen, A., M. Simpson, B.T. Ribesen, et al. 2010. *Anchor Handling and Rig Move for Short Weather Windows During Exploration Drilling*. in *IADC/SPE Drilling Conference and Exhibition*. SPE-128442-MS, New Orleans, Louisiana, USA: Society of Petroleum Engineers. https://doi.org/10.2118/128442-MS.
9. Bai, Y. and Q. Bai. 2010. *Subsea engineering handbook*. USA: Gulf Professional Publishing. 978-1-85617-689-7.
10. Bhattacharyya, S.K., S. Sreekumar, and V.G. Idichandy. 2003. Coupled dynamics of SeaStar mini tension leg platform. *Ocean Engineering* 30 (6): 709–737. https://doi.org/10.1016/S0029-8018(02)00061-6.
11. Zhang, D., Y. Chen, and T. Zhang. 2014. Floating production platforms and their applications in the development of oil and gas fields in the South China Sea. *Journal of Marine Science and Application* 13 (1): 67–75. https://doi.org/10.1007/s11804-014-1233-2.
12. Nielsen, A. 2016. *Unmanned wellhead platforms—UWHP summary report*. Oljedirektoratet Norway: Norway. 27.
13. Alyafei, O. 2014. Well services operations in offshore unmanned platform; challenges and solutions. In *International petroleum technology conference*. IPTC-17231-MS, Doha, Qatar: International Petroleum Technology Conference. https://doi.org/10.2523/IPTC-17231-MS.
14. Flory, J.F., S.J. Banfield, and C. Berryman. 2007. *Polyester Mooring Lines on Platforms and MODUs in Deep Water*. in *Offshore Technology Conference*. OTC-18768-MS, Houston, Texas, U.S.A.: Offshore Technology Conference. https://doi.org/10.4043/18768-MS.
15. Duggal, A., A. Izadparast, and J. Minnebo. 2017. Integrity, monitoring, inspection, and maintenance of FPSO turret mooring systems. In *Offshore Technology Conference*. OTC-27938-MS, Houston, Texas, USA: Offshore Technology Conference. https://doi.org/10.4043/27938-MS.
16. Mack, R.C., R.H. Gruy, and R.A. Hall. 1995. Turret Moorings for extreme design conditions. In *Offshore technology conference*. OTC-7696-MS, Houston, Texas: Offshore Technology Conference. https://doi.org/10.4043/7696-MS.
17. Pollack, J., R.F. Pabers, and P.A. Lunde. 1997. Latest breakthrough in turret moorings for FPSO systems: The forgiving Tan kerrrurret interface. In *Offshore technology conference*. OTC-8442-MS, Houston, Texas: Offshore Technology Conference. https://doi.org/10.4043/8442-MS.

18. Shneider, W.P. 1969. Dynamic positioning systems. In *Offshore technology conference*. OTC-1094-MS, Houston, Texas: Offshore Technology Conference. https://doi.org/10.4043/1094-MS.
19. Hammargren, E. and J. Tornblom. 2012. Effect of the moonpool on the total resistance of a drill-ship. In *Department of Shipping and Marine Technology*. Chalmers University of Technology: Sweden.

Chapter 6
Work Classification and Selection of Working Units

As reviewed in the previous chapter, there are different types of working units employed for permanent P&A operations depending on the well location and type of production facility. When considering onshore wells, the site space is not a concern, except for mountainous areas and swamps. However, site space for offshore wells may be a concern. Selection of working units for offshore wells is a concern as the daily rate of working units highly affect the final cost of permanent P&A operation. Generally speaking, there are some critical factors to be considered for the selection of the optimal working unit for the P&A job including: well location (either onshore or offshore), depth of operation/hole, required horsepower, availability of working unit, type of production facility, type of well completion, unit capacity (offshore wells) and availability of space (offshore wells). Unit capacity is the maximum load which a unit, production facility or platform, can support without collapsing.

When considering offshore wells, more than one type of working unit may be employed when performing different phases of P&A operation, depending on the type of production facility and availability of units. Generally, if a working unit is supposed to fully conduct the permanent P&A operation, the following systems are necessary: power system, fluid-circulating system, hoisting system, rotary system, well control system, well-monitoring system, and special marine equipment. At first glance, the required systems may suggest a drilling rig to be convenient, however, this might not be correct. A drilling rig is designed for the purpose of drilling. It has a high daily rates and needs to have all the mentioned systems at the same time. A P&A working unit may not need all the available systems at the drilling rig. For example: a multipurpose modular working unit could be an example of such a unit whereby different systems can be integrated at the required time. A modular rig may be used when either the existing drilling rig is not properly maintained or when there is no existing drilling rig onboard the platform. A P&A unit requires much lower drilling fluid pit volume compared to a drilling rig. However, it should be mentioned that a modular rig has a significant mobilization and demobilization cost and relies on existing infrastructure such as cementing system, mud system, utility, etc. to a large extent.

M. Khalifeh and A. Saasen, *Introduction to Permanent Plug and Abandonment of Wells*, Ocean Engineering & Oceanography 12,
https://doi.org/10.1007/978-3-030-39970-2_6

6.1 P&A Code System

Consider a situation where you wish to briefly explain the complexity of permanent abandonment of a well or several wells. Perhaps, it would be time consuming to explain the well location and complexity of each abandonment phase. A P&A code system can address their challenge. A P&A code system aims to classify wells for abandonment cost estimation. The P&A code system classifies wells according to three factors [1]:

- Well location
- Abandonment phases
- Abandonment complexity.

The well location is presented by two letters and followed by three digits, where each digit represents the complexity of abandonment operation for each abandonment phases.

6.1.1 *Well Location*

The Well location defines the physical location of well; land, platform, or subsea. So, the first two letters are as following:

- LA—Land well,
- PL—Platform well,
- SS—Subsea well.

6.1.2 *Abandonment Phases*

A P&A operation, can be divided into three different phases: Phase 1—reservoir abandonment, Phase 2—intermediate abandonment, and Phase 3—wellhead and conductor cut and removal. These phases are regardless of well location. The main goal in P&A is to perform the full operation as possible without a rig and removing as little steel as possible. The footprint should also be kept small. Therefore, experienced personnel should be involved with a good knowledge of what is required.

6.1.2.1 Phase 1—Reservoir Abandonment

This phase includes the following activities: wellhead is checked, waste handling system is prepared, wireline investigations are conducted and if possible, cement is squeezed into the reservoir perforations. If the squeezed cement is extended across the cap rock and is qualified, it is counted as a primary permanent barrier. So far, these activities are performed with XMT in place and it is a rigless operation. If the squeezed cement is not qualified, the primary and secondary permanent barriers shall be established to secure the reservoir. This step may be carried out rigless or using a rig. When a rig is required, the well control system needs to be established. There are circumstances which require the use of a rig during the operation of Phase 1. These circumstances include: restricted access through the tubing to the barrier depth, lack of technology to log casing cement through the production tubing, poor cement or no cement behind the production casing, retrieval of production tubing due to presence of control lines at the barrier depth, a permanent packer set above cap rock, presence of an Annulus Safety Valve (ASV), and experiencing SCP due to hydrocarbon or overpressures.

6.1.2.2 Phase 2—Intermediate Abandonment

In this phase, all the identified zones with flow potential in the overburden need to be isolated. All the hydrocarbon flow potentials are secured by primary and secondary permanent barriers. Hydrocarbon zones with no flow potential and water bearing zones are isolated by establishing one permanent barrier. If the water bearing zone is a pressurized zone, then two permanent barriers, primary and secondary, are required. In the last part of Phase 2, a top plug often called and environmental plug, is installed. This phase may be carried out with a rig or rigless. The circumstances which dictate the use of a rig, in Phase 2, include: SCP due to hydrocarbons or overpressure at barrier depth which originates from the reservoir, restricted access to the casing, no isolated fresh water aquifers or zones, and non-isolated shallow gas. In addition, poor cement or uncemented casing at barrier depth, lack of technology to log casing cement behind the second casing string, and the presence of control lines (if not retrieved during Phase 1) can dictate the use of rig.

6.1.2.3 Phase 3—Wellhead and Conductor Cut and Removal

This phase is the last stage of a permanent P&A operation whereby the well control system is dismantled and wellhead and conductor are cut and pulled. When the wellhead is removed, re-entry to the wellbore would be almost impossible as the well control system cannot be installed. The cut and removal can be performed by use of a rig, conductor jack, vessel (subsea wells), or heavy lift vessel (offshore wells). The circumstances which may require the use of a rig, in Phase 3, may include: poor conductor integrity, platform may not be able to support the conductor load during pulling (offshore wells), water depth is beyond the limitation for cutting by anchor handling vessels or LWIV for subsea wells. Poor integrity of conductor may be caused by corrosion, weak connectors or shallow damage.

6.1.3 Abandonment Complexity

The well abandonment complexity is shown by a digit from 0 to 4, which is distinguished for each abandonment phases.

Complexity 0: **No work is required**. The abandonment phase has already been accomplished and no further work is required.
Complexity 1: **Simple rig-less operation**. A wireline unit, pumping, crane and jacks are utilized for the operation. Riser-less LWIV will be employed for subsea wells.
Complexity 2: **Complex rig-less operation**. Wireline unit, coiled tubing unit, hydraulic work-over unit (HWU), crane, and jacks are utilized during the operation. A heavy duty well intervention vessel with riser may be used for subsea wells.
Complexity 3: **Simple rig-based operation**. Operation requires retrieval of tubing and casing.
Complexity 4: **Complex rig-based operation**. Due to limited access to barrier depth, poor casing cement, or no casing cement, tubing and casing are required to be retrieved, and section milling and cement repair are necessary.

To figure out the complexity of each phase, it is recommended to consider some criteria, which are based on experience. These criteria are presented in Tables 6.1, 6.2 and 6.3. Sustained casing pressure related to hydrocarbons or overpressure is an indication of a well integrity issue associated with failure of the primary cement. The cement failure needs to be mitigated at cap rock level and there are risks associated with well control. Therefore, the operation is highly complex and well control equipment is required. Uncemented casing or poor casing cement means that the annulus needs to be accessed and a new annular barrier needs to be established. In a conventional P&A operation, section milling or an alternative technique is required. Therefore, the operation is of high complexity and HSE issues are associated with it. One of the main concerns, especially at the first stage of the P&A operation, is access downhole to below the estimated minimum setting depth where permanent barriers are required to be established. Drift diameter can be limited due to collapse or deposits of downhole minerals or chemicals. Such circumstances may be mitigated by injecting chemicals or it may require retrieval of the production tubing, either by a cut and pull operation or milling. There are situations where due to restricted access, the production tubing is milled from near surface to all the way down to the required depth of barriers. High torque and circulation gives the operation a complexity level of 4. The production tubing can also create a challenge if it is leaking, as the circulation of fluid or cement cannot be done. If a coiled tubing unit cannot be utilized for circulating or pumping cement, then the production tubing needs to be retrieved. Even though tests have shown that zonal isolation is possible, most authorities do not accept the presence of control lines and downhole gauges as a part of the permanent barrier. Because the control lines are attached to the production tubing, the tubing needs to be retrieved which requires a high pulling capacity. It should be noted that retrieval of production tubing means a higher cost of pipe pulling and handling, HSE issues for personnel and transportation to a location for disposal.

Table 6.1 The considered criteria for classifying complexity of Phase 1 of a permanent P&A operation [1]

×: Not feasible, √: Required, O: Optional		Well abandonment complexity			
		Type 1	Type 2	Type 3	Type 4
		Simple rigless	Complex rigless	Simple rig-based	Complex rig-based
1	Sustained Casing Pressure (SCP) due to hydrocarbons or overpressure	×	×	×	√
2	Uncemented casing or liner at barrier depth (cap rock)	×	×	×	√
3	Restricted access to tubing	×	×	√	O
4	Deep electrical or hydraulic lines present at barrier depth	×	×	√	O
5	Annular Safety Valve (ASV) present	×	×	√	O
6	Packer set above cap rock	×	×	√	O
7	Site does not allow CT/HWU pumping operations	×	×	√	O
8	Multiple reservoirs to be isolated	×	√	O	O
9	Tubing leak (e.g. corrosion, accessories)	×	√	O	O
10	Inclination >60° above packer (wireline access)	×	√	O	O
11	Well with good integrity, no limitations	√	O	O	O

Table 6.2 The criteria considered for classifying the complexity of Phase 2 of a permanent P&A operation [1]

×: Not feasible, √: Required, O: Optional		Well abandonment complexity			
		Type 1	Type 2	Type 3	Type 4
		Simple rigless	Complex rigless	Simple rig-based	Complex rig-based
1	Sustained Casing Pressure due to hydrocarbons or overpressure	×	×	×	√
2	Restricted access to tubing	×	×	×	√
3	No isolated fresh water aquifers/zones	×	×	×	√
4	Uncemented casing or liner at barrier depth (cap rock)	×	×	×	√
5	No isolated shallow gas	×	×	×	√
6	Site does not allow CT/HWU pumping operations	×	×	√	O
7	Poor primary casing cement	×	×	√	O
8	No tubing in well	×	√	O	O
9	Inclination >60° above barrier depth (wireline access)	×	√	O	O
10	Well with good integrity, no obstacles, tubing in place	√	O	O	O

An annulus safety valve may represent a restriction and limit the maximum flowrate required when circulating fluids or cement through tubing, which may demand retrieval of the production tubing. When the permanent packer is above the estimated minimum setting depth, and the workstring cannot pass through it

Table 6.3 The criteria considered for classifying the complexity of Phase 3 of a permanent P&A operation [1]

×: Not feasible, √: Required, O: Optional		Well abandonment complexity			
		Type 1	Type 2	Type 3	Type 4
		Simple rigless	Complex rigless	Simple rig-based	Complex rig-based
1	Poor conductor integrity	×	×	×	√
2	Platform unable to manage conductor load during retrieval	×	×	√	O
3	Water depth beyond limitation for cutting by LWIV (Subsea well)	×	×	√	O
4	Conductor cutting/rigless retrieval	√	O	O	O

to establish the primary and secondary permanent barriers, the packer needs to be milled.

As discussed in Chap. 5, that actual offshore facilities may not be able to accommodate crews, store equipment, use cranes or withstand load capacity. A support vessel may therefore be required. A typical situation is the need to isolate multiple reservoirs or multiple high pressure zones which may require a rig to remove the downhole completion and packers. Permanent plug and abandonment of multiple reservoirs or flow potential sections means a higher complexity of the operation.

When the permanent barriers are to be installed in depths with inclinations greater than 60°, a tractor is necessary for conducting wireline operation such as setting wireline plugs. Even at high inclinations, punching casing or running a wireline operation can be almost impossible. Therefore, high inclination at the barrier depth introduces multiple challenges.

Permanent P&A operation of wells with good integrity can be done rigless as internal plugs only need to be installed across the qualified annular barrier. Such operations can be done utilizing a coiled tubing unit.

Fresh water zones, abnormally pressured water bearing zones, and shallow gas zones need to be isolated by installing cross-sectional barriers. If such zones are poorly isolated or the corresponding annular space is uncemented, access to the formation should be achieved to establish permanent barriers. Such an operation may require section milling and well control systems.

Poor integrity of the conductor caused by corrosion, weak connectors, leaking connectors, etc. requires program with contingency plans. There are circumstances where a platform or mobile offshore unit is not able to manage the conductor load during conductor removal. The situation is amplified when the inner casing is cemented. For subsea wells, the wellhead and conductor are usually cut and retrieved using an anchor handling vessel or LWIV. However, if the water depth is beyond the operational water depth for the LWIV, a heavy offshore unit may be required. Water depth can create challenges for the cut and removal of the wellhead. As an example, currently 500 m water depth is the limit for abrasive cutting of the wellhead or conductor due to the limit for compressor capacity.

Example 6.1 A subsea well, which is located in an ultra-deep water area, is going to be permanently plugged and abandoned. The well suffers from sustained casing pressure in A- and B-annulus. Logging data shows a shallow gas zone, which has not been isolated properly. What is the P&A code for the well.

Solution As the well is a subsea well, the first two letters are SS. The well suffers from SCP which means well integrity issue at cap rock level. By refereeing to Table 6.1, the P&A complexity of operation for Phase 1 is 4. The shallow gas zone needs to be secured and as there is uncemented casing at the depth of the gas zone, by referring to Table 6.2, the complexity of operation for Phase 2 is 4. The well is located in ultra-deep water area which is beyond conventional vessels. By referring to Table 6.3, the operation complexity is 4. Therefore, the P&A code system for the well is: SS-4-4-4.

6.2 Time and Cost Estimation of a P&A Operation

As a candidate well for permanent P&A is not going to be profitable, and all the P&A cost associated with it are not going to be recovered, cost estimation is an important process. To understand necessary time and cost of a P&A operation, it is necessary to identify the factors affecting the operation and to quantify their interaction. However, it is impractical to identify all the characteristics of a P&A operation. Therefore, in practice it is important to consider those factors that adequately represent the P&A operation. The contributing factors can be classified as either *observable* or *unobservable* factors. Observable factors are measured and quantified directly such as well characteristics, or may need to be represented by a proxy variable, such as operator experience. Unobservable factors are those kinds of factors which also affect the P&A operation, but are impossible to quantify, such as project management skills, communication skills, and readiness level of personnel. The observable and unobservable factors can be either dependent variables or independent variables. When time for a P&A operation is estimated, then cost of operation is consequently estimated.

6.2.1 Description of Factors

There are many factors and events that impact time and cost associated with P&A operations. The factors can include well characteristics, well complexity, site characteristics, working unit, operator philosophy, local regulations, exogenous events, dependent variables, and unobservable variables.

6.2.1.1 Well Characteristics

In P&A operations, characteristics of the well such as well length, hole diameter, and well inclination contribute to time, cost, and HSE risk. For example, hole diameter and length of the required plug determine the P&A material volume, and the material to be removed from the well.

6.2.1.2 Well Complexity

The well complexity can be increased because of different reasons, including but not limited to: limited access to the desired interval, type of completion, high-pressure and high-temperature condition, and well integrity issues. Consequently, increase of well complexity can increase the duration and cost of P&A operation. Well complexity can also directly influence the type of required working unit.

6.2.1.3 Site Characteristics

Geographical location, distance from the well to the nearest service station, and water depth at the site for offshore well are some of the main site characteristics.

6.2.1.4 Working Unit

Type of working unit and personnel on board, directly contribute to a large part of the total P&A cost. Selection of the working unit depends on other factors such as well complexity, site characteristics, vessel availability for offshore wells, environmental criteria, etc. Therefore, this factor is a dependent factor.

6.2.1.5 Operator Philosophy

The operator decides when to permanently P&A a well, what type of contract is necessary, and how to carry out the operation. In addition, duration, P&A design, job specification (single well or campaign P&A), and strategies are the main parameters, which are based on operator preferences, for determining time and cost of P&A.

6.2.1.6 Local Regulations

As discussed in Chap. 1, different authorities have their own requirements with respect to permanent P&A of wells. Local regulations may impact the time and cost of the operation. Consider the North Sea where the UK regulation requires 30 (m) of a continuous qualified plug whereby the Norwegian regulation requires 50 (m) of a continuous qualified plug.

6.2.1.7 Exogenous Events

There are situations where the P&A operation may be subject to delays. Equipment failure is an example of such circumstances. If a spare part is not available, then activities are delayed. Sometimes, equipment or equipment's parts may be lost in the well. Fishing or a multiple fishing operation may be necessary to retrieve these elements. For offshore P&A, activities may be significantly delayed due to weather. Weather downtime can become an important factor in the total time and cost of an operation. Severe weather conditions can cause delays for supply boats to deliver equipment or material which are of a critical stock level. Weather can also impact anchoring and moving time of floating working units. In some geographical locations, such as the North Sea, the weather can be too severe and cause the operation to be suspended. Therefore, weather conditions and waiting on weather time needs to be considered for P&A.

6.2.1.8 Dependent Variables

The number of days spent to accomplish Phase 1 to Phase 3, is defined as time to P&A a well. It includes, mooring and demooring (if applicable), time to survey the well condition, tripping, time spent on barrier installation and its verification, weather time, and cut and removal of the wellhead.

6.2.1.9 Unobservable Variable

There are many factors known to be difficult to quantify and incorporate directly into time and cost analysis. Of these one can refer to unique P&A design, incidents during preparation, project management and leadership skills, availability of technology and technique, and personnel skills.

P&A design and preparation—Evaluation of well condition and careful planning is required to complete a P&A project successfully. The first step in P&A design is to identify the different permeable zones that needs to be isolated. There are two different approaches with respect to the number of plugs to be installed: traditional approach or risk-based approach. In a traditional approach, each hydrocarbon zone requires its permanent barriers. However, in a risk-based approach, the consequence of combining formations which contain fluid is carefully studied. If the risk of any harm to environment or failure of barrier is low, the two or more formations are grouped and a barrier is installed for them. The installed barrier consists of a primary and a secondary permanent barrier. Each approach has its impact on time and cost. In addition, a multidisciplinary team should design the P&A efficiently to deal with objectives of the operation.

Project management and leadership skills—Appropriate project management and leadership has to have comprehensive and integrated engineering planning, with coordinated skills and well defined contingencies. The project is to be executed in the shortest possible time in collaboration with all team members.

Technology and technique—The impact of technology and technique on performance of a P&A operation is extensive. New technology can be enabling or enhancing or both, and will shift from enabling to enhancing over time, due to learning effects. Generally, new technology is expensive, but if the performance of the operation and safety is improved, then costs will decline. It is difficult to find out the impact of new technologies on estimation of time and costs of an operation. A kind of such tool is perforate, wash and cement technique which reduces the time of P&A operation by eliminating section milling (see Chap. 8).

Personnel skill—Another factor which is part of unobservable variables is skill and experience. During P&A design, experienced engineers can include their learnings from other operations which can significantly influence the time and cost. During the operation, experienced personnel, crossed trained, can implement their experiences to solve the challenges on site instead of awaiting personnel to arrive from another site. Hence, the cost can be significantly reduced by suing properly trained personnel.

6.2.2 Traditional Method for Time Estimation

Traditionally time of a P&A operation is estimated using deterministic values. This statistical approach, also known as the deterministic method, uses a mathematical model to estimate the outcomes precisely. In deterministic methods, a deterministic model governs the outcomes through known relationships among the factors, observable and quantifiable factors, Fig. 6.1. However, there is no room for unobservable and variable factors. In this method, a given input will always produce the same output which means that the model defines an exact relationship between the variables. This defined relationship allows prediction of the impact of one variable on the other. The traditional method assumes certainty in its solution.

The deterministic approach has its advantages and limitations. The advantages include simplicity of approach, clear assumptions, and transparent communication

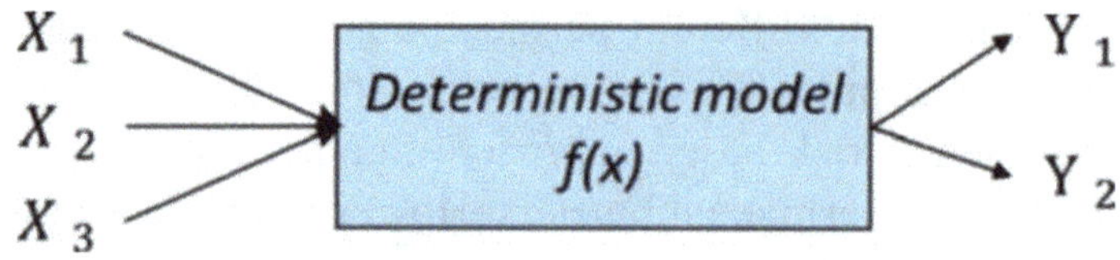

Fig. 6.1 A deterministic model precisely estimates the outcome based on the input factors

of results [2]. The limitations are the prediction's optimistic bias on good results, not presenting the full range of possible outcomes. Uncertainty associated with sub-operations are not included in the final results [2].

6.2.3 Probabilistic Method for Time Estimation

The probabilistic approach, also known as the stochastic method, uses a mathematical model which presents probability of a random phenomenon. In this method, the probability of an event occurring again is estimated based on historical data and governing statistical analysis models. The probabilistic model is likely to produce different outcomes even with the same initial conditions. So, variation and uncertainty of data on each event are considered in the probabilistic model, Fig. 6.2. In fact, a probabilistic model includes both deterministic components and random error components.

The advantages of the probabilistic approach are addressed as: reflecting uncertainties, presenting a range of possible outcomes, including unexpected events, providing the opportunity to apply sensitivity analysis, possible to include learning effects, the interaction between sub-operations can be analyzed, and decision making process of the operation can be improved [2]. Although the probabilistic approach introduces advantageous features, there are also limitations and subjectivity associated with it. A probabilistic model will not and should not be expected to identify and

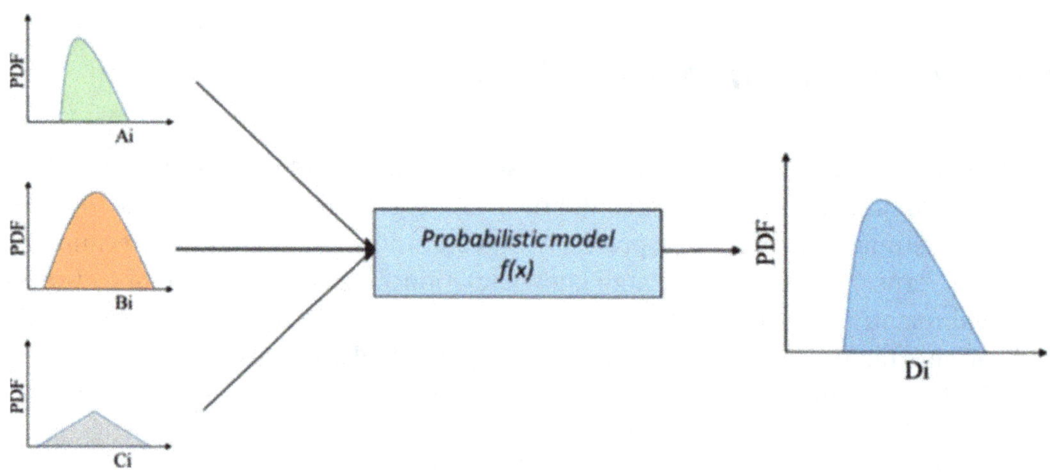

Fig. 6.2 A probabilistic model includes uncertainty of input data and includes uncertainty in the outcome values

capture all risks as the presence of unknowns are always part of the story. Considering the results of a probabilistic approach without the accompanying philosophy behind them has limited value. Defining the relationship of inputs is not straight forward [3].

6.2.3.1 Refreshing Statistics

It is important to avoid any misinterpretation or misconstruction of results obtained by probabilistic estimates. Therefore, there are terms which need to be elaborated and be properly used. The most common terms are *percentiles*, *mode*, *mean* and *median* [4].

Percentiles (also known as the "P" number)—In probabilistic methods, the range of outcomes produced by models are divided into 100 parts and presented by 99 percentiles, P_1–P_{99}. Each percentile contains 1% probability of the outcomes. Of these percentiles, three percentiles are the most common to be used when discussing the results. These include P_{10}, P_{50}, and P_{90} (see Fig. 6.3). P_{10} is the percentile whereby 10% or less of outcomes have the probability to fall in the range of P_1–P_{10}. P_{90} is the percentile whereby 10% or less of outcomes have the probability to fall in the range of P_{90}–P_{99}. In other words, P_1–P_{10} and P_{90}–P_{99} are less likely outcomes. P_{50}, also known as *median*, is the probability of having 50% of the outcomes equal to or exceeding the best estimate. Similarly, 50% of the outcomes equal or less than the best estimate. The distribution curve of outcomes can be symmetric or asymmetric (skewed), Fig. 6.4. For an asymmetric distribution, the P_{50} and mean value are unequal.

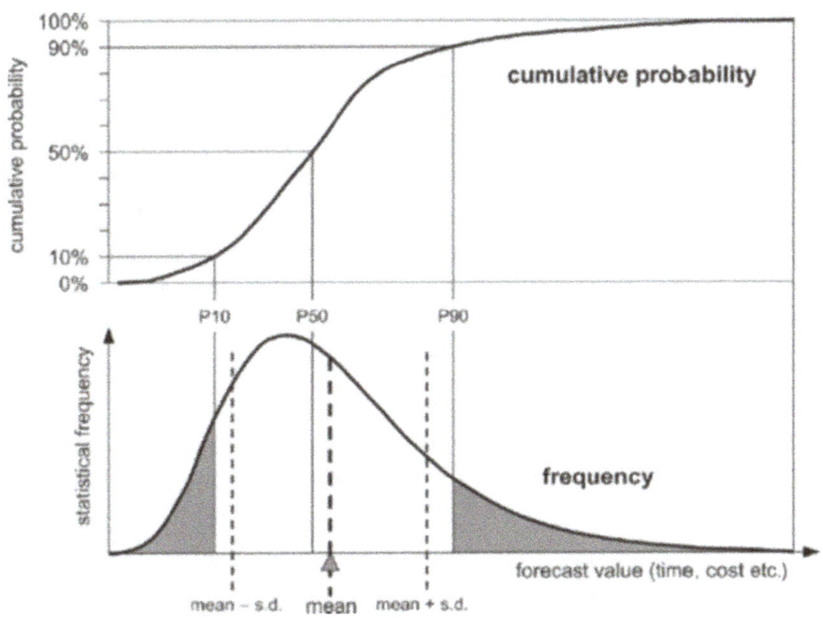

Fig. 6.3 Distribution outcome with statistical terms used in probabilistic method of estimation

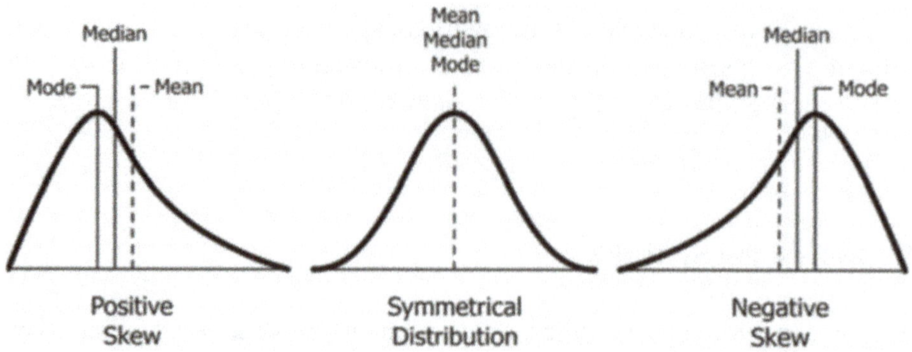

Fig. 6.4 Symmetric and a symmetric distribution with mode, median and mean values

Mode (also known as the Most Likely Value)—This is the most frequent value in the data set which occurs during thousands of iterations of time or cost estimation. On a probability frequency chart, the mode is the value at the highest point, see Fig. 6.4.

Mean (also known as the Expected Value)—This is the arithmetic average, sum of values of a data set divided by number of values, of all the outcomes of simulation iterations. If mean values are summed up together, a meaningful result is obtained which is commonly used for Authority for Expenditure (AFE) or analysis of a single well.

Median (also known as the "P_{50}")—This is the middle value which separates the greater and lesser outcomes of a data set.

6.2.3.2 Probability Distributions

In the probability approach, an offset data is a value which indicates the distance between the beginning of the value and a given point. Probability distributions are used to characterize the behavior of random variables. To fit offset data, there are several probability distributions to choose: normal, triangular, lognormal, and uniform [5]. Of these, the two widely accepted probability distributions are uniform distributions and triangular distributions (the two lower left distributions shown in Fig. 6.2). The uniform distributions are the simplest of all and are described by a minimum and a maximum value. The triangular distributions are an extended form of uniform distributions by adding the mode, the most likely value.

When considering probability distributions three terms are distinguished: Probability Mass Function (PMF), Probability Distribution[1] Function (PDF) and Cumulative Distribution Function (CDF). To clarify the difference between these terms, it is necessary to understand two main types of random variable distribution: *discrete* or *continuous*. A continuous random variable distribution is a curve with an infinite number of values on it which characterizes the distribution of a random variable.

[1]Distribution is also noted as Density in some references.

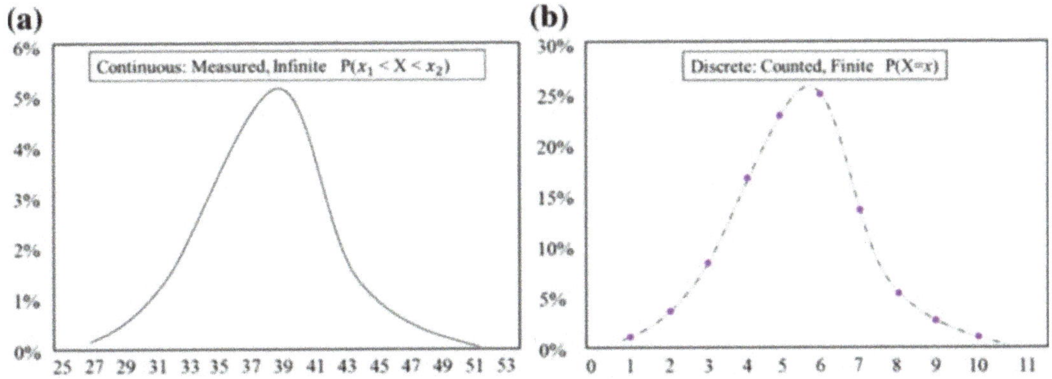

Fig. 6.5 Random variable distributions: **a** continuous, **b** discrete

To specify a local location on a continuous distribution, probability, one exact value cannot be given but an interval is presented. So to specify the interval on a continuous distribution, a PDF represents the interval (see Fig. 6.5a). Examples of continuous random variables are distance, time and asset returns in finance. A discrete random variable distribution is obtained by counting and is a finite measurement (see Fig. 6.5b). So the probability of a random variable is an exact value. Therefore, a PMF presents the outcome. Briefly, continuous random variables are measured, however, discrete distribution random variables are counted.

When considering a continuous distribution of a random variable, to present the probability of outcomes, approximately equally, of an input value, PDF is used. But to present the outcome probability of an input value, CDF is used.

Example 6.2 Figure 6.6 shows Monte Carlo simulation results of time estimation for a subsea single well which is going to plugged and abandoned. In one scenario, the entire operation is supposed to be carried out by a semisubmersible rig while in another scenario the operation is carried out partly by a semisubmersible unit and partly by a vessel.

(a) What will be the most likely value, the mode, for time of this operation when performing the operation entirely by deploying a semisubmersible unit and what will be the occurrence probability of the most likely value?

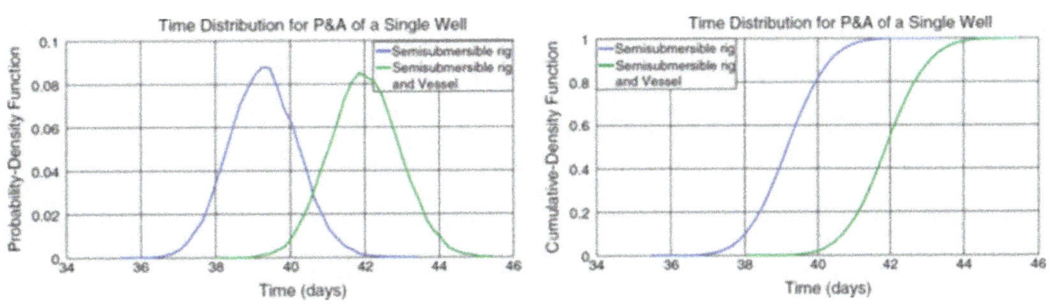

Fig. 6.6 Time distribution for P&A of a single well: **a** PDF, **b** CDF [2]

(b) What will be the most likely value, the mode, for time of this operation when performing the operation by deploying a combination of a semisubmersible unit and a vessel and what will be the occurrence probability of the most likely value?
(c) Interpret the outcomes of PDF and CDF of having 42 days when using a combination of a submersible and a vessel.

Solution The probability is presented by a value from 0 to 1.0 or in percentage, from 0 to 100%.

(a) By referring to the PDF curve, semisubmersible rig, the occurrence probability of the mode is approximately 85.7% and the associated time will be approximately 39 days.
(b) By referring to the PDF curve, semisubmersible rig and vessel, the occurrence probability of the mode is approximately 83% and the associated time will be approximately 42 days.
(c) By considering the probability of having the time of operation 42 days, when combination of semisubmersible and vessel are employed, the CFD is approximately 55%. In other words, having the expected time of the P&A operation being less or equal to 42 days is 55%.

Offset data analysis includes data gathering and analyzing the offset data. Although this planning procedure is time consuming it should be carried out properly as more accurate input results in a better quality of outcomes. Therefore, during data gathering, team members need to discuss and analyze the data. Non-Productive Time (NPT) should be split into predicted and unpredicted offset data and analyzed separately. It is important to document the analysis of offset data.

Considering time estimation of P&A operations, limited or poor offset data requires predominantly expert judgement. However, this might become a challenge as biased positive and negative experiences of experts may lead to an unrealistic understanding of their ability to assess uncertainties. In addition, poor handling of uncertainty may also be rooted in outdated methodology and poorly quantitative ideas about it.

6.2.3.3 Central Limit Theorem

When building a model for time estimation, the Central Limit Theorem (CLT) can be used. According to the CLT, when independent random distributions, of any distribution shape, are added, the sum of probability distribution tends toward a normal distribution.

To reduce the impact of the CLT and avoid unrealistically narrow results of estimations, three main ways have been suggested and addressed. These includes: restricting the number of input variables, avoid using too narrow input ranges and not underestimating uncertainty, and dependencies between input variables to be addressed by use of *correlation* [4].

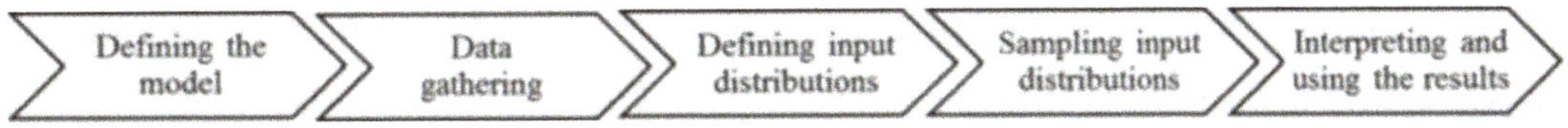

Fig. 6.7 Five general steps of Monte Carlo simulation

6.2.3.4 Monte Carlo Simulation of Time Estimation

The concept of the Monte Carlo was developed by a Polish-American mathematician, Stanislaw Ulam, in the late 1940s. Before Monte Carlo simulation was developed, statistical sampling was used to estimate uncertainties of deterministic simulations. Monte Carlo simulation is a method of estimating the value of an unknown quantity by deploying the principles of inferential statistics. The inferential statistics make inferences and predictions about population (a set of examples) and samples (a proper subset of a population). In other words, Monte Carlo is a numerical technique to forecast the outcomes based on the available evidences [6]. The predicted outcome depends on the size of input variables and the variance. Since the introduction of the Monte Carlo method and with advancement of computers, application of the technique has been adapted in different fields, including a wide variety of problems in the petroleum industry [2, 4, 7–11].

The Monte Carlo simulation method can be divided into five steps (see Fig. 6.7): defining the model, data gathering, defining input distributions, sampling input distributions, and interpreting and using the results.

Defining the model—A Monte Carlo simulation begins with a model. The forecaster needs to clarify the scope of the analysis including the contingencies and possibilities, to be determined. Once done, appropriate input parameters are specified to the model. Then, the output values are calculated. Each of these parameters are viewed as random parameters [5].

Data gathering—In a P&A operation, the assumption is that the exact values of the model inputs are unknown, so offset data are used which means uncertainty in modeling. Data gathering is necessary to quantify this uncertainty. Offset data is the key step in Monte Carlo forecasting.

Defining input distributions—When offset data are ready, the probability distributions are defined and sampling of each uncertain input value to the model is accomplished. This process can be subdivided into two steps: selection of distribution shape (e.g. uniform, normal, lognormal, etc.) and distribution parameters (e.g. minimum, standard, deviation, P_{90} percentile, etc.).

Sampling input distributions—Monte Carlo simulation performs random sampling from input distributions and conducts a large number of trials. A trial is the process of selecting one value for each input and calculating the output or possible results. A simulation is a series of hundreds or thousands of repeated trials for which the outputs are stored. The selection of random numbers from the input distributions can have a significant effect on the outcomes.

A major driver of the final range of outcomes is correlation. A *correlation* is defined as any relationship or dependency between two input quantities which motivates their joint distribution to deviate from statistical independency [4, 11]. Correlation is part of physical reality and as the relationships between input quantities are often not amendable to quantification, correlations are quite subjective and amorphous [12]. Spearman's rank order, Pearson, and Kendall Tau are some common correlations to consider the dependency [13]. The dependency between inputs is assigned with a value between –1 and +1 whereas full independency is shown by –1 and full dependency by +1. Spearman's rank order correlation, also known as Spearman's rho, is a measure of statistical dependency between the rankings of two variables. For two data sets of X and Y, Spearman's rank order correlation coefficient is given by [13].

$$\rho = 1 - \frac{6 \times \sum (\Delta r^2)}{n \times (n^2 - 1)} \tag{6.1}$$

$$\Delta r = x - y \tag{6.2}$$

where n is the number of correlated data pairs between the data sets, and x and y correspond to ranks of the data sets. It should be noted that the Spearman's rank order correlation is independent of distribution of data.

Pearson correlation coefficient for two data sets is given by:

$$P = \frac{\sum_{i=1}^{n} (X_i - X')(Y_i - Y')}{\sqrt{\sum_{i=1}^{n} (X_i - X')^2} \sqrt{\sum_{i=1}^{n} (Y_i - Y')^2}} \tag{6.3}$$

where X_i and Y_i are a pair of the correlated data sets, X' and Y' are their mean values, and n is the number of correlated data pairs between the data sets. One of the main drawbacks of this correlation is that a non-linear transformation between two variables is not preserved. In addition, this correlation does not capture a non-linear relationship between the two variables.

Unlike the Spearman's rank order and the Pearson's correlations, the Kendall Tau rank correlation coefficient captures the dependency pattern between two variables. The Kendall Tau rank correlation coefficient is given by:

$$\tau = \frac{(number\ of\ concordant\ pairs) - (number\ of\ discordant\ pairs)}{\frac{n(n-1)}{2}} \tag{6.4}$$

where n is the size of samples, concordant pairs are the pairs that are moving in the same direction, and discordant pairs are the ones that are moving in the opposite direction. When the correlations are generated, they will be included in the model.

Interpreting and using the results—PDF and CDF are the basic outcomes of a Monte Carlo simulation, for each quantity [2]. In order to avoid any false outcome or misinterpretation, these must pass some quality-assurance or a sensitivity analysis prior to being used. When the process is successfully completed, the distribution curves are ready to be used in a variety of processes of a P&A operation, including risk analysis and decision making, budget allocation, setting targets and expectations [4, 11].

6.2.4 Regression Method for Time Estimation

A regression approach uses a model in which deterministic and probabilistic models are applied. The contribution of the deterministic part in the regression model provides the opportunity that output results depend on input values. But the contribution of the probabilistic part in the regression model does not allow to produce an exact value for output values.

References

1. Oil and Gas UK. 2011. *Guideline on well abandonment cost estimation.* The United Kingdom Offshore Oil and Gas Industry Association: United Kingdom.
2. Moeinikia, F., K.K. Fjelde, A. Saasen, et al. 2015. A probabilistic methodology to evaluate the cost efficiency of rigless technology for subsea multiwell abandonment. *SPE Production and Operations* 30 (04): 270–282. https://doi.org/10.2118/167923-PA.
3. Moeinikia, F., E.P. Ford, H.P. Lohne, et al. 2018. Leakage calculator for plugged-and-abandoned wells. *SPE Production and Operations* 33 (04): 790–801. https://doi.org/10.2118/185890-PA.
4. Akins, W.M., M.P. Abell, and E.M. Diggins. 2005. Enhancing drilling risk and performance management through the use of probabilistic time and cost estimating. In *SPE/IADC drilling conference*. SPE-92340-MS, Amsterdam, Netherlands: Society of Petroleum Engineers. https://doi.org/10.2118/92340-MS.
5. Murtha, J.A. 1994. Incorporating historical data into Monte Carlo simulation. *SPE Computer Applications* 06(02). https://doi.org/10.2118/26245-PA.
6. Stoian, E. 1965. Fundamentals and applications of the Monte Carlo method. *Journal of Canadian Petroleum Technology* 04(03). https://doi.org/10.2118/65-03-02.
7. Adams, A., C. Gibson, and R.G. Smith. 2010. Probabilistic well-time estimation revisited. *SPE Drilling and Completion* 25(04). https://doi.org/10.2118/119287-PA.
8. Banon, H., D.V. Johnson, and L.B. Hilbert. 1991. Reliability considerations in design of steel and CRA production tubing strings. In *SPE health, safety and environment in oil and gas exploration and production conference*. SPE-23483-MS, The Hague, Netherlands: Society of Petroleum Engineers. https://doi.org/10.2118/23483-MS.
9. Sawaryn, S.J., K.N. Grames, and O.P. Whelehan. 2002. The analysis and prediction of electric submersible pump failures in the milne point field, Alaska. *SPE Production and Facilities* 17(01). https://doi.org/10.2118/74685-PA.
10. Sawaryn, S.J. and E. Ziegel. 2003. Statistical assessment and management of uncertainty in the number of electrical submersible pump failures in a field. *SPE Production and Facilities* 18(03). https://doi.org/10.2118/85088-PA.
11. Williamson, H.S., S.J. Sawaryn, and J.W. Morrison. 2006. Monte Carlo techniques applied to well forecasting: some pitfalls. *SPE Drilling and Completion* 21(03). https://doi.org/10.2118/89984-PA.

12. Purvis, D.C. 2003. Judgment in probabilistic analysis. In *SPE hydrocarbon economics and evaluation symposium*. SPE-81996-MS, Dallas, Texas: Society of Petroleum Engineers. https://doi.org/10.2118/81996-MS.
13. Moeinikia, F., K.K. Fjelde, A. Saasen, et al. 2015. Essential aspects in probabilistic cost and duration forecasting for subsea multi-well abandonment: simplicity, industrial applicability and accuracy. In *SPE Bergen one day seminar*. SPE-173850-MS, Bergen, Norway: Society of Petroleum Engineers. https://doi.org/10.2118/173850-MS.

Chapter 7
Fundamentals of Plug Placement

The best practice of permanent plug and abandonment requires a cross sectional barrier, which is known as *rock-to-rock barrier*. The barrier is placed at the right depth where formation is capable to hold the maximum anticipated pressure. To fulfill the requirement, two general situations could be encountered: openhole plug placement or cased hole plug placement.

7.1 Openhole Plug Placement

To place a permanent cement plug in an openhole, the fluid in the well needs to be replaced with cement. As the compositions and properties of drilling (or milling) fluids and cement slurries vary widely, severe contamination can occur at the interface of drilling fluid and cement slurries due to incompatibility. Therefore, fluid removal during cement plug placement is a crucial task.

7.1.1 Fluid Removal

Fluid removal has been an interest for cement engineers for many years. To achieve the objectives, drilling fluid and pre-flushes must be fully removed from the openhole interval and be exchanged fully with cement or any plugging material. The fluid removal process is a function of borehole quality, circulation and displacement efficiency, fluid conditioning and properties of drilling fluids, spacers and washes [1–6]. Fluid removal process can be carried out in two main different ways: hydraulically or mechanically. In the hydraulic process, spacer fluids with specific viscous behavior are pumped ahead of cement slurry to displace drilling or milling fluid. The contamination effect of these spacer fluids on cement is less compared to drilling or milling fluids.

M. Khalifeh and A. Saasen, *Introduction to Permanent Plug and Abandonment of Wells*, Ocean Engineering & Oceanography 12,
https://doi.org/10.1007/978-3-030-39970-2_7

One major difference, when considering milling operation during permanent P&A, is that a window of casing is milled to reach the formation and therefore, milling fluid is used instead of drilling fluid. So, compatibility of cement and milling fluid is strongly dependent on the chemistry and properties of milling fluid; therefore, milling fluids will be reviewed briefly.

7.1.2 Milling Fluid

When casing is milled away, the generated debris known as *swarf* (Fig. 7.1) needs to be transported to surface or left behind in the bottom of the well as will be discussed in Chap. 8. As drilling fluids do not have the transportation capacity of swarf, special fluids known as milling fluids are used. Milling fluids are usually water based. Of milling fluids one can list: bentonite/bicarbonate mud, bentonite/MMH (Mixed Metal Hydroxide) mud, xanthan gum/sea water mud, and potassium formate milling fluid [7–9]. Considering the geometry of the circulation system and non-Newtonian behavior of milling fluids, the hydrodynamics of swarf transportation and hole cleaning are identical to cutting transportation and hole cleaning during drilling. However, given the fact that swarf are much larger (see Fig. 7.1), having higher density compared to rocks and having irregular shapes, the problems are different [7]. A desired milling fluid should have high transportation capacity with low shear rate viscosity.

When considering the transportation of swarf by milling fluid, settling velocity of swarf in static and dynamic fluids and transportation velocity of debris are important. Experimental studies show that in static fluids the gel strength and effective viscosity of milling fluid are critical factors besides, the shape, surface area and the settlement orientation of swarf. The gel strength causes suspension of swarf but when the swarf is sharp, the gel strength can be overcome.

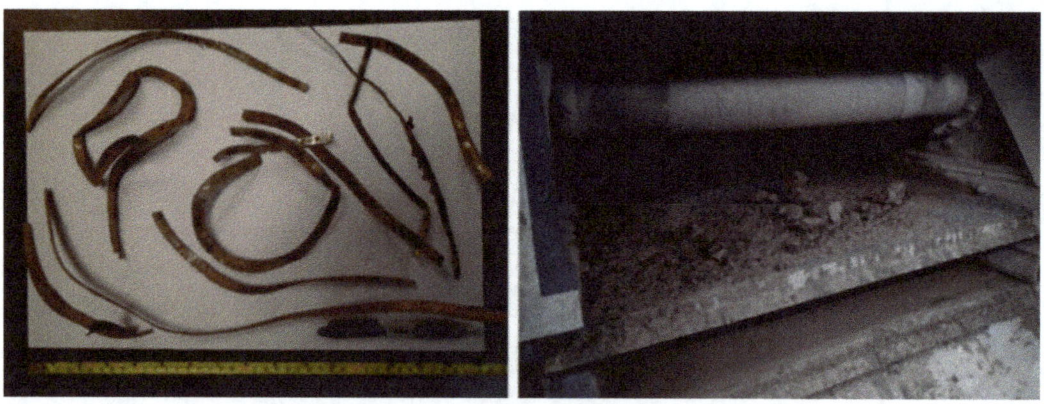

Fig. 7.1 Swarf from a milling operation in the North Sea (Courtesy of equinor)

When swarf is in dynamic conditions, flowing fluid, the mean circulating velocity required to prevent the swarf from settling is significantly higher. In dynamic conditions, there are areas close to the wall where the flow velocity is near zero and large volumes of swarf will be located in an un-sheared zone [7].

7.1.3 Hydraulic Mud Removal

Spacer or *displacement fluids* are type of fluids which minimize the cement contamination and improve the fluid removal efficiency. Any liquid which physically separates a special liquid from another is known as spacer fluid. In most practical operations, a cement slurry should have turbulent flow conditions to displace drilling fluid. But the flow regime cannot be achieved because of operational restrictions. So, spacer fluid needs to be selected to reach a turbulent or pseudo-laminar flow to remove the left fluids. Displacement fluid is usually used to force a cement slurry out of the workstring or into the annulus behind casing.

A spacer should have the following characteristics: compatible with a given type of drilling fluid or milling fluid, including bentonite muds and polymer based muds. The spacer properties should not affect the cement slurry viscosity nor changing the pumping time; to tolerate high solids and mud cake; to tolerate addition of wetting agents, dispersants, friction reducers, and retarders; low-fluid-loss properties; and permitting turbulence flow regime at low pumping rates for efficient mud removal [10–14]. Although spacers are used to remove drilling fluid and mud cake but it is unlikely to remove the mud cake without using mechanical aids.

7.1.4 Mechanical Filter Cake Removal

To clean the formation interface for achieving better bonding between the plugging material and formation, mechanical devices known as *wall cleaners* can be utilized. *Mud cleaners* or *scratchers* (sometimes called *mud stirrers*) are mechanical devices used to remove the mud or condition the drilling fluid filter cakes off of openhole wall for achievement of a better shear bond strength and hydraulic bond strength. The wall cleaning operation is different from *reaming* and *under-reaming*. *Reaming* operation is for enlarging wellbore by utilization of a mechanical device. However, enlarging hole is avoided during plug placement as it creates challenges during cement placement. Mechanical cleaners are fastened on the outside of workstring to agitate the mud and make it easier to displace it. The introduced motion breaks the gel strength of the mud filtercake and with help of wash fluid, the drilling fluid is displaced easier. The *rotational type* and the *reciprocation type* scratchers are the two commonly used types, see Fig. 7.2.

The rotational type scratcher cleans the formation when workstring is rotated. A continues length of scratcher is fastened on the workstring, Fig. 7.2a. The steel spike

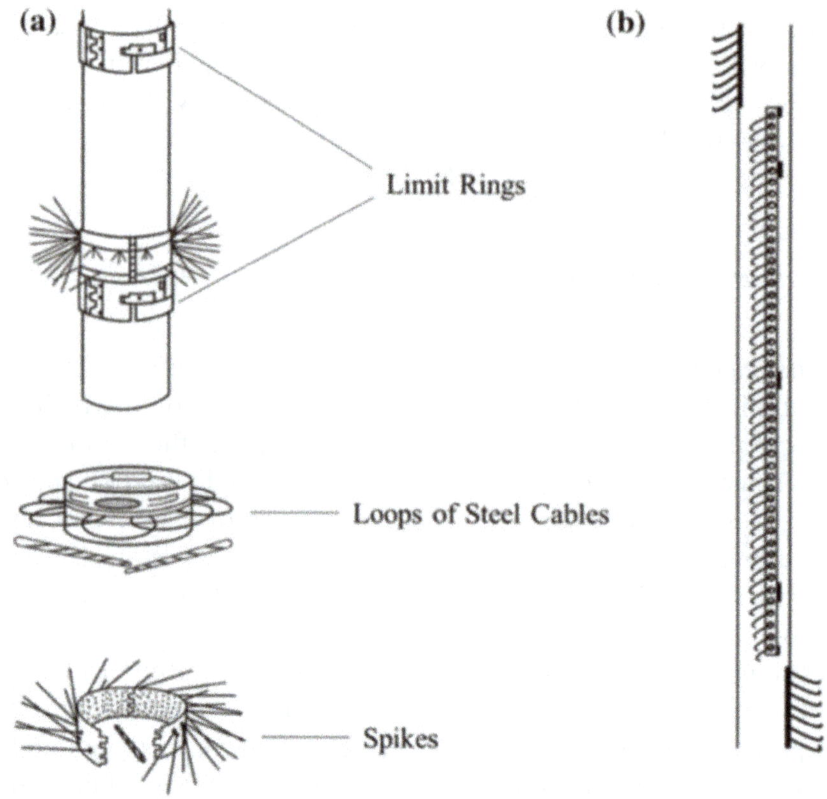

Fig. 7.2 Two commonly used scratchers: **a** rotational type scratcher, **b** reciprocation type scratcher

or steel cable sets are installed on workstring with different phasing to improve the cleaning efficiency.

The reciprocation type scratcher cleans the formation has either steel spikes or steel cables. Depending on the length of zone to be cleaned, one or more scratchers are attached to outside of workstring. Each scratcher is limited by two rings or clamps in the desired interval: one above and one below, Fig. 7.2b. These types of scratchers clean the formation when workstring is moved upwards and downwards.

During the mechanical cleaning operation, a wash fluid is pumped to displace and wash the mud and filter cakes. If the plug is off-bottom plug, when the interval is clean, a viscous reactive pill is pumped to create a base for cement plug and keeping plug in position, Fig. 7.3. *Viscous reactive pill* is a special blend of drilling fluid containing silicate component, which has higher density than cement slurry. When the calcium in the cement reacts with the reactive pill, a gel forms that prevents flow between cement and the pill. The reactive viscous pill is compatible with cement slurry and its high yield stress provides base functionality while cement sets. When the reactive viscous pill is in place, the cement slurry is placed on top of it which is across the cleaned formation.

If the drilling fluid present in bore is an oil-base mud, a viscous spacer is necessary before and after pill to minimize slurry contamination. The failure roots of plugs in openholes have been investigated by different authors and include the following [16]:

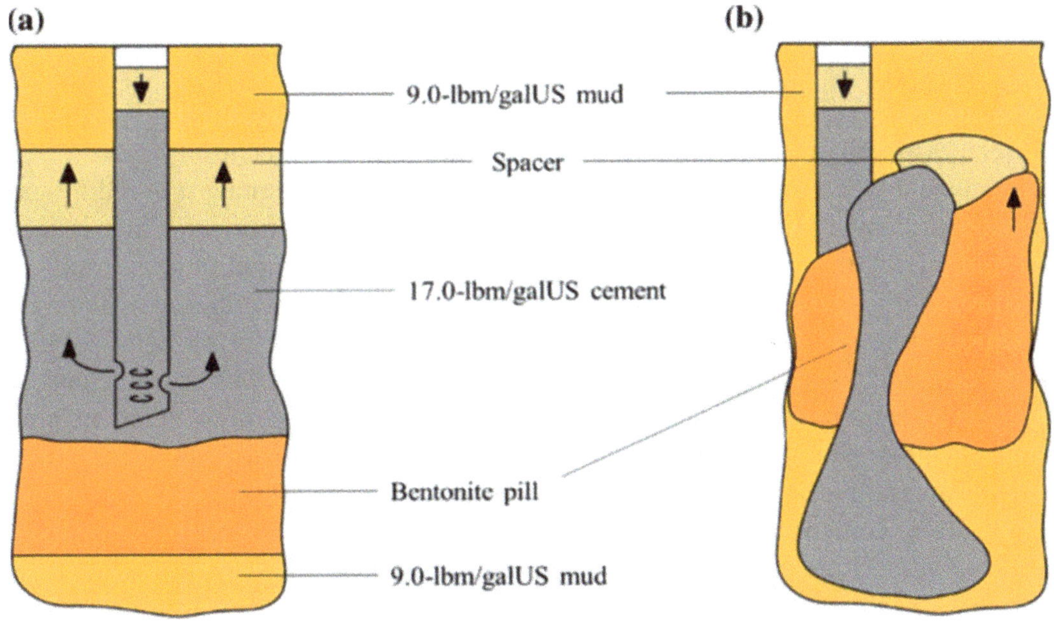

Fig. 7.3 Cement plug is placed in an openhole on a viscous pill, **a** ideal cement set, **b** unsatisfactory results for a cement plug placed on a lighter fluid (Taken from well cementing) [15]

- Poor mud removal
- Poorly designed slurry properties
- Incorrect estimation of slurry volume
- Poor downhole temperature estimation
- Poor job execution and placement
- Instability of the interfaces and swapping.

When cement plug is placed, it is left undisturbed until it develops high enough strength. When cement plug solidified sufficiently, top of cement is *dressed off* (top of cement is drilled out until hard cement is reached). As the plug is in an openhole, pressure testing is meaningless. Therefore, the TOC is tagged and a certain weight is applied on cement. If position of cement did not change, the cement plug is regarded as qualified. However, if the plug is not capable to hold the weight or the tagged TOC is not at the right depth, the plug is regarded as disqualified and a new cement plug needs to be established. In case of wireline and coiled tubing utilization, the maximum weight availability is limited compared to use of drillpipe.

7.2 Cased Hole Plug Placement

When considering plug placement for a cased hole, two different scenarios can be considered: either qualified annular barrier is proven or annular barrier is disqualified. Each case can dictate different operations.

7.2.1 *Qualified Annular Barrier*

If the annular barrier behind casing is qualified, then a mechanical plug is installed to create a foundation for cement plug. The mechanical foundation is not a part of permanent well barrier envelope but has the following advantages: avoiding gas invasion of cement while it sets, avoiding dispositioning of cement while it sets, and minimizing the cement contamination. When the mechanical plug is installed, it is pressure tested. If it successfully passes the pressure test, then cement plug is poured on top of it and left undisturbed until it develops high enough strength. When cement is solidified, cement is dressed off and tagged. As the mechanical plug has already passed the pressure test, the pressure testing of the cement plug is meaningless. However, if mechanical plug has not tested or did not successfully pass the pressure test, the cement plug is pressure tested and documented. Pressure test failure of cement plug means that another cement plug needs to be established. Different authorities require different plug length and different rate of pressure test.

7.2.2 *Disqualified Annular Barrier*

Wherever the quality of casing cement is not qualified or there is no annular cement, access to the annular space behind casing should be established to place a qualified barrier both inside and outside the casing. The conventional approach is section milling. The operation of removing a part of casing by milling or machining the casing is called *section milling*. To mill out casing steel, special knives are employed. Section milling is explained in Chap. 8. New methods exist called Perforate, Wash and Cement (PWC). PWC is described in the next chapter.

7.3 Plug Placement Techniques

7.3.1 *Balanced-Plug Method*

This is the most common plug placement technique used for placing permanent plugs. A work string is run into the hole to the desired depth for the plug base. As the work string is surrounded by mud, spacer and chemical wash are pumped ahead and behind the slurry to avoid mud contamination and ensure wetting of the surface of casing or formation. Cement slurry is pumped down through the work string and up in the annulus between the work string and casing or formation. The volumes of spacer ahead and behind the slurry are calculated so that the spacer height inside and outside the work string end up at the same level, Fig. 7.4.

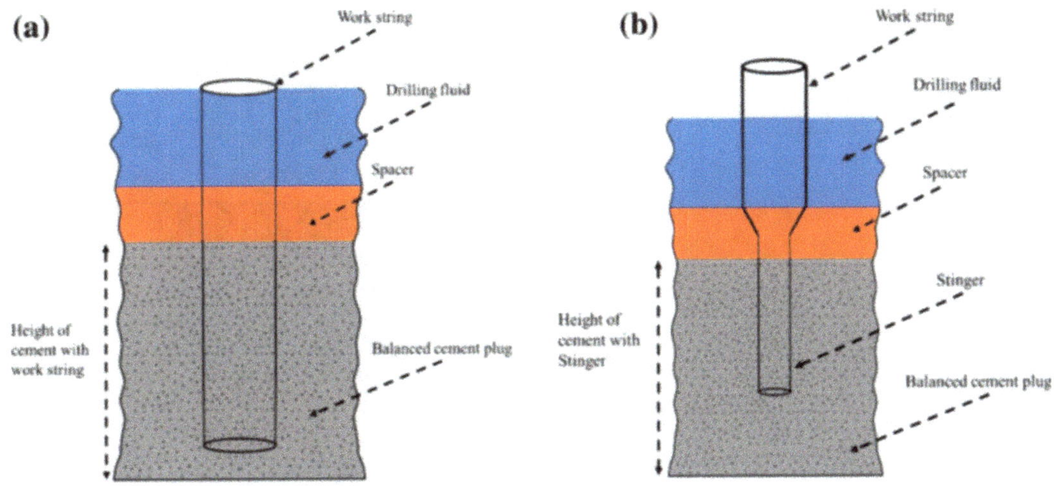

Fig. 7.4 Balanced-plug placement technique, **a** common work string, **b** deploying a stinger to minimize the agitation of slurry during pulling out of slurry

Example 7.1 You are asked to install a balanced plug across a suitable formation whereby the plug base is supposed to be at 10,000 ft measured depth. For this job, a 4½-in. drillpipe will be used as a workstring in an openhole with 8¾-in. diameter. The plug length is expected to be 200 ft and 24 bbl of fresh water will be pumped ahead of cement as spacer. Additional information: string capacity = 0.01422 bbl/ft, annular capacity = 0.0547 bbl/ft. Assume the wellbore is vertical.

(a) Calculate the required volume of cement.
(b) Calculate the height of cement plug with workstring in.
(c) Calculate the required volume of spacer behind slurry.
(d) Calculate the volume of displacement fluid.

Solution The goal of balanced plug placement technique is to have an equal drillpipe pressure and annular pressure, at the plug base (see Fig. 7.5). It can be written as:

$$\begin{aligned} \Delta P_{CD} + \Delta P_{WD} + \Delta P_{MD} &= \Delta P_{CA} + \Delta P_{WA} + \Delta P_{MA} \\ PD &= PA \end{aligned} \tag{7.1}$$

whereas ΔP_{CD} is the hydrostatic pressure exerted by cement inside workstring, ΔP_{WD} is the hydrostatic pressure exerted by spacer inside workstring, ΔP_{MD} is the hydrostatic pressure exerted by mud inside workstring, ΔP_{CA} is the hydrostatic pressure exerted by annular cement, ΔP_{MA} is the hydrostatic pressure exerted by annular spacer, and ΔP_{MA} is the hydrostatic pressure exerted by annular mud.

In this example, the spacer ahead and behind the slurry are the same type with the same characteristics. However, there are circumstances where spacer ahead and behind the slurry are different. The latter case is given as problem, at the end of this chapter.

(a) Volume of cement assuming no washout:

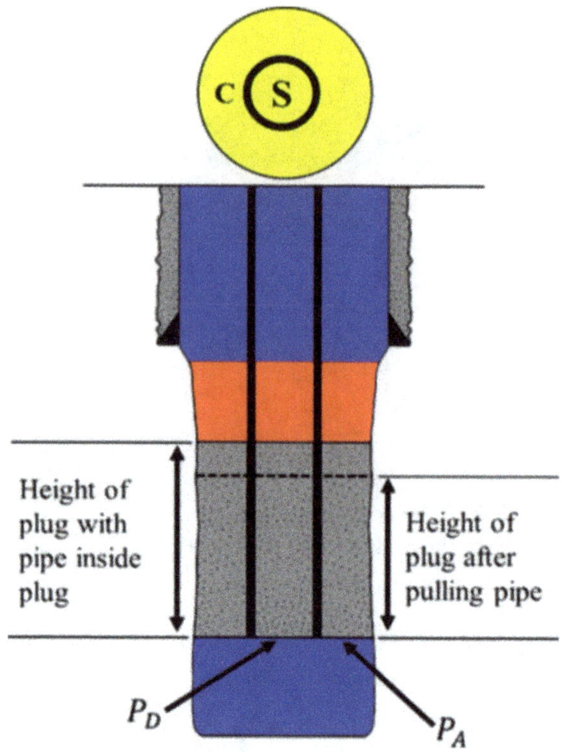

Fig. 7.5 A balanced plug whereas the spacer inside and outside workstring are in the same level and the same chemistry

$$V = \frac{\pi D^2}{4} * h \tag{7.2}$$

where D is wellbore diameter (ft), and h is plug length with no workstring inside plug.

$$V = \frac{\pi \times 8.75^2}{4} \times 200 \times \frac{(1 \ \text{ft})^2}{(12 \ \text{in.})^2} = 83.517 \left(\text{ft}^3\right)$$

(b) Height of cement plug when workstring is inside plug is given by equation [15]:

$$H = \frac{V}{C + S} \tag{7.3}$$

where C is annular capacity (bbl/ft), S is workstring capacity (bbl/ft), V is volume of slurry (bbl), and H is height of plug with pipe in place (ft).

$$H = \frac{83.517 \left(\text{ft}^3\right) \times \frac{1 \,\text{bbl}}{5.615 \,\text{ft}^3}}{(0.01422 + 0.0547) \frac{\text{bbl}}{\text{ft}}} = 215.814 \,\text{ft}$$

(c) The required volume of spacer behind slurry is supposed to be at the same height as spacer ahead of slurry. It means:

$$L_{sp2} = L_{sp1} \tag{7.4}$$

Then, it can be written as:

$$\frac{V_{sp2}}{S} = \frac{V_{sp1}}{C} \tag{7.5}$$

where V_{sp1} is the spacer volume ahead of slurry (bbl) and V_{sp2} is the spacer volume behind slurry (bbl).

$$\frac{V_{sp2}}{0.01422} = \frac{24}{0.0547}$$
$$V_{sp2} = 6.24\ \text{(bbl)}$$

(d) Volume of displacement fluid is the amount of fluid which needs to be pumped behind spacer to level the heights and keep the pressure equal, at the base of plug. The displacement volume in bbl is given by Eq. (7.6):

$$V_{dis} = S \times \left[L_{dis} - \left(H + L_{sp2}\right)\right] \tag{7.6}$$

whereas L_{dis} is length to be displaced which is measured depth (ft), and L_{sp2} is length of spacer behind slurry.

$$V_{dis} = 0.01422 \times [10{,}000 - (215.814 + 438.7)] = 132.89\ \text{bbl}$$

Cement plug contamination is one of the main challenges associated with the balanced plugs and can occur in three different ways: mud contamination during pumping, contamination caused by cement agitation while pulling the work string out of the plug, and plug displacement while it sets. Contamination during pumping the slurry may occur in the slurry-spacer interface and due to poor mud removal from formation or casing surface. The best practice to minimize the effect is to properly design the type, volume and flowrate of spacer and chemical wash or use a two-plug method.

7.3.2 *Two-Plug Method*

To minimize the contamination of the cement plug with the fluid ahead and behind, the two-plug method is used (see Fig. 7.6). In this technique, a wiper dart is run ahead of the cement plug (between the lead cement slurry and spacer) and another wiper dart behind the slurry (between the tail cement slurry and spacer). Thus, from surface down to a depth close to the tailpipe or stinger, the slurry is fully separated from the spacer and consequently, the risk of contamination is decreased. Each wiper dart has a diaphragm which holds pressure up to a certain point and ruptures at a

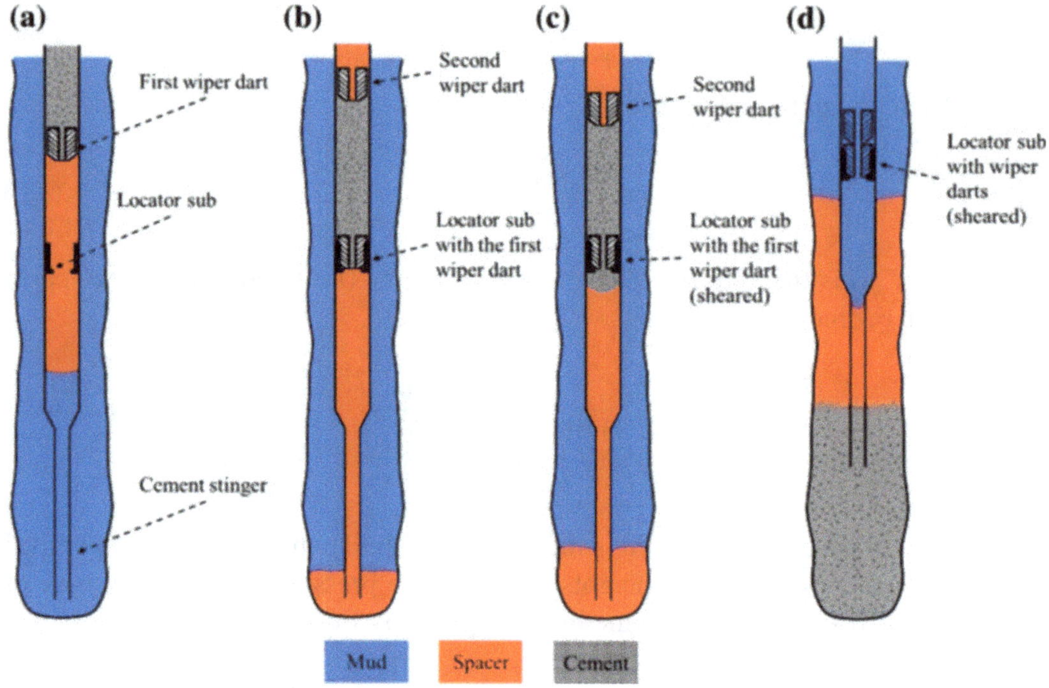

Fig. 7.6 Two-plug method; **a** first wiper dart separates cement from spacer until it lands on the locator sub, **b** second wiper dart separates cement from spacer behind cement, **c** the diaphragm of the first wiper dart is sheared due to the increased pressure and cement slurry passes through it, **d** second wiper dart seats on the first wiper dart and its diaphragm is sheared due to the increased pressure and the spacer passes through it [15]

higher pressure. The work string is equipped with a locator sub close to the stinger or tailpipe. When the first wiper dart seats on the locator sub, the pressure increases until the diaphragm ruptures, and the cement passes through the first wiper dart. Afterwards, the second wiper dart seats on the first wiper dart and causes a pressure increase. Due to the increased pressure, its diaphragm is ruptured and spacer passes through.

7.3.3 Dump Bailer Method

It is a wireline tool for placing small volumes of slurries at the desired depth with a minimal contamination. It is normally used only for onshore wells. The bailer is filled with cement and run into the wellbore. When it reaches the desired depth, the bailer cap is opened electronically via a signal or mechanically via touching a mechanical foundation. It is a common practice to use a mechanical foundation when a dump bailer is going to be used, Fig. 7.7. Some of its advantages are: minimizing the effect of contamination, it is inexpensive, drilling rig is not necessary for the operation, plug depth is easily controlled and operational time is significantly less compared to other methods. Low capacity of bailers and multi runs may be necessary, cement

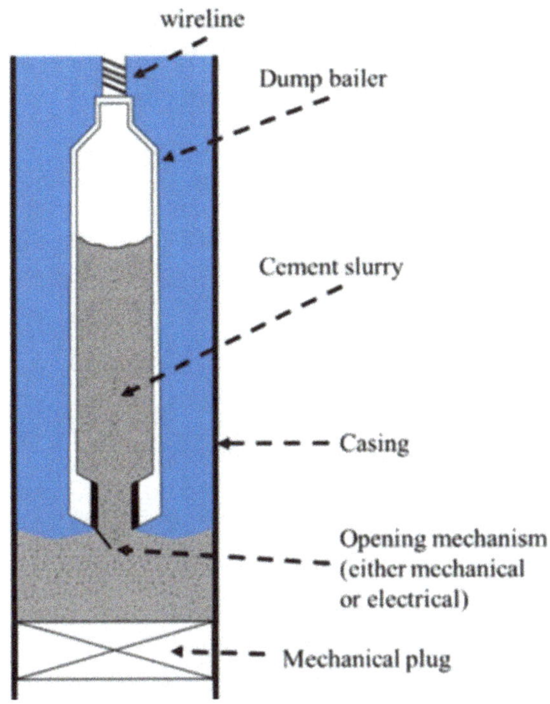

Fig. 7.7 Dump bailer method for plug placement

may set inside the bailer due to static conditions inside the bailer while running it down to the desired depth, and uncertainties associated with mud or spacer removal are some of the limitations for this method. It should be noted that slurry gelation or instability must be avoided to ensure that the slurry can exit the dump bailer.

7.3.4 Coiled Tubing Method

Coiled tubing is a long continuous pipe wound on a spool. The pipe is straightened prior to being pushed into the wellbore and rewound to recoil the pipe back onto the transport and storage spool. Depending on the pipe diameter and the spool size, coiled tubing can range from 2000 to 15,000 ft or greater lengths. Table 7.1 presents typical coiled tubing sizes.

Utilization of coiled tubing for remedial cementing began in the early 80's. Since then, the technique has received considerable attention. This technique has proved to be very economical to place small volumes of cement slurries required in curing channeling behind tubulars, blocking off perforations, squeezing cement into perforations, curing lost circulation zones during drilling, and placing cement whipstocks [17]. As the pipe is continuous, challenges associated with making connections and the need for a conventional rig are minimized which means it is a cost effective technique. However, there are some concerns limiting the use of coiled tubing for cement plug placement including fatigue problems, hole cleaning, special cement slurry design, unit space and capacity, crane capacity, and local regulations.

Table 7.1 Industry coiled tubing sizes available to date for material grades GT-80 (courtesy of global-tubing)

Specified dimensions				Axial load capacity		Pressure capacity		Torsional strength		External displacement	Internal capacity per 1000 ft
Outside diameter (in.)	Wall thickness (in.)	Inside diameter d (in.)	Nominal weight (lb/ft)	Yield load (lb) t_{nom}	Tensile load (lb) t_{nom}	Yield pressure (psi)	Hydrotest pressure 90% (psi)	Yield (ft/lb) t_{min}	Ultimate (ft/lb) t_{min}	Barrels	Barrels
1.250	0.190	0.870	2.16	50,620	55,680	23,040	15,000	1,097	1,206	1.52	0.74
1.500	0.087	1.326	1.32	30,900	33,990	8,850	7,970	955	1,051	2.19	1.71
1.750	0.109	1.532	1.91	44,950	49,950	9,510	8,560	1,609	1,770	2.97	2.28
2.000	0.109	1.782	2.21	51,800	56,980	8,320	7,490	2,149	2,364	3.89	3.08
2.375	0.125	2.125	3.01	70,690	77,750	7,950	7,160	3,463	3,809	5.48	4.39
2.625	0.134	2.357	3.57	83,890	92, 280	7, 740	6,970	4,571	5,028	6.69	5.40
2.875	0.156	2.563	4.54	106,600	117,260	8,240	7,420	6,330	6,963	8.03	6.38
3.250	0.156	2.938	5.17	121,310	133,440	7,290	6,560	8,236	9,060	10.26	8.39
3.500	0.175	3.150	6.23	146,240	160,870	7,630	6,870	10,708	11,778	11.90	9.64
4.500	0.224	4.052	10.25	240,730	264,800	7,610	6,850	22,639	24,962	19.67	15.95
5.000	0.276	4.448	13.96	327,690	360,460	8,510	7,660	34,231	37,654	24.29	19.22

Fatigue problems—Coiled tubing fatigue life is a major area of concern as the coiled tubing diameter increases for cementing applications. Each time that coiled tubing is spooled on and off the reel and over the gooseneck of the coiled tubing unit, it is stressed. This concern is greater in coiled tubing with larger diameters. Another cause is the internal pressure in the coiled tubing during bending and straightening [18]. As there is no practical non-destructive means of measuring the amount of damage accumulation, coiled tubing lifetime prediction models have been developed to predict the coiled tubing properties.

Hole cleaning—Limited flow capacity due to the size of the coiled tubing and lack of mechanical agitation effects through pipe rotation reduces the hole cleaning efficiency in large hole sizes [19].

Unit space and capacity—When considering the feasibility of coiled tubing for cementing, the unit deck area for placing the coiled tubing equipment such as reel, injector, pumping equipment, cementing equipment, and testing equipment need to be studied. In addition, the unit structure should have the capacity to hold the weight of equipment without introducing a risk of failure. For onshore wells the soil and the area should be able to hold the weight and also for offshore wells platform, drillship, vessel, semi-submersible or other working units the weight must be considered. As cementing utilizing a coiled tubing unit requires larger pipe diameters, the size and capacity of pipe handling equipment (e.g. injector heads, reels, well control equipment, etc.) have to be increased. Therefore, the unit space and capacity need particular consideration.

Crane capacity—In case of offshore activities, platform cranes must be able to lift up equipment from a supply boat to platform or any other offshore working unit [20]. The increase in weight and dimensions created by larger pipe diameters require higher crane capacity and introduces additional hazards.

Local regulations—Regulations aim for performing a safe coiled tubing operation which requires quality control of coiled tubing, wellsite safety standards, and safe deployment of tools in and out of the well. Well control equipment (e.g. BOPs), pressure rating of coiled tubing, fatigue prediction, unit capacity, and crane capacity are some of the main concerns focused on by local regulators. However, different regulatory authorities have different criteria.

Cement slurry design—As coiled tubing has a lower flow capacity compared to drillpipe, a standard coiled tubing cement recipe is not the same as a standard primary cement recipe. Due to the mixing energy introduced by coiled tubing on the cement slurry, a mechanical acceleration result. Therefore, a typical cement slurry designed for placement with coiled tubing has a longer thickening time, and lower viscosity and yield stress [21].

7.4 Mud Displacement During Cementing

The replacement of drilling fluids with cement to establish a barrier and to seal formation pressures hydraulically is the main task to be achieved during plug placement, besides the prime physical properties of the cured cement. Several parameters influencing mud displacement efficiency during plug placement include: hole geometry and inclination, flow rate, degree of turbulence, ECD, cement or mud and spacer design, hole conditioning, rheological behavior, buoyancy and plug stability, pulling out of plug, size of work string, and centralization of work string [3, 22–26]. Obviously, no single technique, will magically make mud displacement and cementing a success.

Hole geometry and inclination—The geometry of openhole where the cement plug is to be placed is very important for mud displacement and pumping of the correct volume of cement. When the milled section (openhole) has a constant diameter, it is referred to as in-gauge. An *in-gauge hole* has a round cross section but as the cross section starts to deviate from the round shape, it is referred to as an oval hole, Fig. 7.8. If the milled section has variations in diameter, it is called an *irregular wellbore geometry*, and has resulted from washouts.

When washouts exist, the annular flow velocity is less than for in-gauge portions of the hole. If the annular velocity is low enough, the mud will be left in the washout in a gelled state and the mud removal by cement becomes very difficult. Another challenge introduced by washouts is that if there is a large uncertainty in hole size, the cement volume will be underestimated and the plug length will be less than required. Therefore, the hole is usually callipered to better describe the wellbore geometry.

In deviated holes, unstable fluid interfaces with regards to gravitational forces, and fluid contamination introduce complications to balance the fluids during plug placement. A properly designed plug can be contaminated during pulling the tailpipe of the work string out of the cement plug, especially in deviated sections [27]. In addition, deviated boreholes intensify challenges related to free fluid and particle segregation [28].

Flow rate—Another major parameter which affects the displacement process is the flow rate. As drilling fluids and cement slurries are non-Newtonian fluids, they

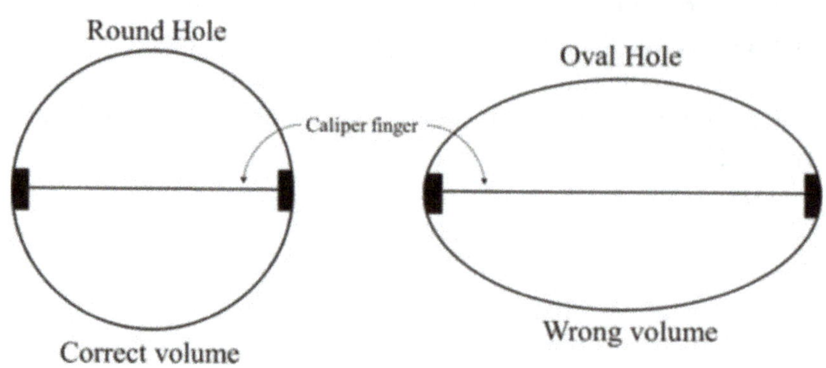

Fig. 7.8 Caliper log may not be able to read the oval-hole diameter

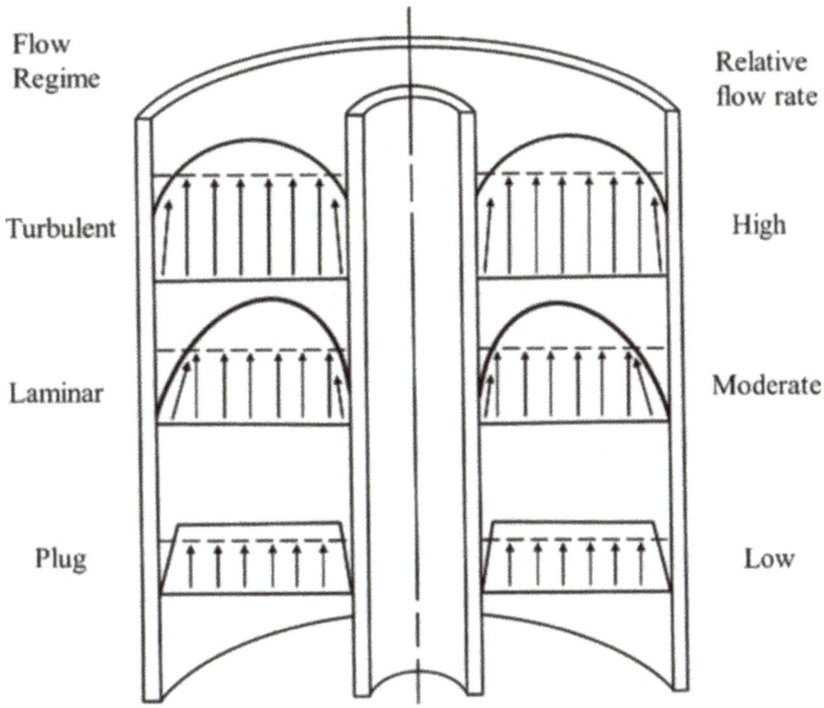

Fig. 7.9 Different flow regimes in which a non-Newtonian may exist (balanced-plug placement technique)

require a certain pressure drop to establish a significant flowrate. There are two possible flow regimes that a non-Newtonian fluid may have (see Fig. 7.9); laminar flow and turbulent flow. Sometimes plug flow regime is also defined as another flow regime but it is a pattern of laminar flow. As shown in Fig. 7.9, the bulk annular velocity profile (dashed lines) and the actual velocity profile (solid lines) are not equal, and the axial velocity (arrows) in the laminar flow regime is not as uniform across the annulus as in the turbulent flow regime. Axial velocity distribution is a maximum in the center of each flow regime and higher than the axial velocity of fluid adjacent to the boundaries [29]. Therefore, mud removal from boundaries might be complex and ineffective.

Cement contamination by drilling fluid is more prone when the drilling fluid removal is inefficient. Haut and Crook [29] showed that the contamination is due to instabilities occurring at the cement-mud interface where the velocities are not strictly axial. The formation of instabilities are a result of nonlinear coupling of changes in shear rate and shear stress at the interface of the fluids, and lead to mud channeling.

Degree of turbulence—In order to achieve a turbulent flow regime for cement, a high flow rate is required; however, it may be unachievable if the slurry has a high viscosity. When a shear force is applied on a non-Newtonian fluid, the fluid resists to flow and undergoes an elastic deformation until the elastic structure breaks down (yields) at some point and the material begins to flow [30]. In practice, turbulent flow of cement during plug placement is less likely to be achieved because of operational limitations. Nevertheless, from a practical point of view, it is important that the frictional pressure drop of cement to be higher than the frictional pressure drop of drilling fluid.

Equivalent circulating density—Long-term zonal isolation requires effective mud displacement which requires the use of high pumping rates during cementing. However, when considering a depleted formation or a subsided field, the formation fracture pressure is lower than the original formation fracture pressure and consequently a narrow pressure window should be expected and tight ECD management is a priority. Therefore, pumping rates during cementing operations are limited. High flowrate results in high frictional pressure, which may exceed the fracture pressure of the formation. This scenario gets even more complex in depleted long horizontal wells. Modifying the rheological behavior of cement and optimizing the pumping rates are to be considered for maintaining low ECDs and to help ensure effective cementing operations [31].

Cement/mud and spacer design—There are several types of spacer systems available including: flushes, gels, water based, oil based, and emulsions (water in oil emulsion and oil in water emulsion). Among these, flushes are mainly used to achieve turbulent flow for improved mud removal [32]. Spacers are designed to improve cement bonds by water-wetting the cement-pipe or cement-formation interfaces while not destabilizing any sensitive zones and not adversely affecting the mud or cement properties [33]. In order to obtain an improved mud removal, studies show that density of displacing fluid should be at least 10% heavier than the displaced fluid, and the friction pressure of the displacing fluid should be greater by at least 20% than the displaced fluid [13]. The maximum mud removal occurs when the viscosity profile of spacer systems is higher than the viscosity profile of the drilling fluid and lower than the cement slurry.

The analysis of removing drilling fluid and replacing it with cement can be performed properly by simulating multiphase flow. In a multiphase flow simulation, interfaces between cement and space fluid, and spacer fluid and drilling mud are presented by solutions of the governing equations [2, 34–37].

In order to minimize the poor mud displacement, a cementing checklist is prepared as guideline [24, 38]:

1. Determine the displacement rates for cement plug on the basis of the mixing and pump capabilities, and ECDs during cementing for typical spacer rheology.
2. Select the spacer and check its compatibility with the mud system.
3. Once the spacer has been selected, determine its viscous properties at bottomhole circulating temperature (BHCT) and bottomhole pressure.
4. Recalculate ECDs during cementing by using viscosity to select the mix, pump, and displacement rate.
5. Calculate cement volumes and annular velocities on the basis of a multi-finger caliper log.
6. Condition the drilling fluid to obtain lower viscosity.
7. Keep solids loading down, especially in high-angle holes.

8. Calculate the surge pressures while running the work string, and run at a speed slow enough to minimize the risk of breaking the formation.
9. Once the work string is at the desired depth, start circulation at the calculated flowrate.

Hole conditioning—Due to high viscosity and gel strength, drilling fluids are not suitable for cement plug operations. Hence mud and hole are conditioned prior to placing cement plugs in open holes. Proper hole conditioning means to establish a hole free of swarf, cuttings, gels, etc., whereas the hole has a mud in a fully displaceable or circulatable condition. This allows the spacer and cement slurry to effectively displace the mud in the desired hole interval. The circulatable hole condition should be established before the first barrel of cement-mud spacer is pumped down [39]. In addition, hole conditioning results in mud conditioning which reduces the yield point of mud and consequently enables more efficient mud removal during cement placement.

Rheological behavior—In order to improve the mud removal efficiency and avoid fracturing formations, modification of cement slurries may be required (role of sophisticated tools and techniques and skilled personnel are inseparable); density changes or rheological behavior modification may be necessary. If pore pressure restrictions do not allow density changes, then modification of slurry rheological behavior is recommended. As an example, rheology of cement slurry can be modified by improving its thixotropic behavior for a better mud displacement [40]. Thixotropy is the characteristic of fluids which have time-dependent shear thinning properties. In other words, when the fluid is in stationary conditions, it forms a gelled structure. But when the fluid is under constant shear rate, the viscosity is decreased over time until it reaches an equilibrium condition. A thixotropic slurry can create a plug flow regime during placement and improve the mud displacement efficiency [41]. However, thixotropic behavior may challenge the plug stability when cement slurry is placed on a pill and while pulling the work string out of the plug.

Buoyancy forces and plug stability—When a cement slurry is placed at the required depth on a less dense drilling fluid, it should resist falling down while setting. Studies performed on the physics of buoyancy driven failure modes of cement plugs placed on drilling fluids, show that a minimum yield stress is required to achieve plug stability. The stability of the interface between drilling fluid and cement is governed by well inclination from vertical, fluids yield stress, the density differences between drilling fluid and cement, the gravity force, and hole diameter [42, 43]. The instability occurring at the interface between two fluids with different densities is known as the *Rayleigh-Taylor instability*. When a cement slurry is placed on a fluid with lower density than the cement in an inclined hole, instabilities in the interface between cement and fluid creates three distinct zones (see Fig. 7.10b); transition zone intruding the mud, exchange flow zone, and the transition zone in the base of cement. The movement of fluids which resulted by the instabilities in the interfaces of cement and slurry caused by buoyancy force is termed slumping motion. In the slumping motion of a cement plug, it is assumed that the bulk of the two fluids moves axially at a slow rate, but in the transition zone, a three-dimensional flow exists.

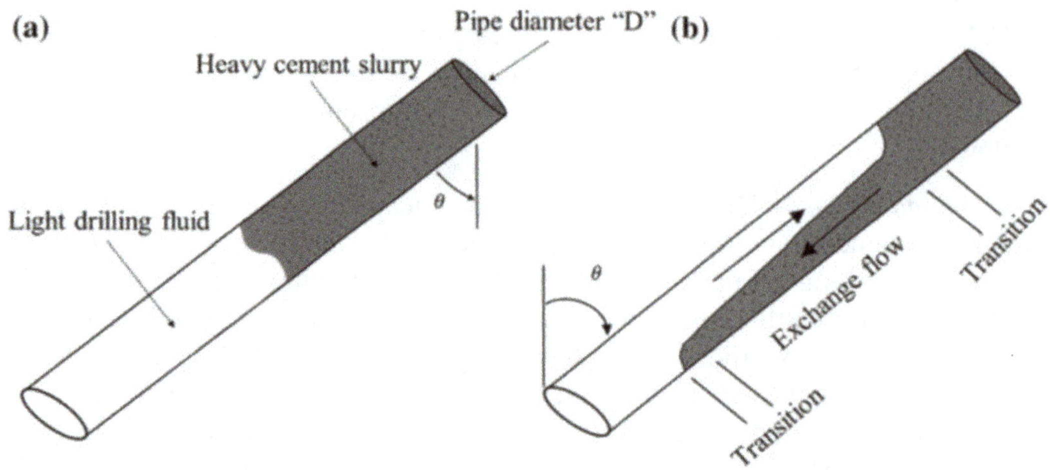

Fig. 7.10 Schematic of stratified axial exchange flow; **a** cement plug placed on a drilling fluid with lower density than cement, **b** buoyancy force is compromising the plug stability and creates three distinct zones [16]

There are some recommendations to minimize the contamination introduced by buoyancy and to achieve a stable plug, including reducing the density differences between cement and drilling fluid (viscous pill), increasing the yield stress or gel strength of the drilling fluid below the intended cement plug, placing a reactive gelled pill between cement and drilling fluid, and avoiding thixotropic cement slurries for balanced plugs [38, 44, 45]. If the induced agitation passes the YP of slurry, the buoyancy and gravity forces will be activated and the contamination effect will be intensified at the interface between the slurry and the drilling fluid and subsequently, the plug stability. It is believed that during balanced-plug placement when a thixotropic slurry is used, the slurry tends to stay in the end of the tailpipe when the intended cement plug is placed. Pulling the tailpipe out of the static thixotropic slurry creates a drag force on the drilling fluid below the cement, and leads to intrusion of the drilling fluid into the slurry, therefore the slurry is contaminated [45]. However, use of a thixotropic cement slurry, which develops gel strength rapidly, improves the plug stability while cement sets [44].

One solution to minimize the effect of buoyancy forces while cement sets is to install a mechanical foundation and place the slurry on top of it [46]. Then, plug stability is provided while cement sets and the gas invasion is minimized. One limitation for the utilization of mechanical foundations is that they cannot be used in openhole sections.

Centralization of work string—An eccentric annulus between work string and formation or casing can channel the displacing fluid to the wide side of the annulus and leave remaining drilling fluid on the narrow side, Fig. 7.11. However, the difference between the density of displacing fluid and displaced fluid creates a hydrostatic pressure imbalance between the narrow side and wide side. On the one hand, the created imbalance pushes the heavier fluid to the narrow side and displacement efficiency is increased. On the other hand, this phenomenon may intensify mud contamination.

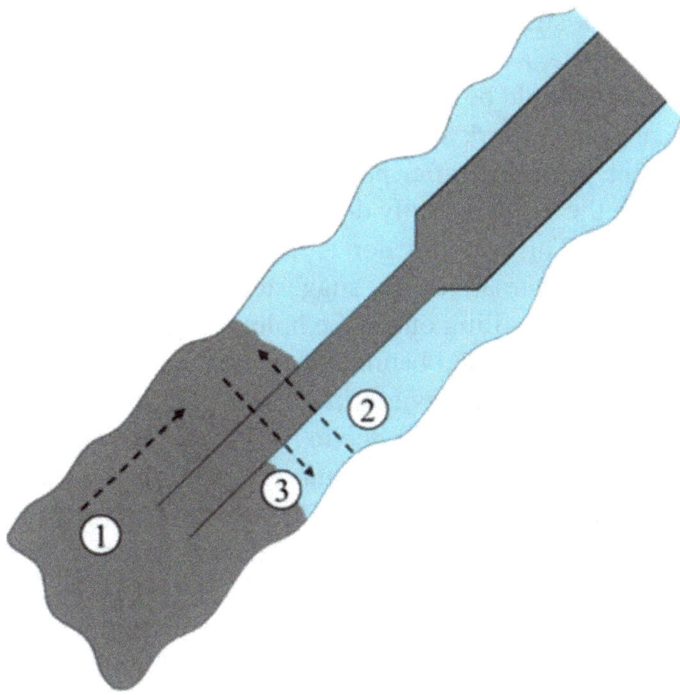

Fig. 7.11 Improper centralization of stinger guides the cement through the large space (1), and gravitational force replaces mud and slurry (2) and (3)

Tehrani et al. [47] studied the effect of eccentricity of pipe on displacement efficiency of mud in inclined wells. Their assumptions included a laminar displacement in the annulus for non-Newtonian fluids in a three-dimensional wellbore. According to their work, good pipe centralization, a high density contrast between drilling fluid and slurry, and positive rheological hierarchy are important factors which improve mud displacement.

Jakobsen et al. [23] considered displacement of fluids with different densities, in eccentric annulus. They concluded that when the displacing fluid is 5% heavier than the fluid to be displaced, the lighter fluid in the narrower part moves to the upper part which is wider. This mechanism, buoyancy-induced, strongly improves the displacement efficiency. This process is recommended when turbulent flow or effective laminar displacement is difficult [48].

Pulling out of plug—The assumption behind the balanced-plug calculation method is that the fluid is going to remain in place while the work string is pulled out of plug with minimal falling of the fluid due to the void caused by metal displacement. However, this assumption is correct only when neglecting the role of drag forces between fluids and work string and the volumes attached to the work string surface, and where a mechanical foundation is used as a base. In order to minimize the agitation effect, a tailpipe or stinger[1] with a smaller diameter and wall thickness is deployed. Because of a thinner wall thickness and a smaller diameter, the fluid volumes involved are smaller and consequently contamination is supposed to be minimized [49]. However, Roye and Pickett [50] showed that the initially balanced plug becomes unbalanced as a dynamic condition is imposed when pulling

[1]The stingers are usually made of fiberglass or aluminum pipe.

the work string out of the plug with a pill as a base. When pulling the work string out of hole, a volume of fluid inside the near surface pipe is displaced, as the same volume should be displaced inside the stinger (with a smaller diameter compared to near surface work string pipe), the height of the displaced fluid inside the stinger is higher. Therefore after pulling a few work string joints out of the hole, the cement inside the stinger is fully displaced with spacer while cement slurry still is left in the annulus. This phenomenon is shown step by step in Fig. 7.12.

Some alternatives are suggested to minimize the effect of dynamic conditions imposed by pulling out of the hole: using a model to correctly calculate the volume of spacer ahead and behind the cement slurry, eliminating the use of stinger, and/or using mechanical devices in the drillpipe just above the stinger [50]. To minimize the plug contamination due to its movement while it sets, it is recommended to use a gelled fluid pill or a mechanical foundation.

Cement job monitoring—The recording of pressure, slurry rate, density, and integrated volume (e.g. mud return rate compared with pump rate) in real time gives a better understanding of the job execution [51, 52]. These data can be analyzed and used for other jobs especially in a campaign plug placement. Figure 7.13 shows the framework of a process control loop for eliminating future plug failures.

7.5 Verification of Placement Operation

Position verification—When a cement plug is placed at the desired interval, its depth and sealability need to be verified. When cement is set, it is *dressed off* and the top of cement is identified by tagging. Cement plugs placed on a mechanical foundations are not tagged when the depth of plug has been verified.

Sealing verification—The evaluation of plug sealing capability is conducted by either pressure testing, or weight testing based on the elements of the well barrier envelope.

Pressure testing—Plugs installed inside casing are placed on either mechanical plugs or viscous pills, Fig. 7.14. When the mechanical plug is used as foundation and it passes the pressure test, usually the cement installed on top of it is not pressure tested. However, if the mechanical plug is not pressure tested or fails to pass the pressure test, the installed plug is pressure tested.

When a cement plug is installed on a viscous pill, its sealability is evaluated by performing either positive or negative pressure testing. In a positive pressure test the well is subjected to a given pressure and the pressure changes are recorded. The given pressure is higher than the pressure below the plug, Fig. 7.15a. When considering a positive test, the primary cement (cement behind casing), cement-pipe bonding, and casing should not be damaged. To avoid this issue, the test pressure is selected to not exceed the casing strength minus wear allowance. Another factor to be considered during positive pressure testing is the effect of ballooning uncemented casing. It happens when the test pressure exceeds the casing mechanical limit and casing is expanded in the intervals where liquid fills the annulus behind the casing. In this case

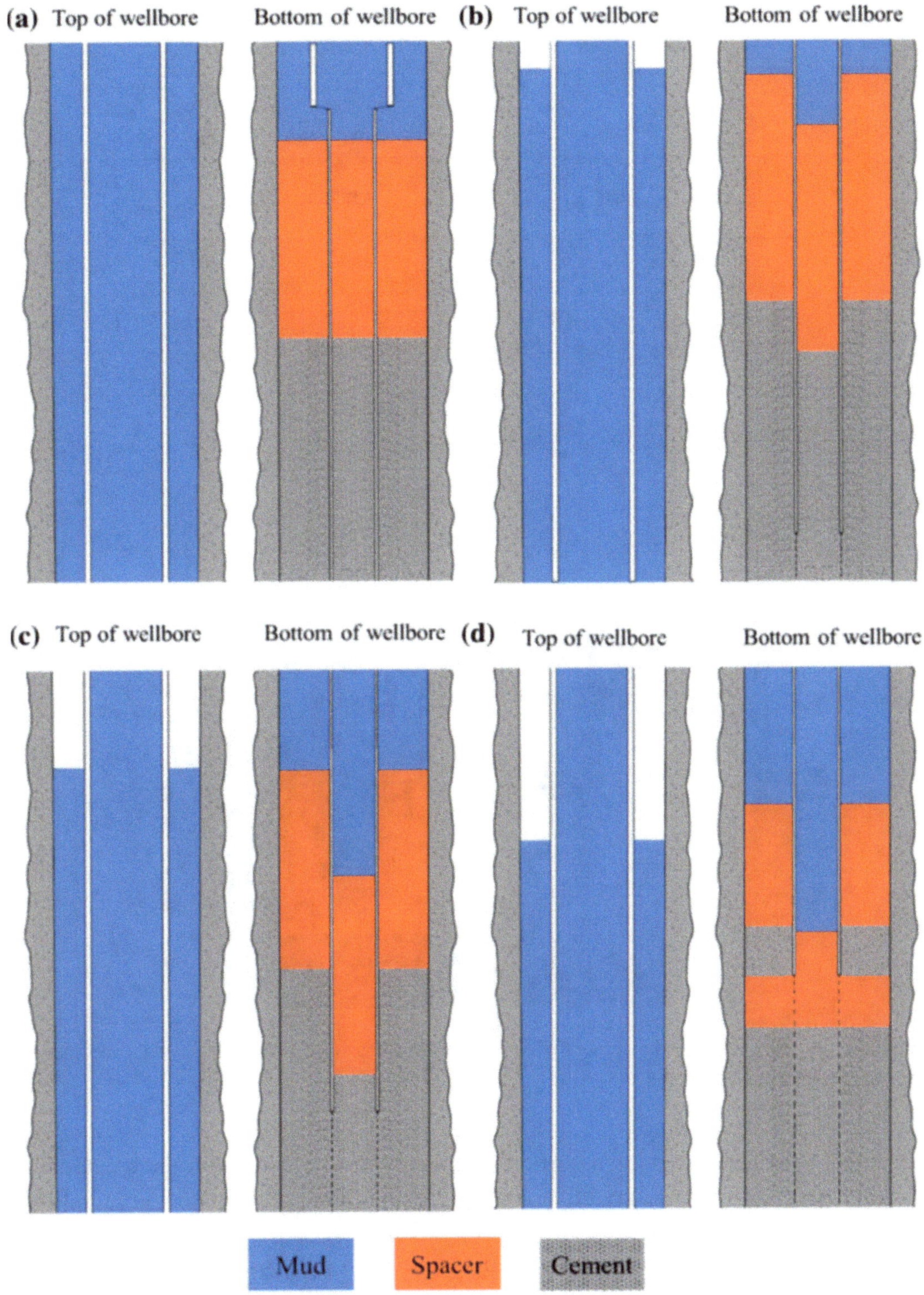

Fig. 7.12 An unbalanced condition during pulling the work string equipped with stinger out of hole: **a** Balanced-plug is established while workstring is inside plug, **b** workstring is removed slowly from the plug but due to the removed volume of workstring, the height of spacer inside and outside is not in the same level, **c** the more the workstring is run out of hole, the higher the differences of the fluid levels **d** spacer reaches the tailpipe while the annular cement slurry is still left [50]

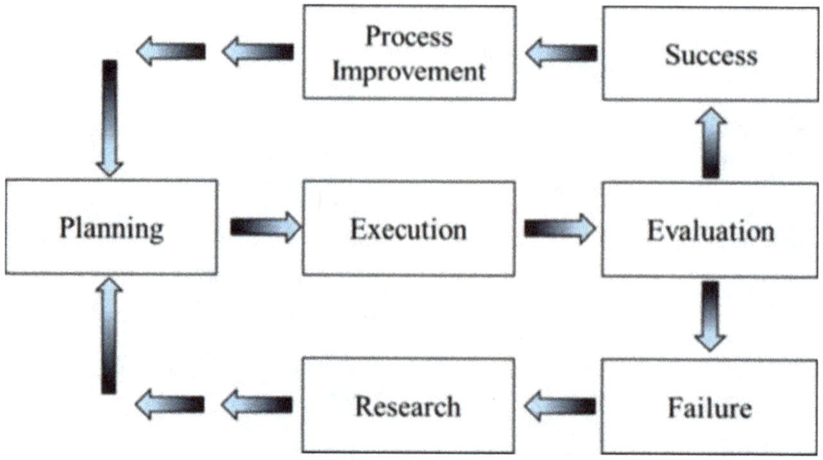

Fig. 7.13 Process control loop for a plug placement operation [53]

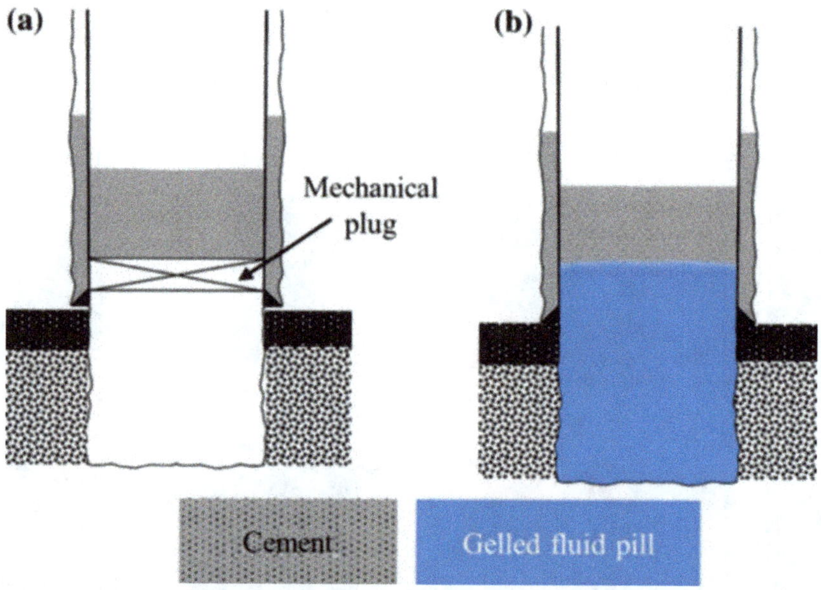

Fig. 7.14 Pressure testing of a cement plug placed inside casing; **a** plug placed on a mechanical foundation, **b** plug placed on a viscous pill

as the test-pressure increases, a portion of injected fluid fills the volume created due to ballooning however, it may be misinterpreted and lead to disqualifying the plug. Pressure testing of cement plug is reviewed in Chap. 9.

In a negative pressure test, the well pressure is dropped and the pressure build-up is recorded. In other words, pressure below the installed plug is higher than the pressure above the plug, Fig. 7.15b. The negative pressure test is also known by other names such as *inflow test* or *drawdown test*. Table 7.2 summarizes requirements for pressure testing of different regulatory authorities.

Weight testing—This method is used for plugs installed in open holes as pressure testing in open holes is meaningless. In this method, top of cement is dressed off

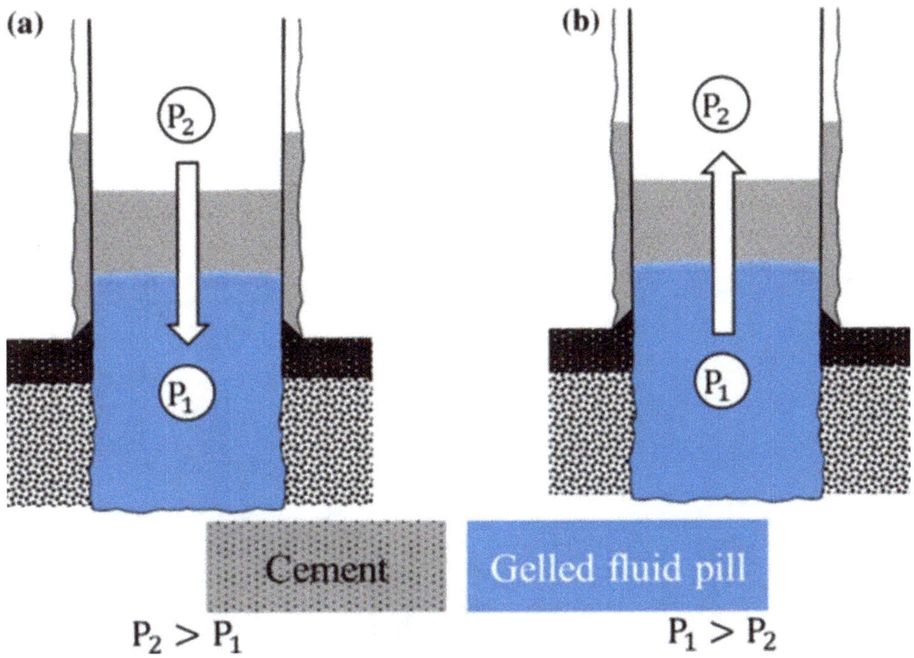

Fig. 7.15 Pressure testing of an installed plug inside casing; **a** positive pressure testing, **b** negative pressure testing

Table 7.2 Requirements for pressure testing and weight testing variation for some countries

Country	Pressure testing requirement	Weight testing requirement
Norway [54]	• The positive pressure test requirement is 1000 psi above the estimated leak off pressure (below casing/potential leak path) • The positive pressure test requirement for a surface casing plug is 500 psi above the estimated leak off pressure • Cement plug installed on a pressure tested foundation need not to be pressure tested	• Cement plug installed in an openhole should be weight tested
United Kingdom [55]	• The positive pressure test requirement is minimum 500 psi above the source pressure • Inflow test requirement at least the maximum pressure differential which barrier will experience after permanent abandonment	• Cement plug installed in open hole is weight tested by drillpipe with typically 10–15 klbs • When cement plug installed in open hole is weight tested by wireline, coiled tubing or stinger then the weight is limited by tools and geometry

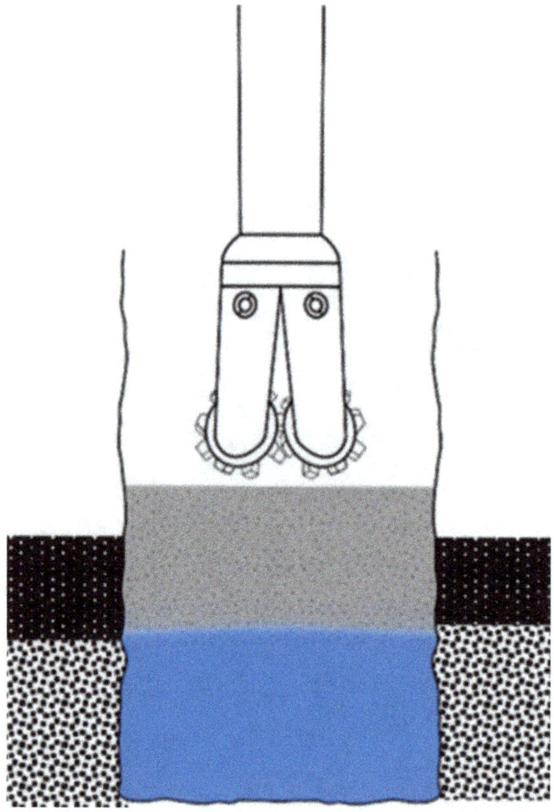

Fig. 7.16 Weight testing of a plug installed in an openhole

and a weight is applied on the plug, Fig. 7.16. Drillpipe, coiled tubing, stinger, and wireline can be used for weight testing however, the application of stinger, wireline, and coiled tubing is limited by weight of the tools or geometry. Different regulators have different weight requirements. Table 7.2 summarizes some specific requirements for weight testing of different regulatory authorities.

The cement plug installed inside tubing is usually verified by pressure testing and tagging. If the plug is installed on a pressure tested bridge plug, then it is not pressure tested. As the plug is installed on a tested bridge plug, its position verification and pressure testing are not possible. Sealing capability of cement plugs installed between tubing and production casing is verified by pressure testing and position verification is done by bond logging.

References

1. Aranha, P.E., C.R. Miranda, J.V.M. Magalhães, et al. 2011. Dynamic aspects governing cement-plug placement in deepwater wells. *SPE Drilling & Completion* 26 (03): 341–351. https://doi.org/10.2118/140144-PA.
2. Chen, Z., S. Chaudhary, and J. Shine. 2014. Intermixing of cementing fluids: Understanding mud displacement and cement placement. in IADC/SPE drilling conference and exhibition. SPE-167922-MS. Fort Worth. Texas, USA: Society of Petroleum Engineers. https://doi.org/10.2118/167922-MS.

3. Clark, C.R., and G.L. Carter. 1973. Mud displacement with cement slurries. *Journal of Petroleum Technology* 25 (07): 775–783. https://doi.org/10.2118/4090-PA.
4. Engelke, B., D. Petersen, and F. Moretti, et al. 2017. New fiber technology to improve mud removal. In *Offshore Technology Conference, OTC Brasil*. OTC-28025-MS. Rio de Janeiro, Brazil. https://doi.org/10.4043/28025-MS.
5. Guzman Araiza, G., H.E. Rogers, and L. Pena, et al. 2007. Successful placement technique of openhole plugs in adverse conditions. In *Asia pacific oil and gas conference and exhibition*. SPE-109649-MS. Jakarta, Indonesia: Society of Petroleum Engineers. https://doi.org/10.2118/109649-MS.
6. Kelessidis, V.C., D.J. Guillot, R. Rafferty, et al. 1996. Field data demonstrate improved mud removal techniques lead to successful cement jobs. *SPE Advanced Technology Series* 4 (01): 53–58. https://doi.org/10.2118/26982-PA.
7. Ford, J.T., M.B. Oyeneyin, and E. Gao, et al. 1994. The formulation of milling fluids for efficient hole cleaning: An experimental investigation. In *European petroleum conference*. SPE-28819-MS. London, United Kingdom: Society of Petroleum Engineers. https://doi.org/10.2118/28819-MS.
8. Messler, D., D. Kippie, and M. Broach, et al. 2004. A potassium formate milling fluid breaks the 400° fahrenheit barrier in a deep Tuscaloosa coiled tubing clean-out. In *SPE international symposium and exhibition on formation damage control*. SPE-86503-MS. Lafayette, Louisiana: Society of Petroleum Engineers. https://doi.org/10.2118/86503-MS.
9. Offenbacher, M., N. Erick, and M. Christiansen, et al. 2018. Robust MMH drilling fluid mitigates losses, eliminates casing interval on 200+ wells in the permian basin. In *IADC/SPE drilling conference and exhibition*. SPE-189628-MS. Fort Worth, Texas, USA: Society of Petroleum Engineers. https://doi.org/10.2118/189628-MS.
10. Carney, L. 1974. Cement spacer fluid. *Journal of Petroleum Technology* 26 (08): 856–858. https://doi.org/10.2118/4784-PA.
11. Labarca, R.A., and J.C. Guabloche. 1992. New spacers in Latin America. In *SPE Latin America petroleum engineering conference*. Caracas, Venezuela: Society of Petroleum Engineers. https://doi.org/10.2118/23733-MS.
12. Moran, L.K., and K.O. Lindstrom. 1990. Cement spacer fluid solids settling. In *SPE/IADC drilling conference*. Houston, Texas: Society of Petroleum Engineers. https://doi.org/10.2118/19936-MS.
13. Shadravan, A., G. Narvaez, and A. Alegria, et al. 2015. Engineering the mud-spacer-cement rheological hierarchy improves wellbore integrity. In *SPE E&P health, safety, security and environmental conference-Americas*. SPE-173534-MS. Denver, Colorado, USA: Society of Petroleum Engineers. https://doi.org/10.2118/173534-MS.
14. Shadravan, A., M. Tarrahi, and M. Amani. 2017. Intelligent tool to design drilling, spacer, cement slurry, and fracturing fluids by use of machine-learning algorithms. *SPE Drilling & Completion* 32 (02): 131–140. https://doi.org/10.2118/175238-PA.
15. Nelson, E.B., and D. Guillot. 2006. *Well cementing*, 2nd ed. Sugar Land, Texas: Schlumberger. ISBN-13: 978-097885300-6.
16. Fosso, S.W., M. Tina, and I.A. Frigaard, et al. 2000. Viscous-pill design methodology leads to increased cement plug success rates; application and case studies from Southern Algeria. In *IADC/SPE Asia Pacific drilling technology*. SPE-62752-MS. Kuala Lumpur, Malaysia: Society of Petroleum Engineers. https://doi.org/10.2118/62752-MS.
17. Portman, L. 2004. Cementing through coiled tubing: Common errors and correct procedures. in *SPE/ICoTA Coiled Tubing Conference and Exhibition*. SPE-89599-MS, Houston, Texas: Society of Petroleum Engineers. https://doi.org/10.2118/89599-MS.
18. Newman, K.R. and P.A. Brown. 1993. Development of a standard coiled-tubing fatigue test. In *SPE annual technical conference and exhibition*. SPE-26539-MS. Houston, Texas: Society of Petroleum Engineers. https://doi.org/10.2118/26539-MS.
19. Elsborg, C.C., R.A. Graham, and R.J. Cox. 1996. Large diameter coiled tubing drilling. In *International conference on horizontal well technology*. SPE-37053-MS. Calgary, Alberta, Canada: Society of Petroleum Engineers. https://doi.org/10.2118/37053-MS.

20. Nick, L., R. Raj, and S. Srisa-ard, et al. 2011. Coiled tubing operations from a work boat. In *SPE/ICoTA coiled tubing & well intervention conference and exhibition*. SPE-141234-MS. The Woodlands, Texas, USA: Society of Petroleum Engineers. https://doi.org/10.2118/141234-MS.
21. Bybee, K. 2011. Cementing, perforating, and fracturing using coiled tubing. *Journal of Petroleum Technology* 63 (06). https://doi.org/10.2118/0611-0054-JPT.
22. Denney, D. 2001. Rheological targets for mud removal and cement-slurry design. *Journal of Petroleum Technology* 53 (08): 65–66. https://doi.org/10.2118/0801-0065-JPT.
23. Jakobsen, J., N. Sterri, and A. Saasen, et al. 1991. Displacements in eccentric annuli during primary cementing in deviated wells. In *SPE production operations symposium*. SPE-21686-MS. Oklahoma City, Oklahoma, USA: Society of Petroleum Engineers. https://doi.org/10.2118/21686-MS.
24. Sauer, C.W. 1987. Mud displacement during cementing state of the art. *Journal of Petroleum Technology* 39 (09): 1091–1101. https://doi.org/10.2118/14197-PA.
25. Smith, T.R. 1989. Cementing displacement practices: application in the field. In *SPE/IADC drilling conference*. SPE-18617-MS. New Orleans, Louisiana: Society of Petroleum Engineers. https://doi.org/10.2118/18617-MS.
26. Smith, T.R. 1990. Cementing displacement practices field applications. *Journal of Petroleum Technology* 42 (05): 564–629. https://doi.org/10.2118/18617-PA.
27. Isgenderov, I., S. Taoutaou, and I. Kurawle, et al. 2015. Modified approach leads to successful off-bottom cementing plugs in highly deviated wells in the Caspian Sea. In *SPE/IATMI Asia Pacific oil & gas conference and exhibition*. SPE-176316-MS. Nusa Dua, Bali, Indonesia: Society of Petroleum Engineers. https://doi.org/10.2118/176316-MS.
28. Webster, W.W., and J.V. Eikerts. 1979. Flow after cementing: A field and laboratory study. In *SPE annual technical conference and exhibition*. SPE-8259-MS. Las Vegas, Nevada: Society of Petroleum Engineers. https://doi.org/10.2118/8259-MS.
29. Haut, R.C., and R.J. Crook. 1979. Primary cementing: the mud displacement process. In *SPE Annual Technical Conference and Exhibition*. SPE-8253-MS. Las Vegas, Nevada: Society of Petroleum Engineers. https://doi.org/10.2118/8253-MS.
30. Barnes, H.A., J.F. Hutton, and K. Walters. 1989. *An introduction to rheology*. Amsterdam, The Netherlands: Elsevier. 0-444-87469-0.
31. Regan, S., J. Vahman, and R. Ricky. 2003. *Challenging the limits: Setting long cement plugs*. In *SPE Latin American and Caribbean petroleum engineering conference*. SPE-81182-MS. Port-of-Spain, Trinidad and Tobago: Society of Petroleum Engineers. https://doi.org/10.2118/81182-MS.
32. Beirute, R.M. 1976. All purpose cement-mud spacer. In *SPE symposium on formation damage control*. SPE-5691-MS. Houston, Texas: Society of Petroleum Engineers. https://doi.org/10.2118/5691-MS.
33. Farahani, H., A. Brandl, and R. Durachman. 2014. Unique cement and spacer design for setting horizontal cement plugs in SBM environment: deepwater Indonesia case history. In *Offshore technology conference-Asia*. OTC-24768-MS. Kuala Lumpur, Malaysia: Offshore Technology Conference. https://doi.org/10.4043/24768-MS.
34. Enayatpour, S., and E. van Oort. 2017. Advanced modeling of cement displacement complexities. In *SPE/IADC drilling conference and exhibition*. SPE-184702-MS. The Hague, The Netherlands: Society of Petroleum Engineers. https://doi.org/10.2118/184702-MS.
35. Frigaard, I.A., M. Allouche, and C. Gabard-Cuoq. 2001. Setting rheological targets for chemical solutions in mud removal and cement slurry design. In *SPE international symposium on oilfield chemistry*. SPE-64998-MS. Houston, Texas: Society of Petroleum Engineers. https://doi.org/10.2118/64998-MS.
36. Frigaard, I.A., and S. Pelipenko. 2003. Effective and ineffective strategies for mud removal and cement slurry design. In *SPE Latin American and Caribbean petroleum engineering conference*. SPE-80999-MS. Port-of-Spain, Trinidad and Tobago: Society of Petroleum Engineers. https://doi.org/10.2118/80999-MS.
37. Li, X., and R.J. Novotny. 2006. Study on cement displacement by lattice-Boltzmann method. In *SPE annual technical conference and exhibition*. SPE-102979-MS. San Antonio, Texas, USA: Society of Petroleum Engineers. https://doi.org/10.2118/102979-MS.

38. Smith, R.C., R.M. Beirute, and G.B. Holman. 1984. Improved method of setting successful cement plugs. *Journal of Petroleum Technology* 36 (11): 1897–1904. https://doi.org/10.2118/11415-PA.
39. Beirute, R.M., F.L. Sabins, and K.V. Ravi. 1991. Large-scale experiments show proper hole conditioning: A critical requirement for successful cementing operations. In *SPE annual technical conference and exhibition*. SPE-22774-MS. Dallas, Texas: Society of Petroleum Engineers. https://doi.org/10.2118/22774-MS.
40. Barnes, H.A. 1997. Thixotropy—A review. *Journal of Non-Newtonian Fluid Mechanics* 70 (1): 1–33. https://doi.org/10.1016/S0377-0257(97)00004-9.
41. Gahlawat, R., S.R.K. Jandhyala, and V. Mishra, et al. 2016. Rheology modification for safe cementing of low-ECD zones. In *IADC/SPE Asia Pacific drilling technology conference*. SPE-180523-MS. Singapore: Society of Petroleum Engineers. https://doi.org/10.2118/180523-MS.
42. Crawshaw, J.P., and I. Frigaard. 1999. Cement plugs: Stability and failure by buoyancy-driven mechanism. In *Offshore Europe oil and gas exhibition and conference*. SPE-56959-MS. Aberdeen, United Kingdom: Society of Petroleum Engineers. https://doi.org/10.2118/56959-MS.
43. Frigaard, I.A., and J.P. Crawshaw. 1999. Preventing buoyancy-driven flows of two Bingham fluids in a closed pipe – fluid rheology design for oilfield plug cementing. *Journal of Engineering Mathematics* 36 (4): 327–348. https://link.springer.com/article/10.1023/A:1004511113745.
44. Bour, D.L., D.L. Sutton, and P.G. Creel. 1986. Development of effective methods for placing competent cement plugs. In *Permian basin oil and gas recovery conference*. SPE-15008-MS. Midland, Texas: Society of Petroleum Engineers. https://doi.org/10.2118/15008-MS.
45. Heathman, J.F. 1996. Advances in cement-plug procedures. *Journal of Petroleum Technology* 48 (09): 825–831. https://doi.org/10.2118/36351-JPT.
46. Harestad, K., T.P. Herigstad, A. Torsvoll, et al. 1997. Optimization of balanced-plug cementing. *SPE Drilling & Completion* 12 (03). https://doi.org/10.2118/35084-PA.
47. Tehrani, A., J. Ferguson, and S.H. Bittleston. 1992. Laminar displacement in annuli: A combined experimental and theoretical study. In *SPE annual technical conference and exhibition*. SPE-24569-MS. Washington, D.C.: Society of Petroleum Engineers. https://doi.org/10.2118/24569-MS.
48. Kroken, W., A.J. Sjaholm, and A.S. Olsen. 1996. Tide flow: A low rate density driven cementing technique for highly deviated wells. In *SPE/IADC drilling conference*. SPE-35082-MS. New Orleans, Louisiana: Society of Petroleum Engineers. https://doi.org/10.2118/35082-MS.
49. Carpenter, C. 2014. Stinger or tailpipe placement of cement plugs. *Journal of Petroleum Technology* 65 (05): 3. https://doi.org/10.2118/0514-0147-JPT.
50. Roye, J., and S. Pickett. 2014. Don't get stung setting balanced cement plugs: a look at current industry practices for placing cement plugs in a wellbore using a stinger or tail-pipe. In *IADC/SPE drilling conference and exhibition*. Fort Worth, Texas, USA: Society of Petroleum Engineers. https://doi.org/10.2118/168005-MS.
51. Marriott, T., H. Rogers, and S. Lloyd, et al. 2006. Innovative cement plug setting process reduces risk and lowers NPT. In *Canadian international petroleum conference*. PETSOC-2006-015. Calgary, Alberta: Petroleum Society of Canada. https://doi.org/10.2118/2006-015.
52. Smith, R.C. 1986. Improved cementing success through real-time job monitoring. *Journal of Petroleum Technology* 38 (06). https://doi.org/10.2118/15280-PA.
53. Heathman, J. and R. Carpenter. 1994. Quality management alliance eliminates plug failures. In *SPE annual technical conference and exhibition*. New Orleans, Louisiana: Society of Petroleum Engineers. https://doi.org/10.2118/28321-MS.
54. NORSOK Standard D-010. 2013. *Well integrity in drilling and well operations*. Standards Norway.
55. Oil & Gas UK. 2012. Guidelines for the suspension and abandonment of wells.

Chapter 8
Tools and Techniques for Plug and Abandonment

8.1 Casing Cut and Removal Techniques

In permanent P&A, establishment of a rock-to-rock barrier is a requirement. There are situations where the annular barrier behind casing is not qualified or there is no annular barrier. Therefore, full access to the formation shall be obtained. Different techniques have been utilized by the petroleum industry such as cut-and-pull, casing milling, and section milling. Some new techniques have been suggested some of which are in use and others are in development. Such techniques include perforate-wash-cement, upward section milling, melting downhole completion, and plasma-based milling. This chapter will present these techniques, briefly.

8.1.1 Cut-and-Pull Casing

In permanent P&A operations, there are situations where there is only a poor annular barrier or no annular barrier at all. When there is a long length of uncemented casing, a cut-and-pull operation can be the necessary option. In this operation, a circumferential cut is made of the casing, above a casing coupling, and then a spear is engaged inside the casing to pull the casing out of hole. The spear can be engaged hydraulically. For the traditional method, the pulling force is provided by the working unit through the workstring to the bottom hole assembly. However, the advancement of cut-and-pull techniques provides a new generation of tools, downhole hydraulic pulling tool anchors, to create large amounts of pulling force without fully engaging the working unit pulling capacity. As an example, by use of 1 psi hydraulic power, 300 psi is generated by the downhole hydraulic pulling tool anchors [1].

M. Khalifeh and A. Saasen, *Introduction to Permanent Plug and Abandonment of Wells*, Ocean Engineering & Oceanography 12,
https://doi.org/10.1007/978-3-030-39970-2_8

Ideally, the cut-and-pull operation is a single trip. However, there are challenges associated with it, which may require multiple trips. Such challenges are settled barite behind casing, scale depositions, collapsed formation, or unknown bond strength of the poor casing cement. Therefore, pipe retrieval requires high pulling capacity. The pulling capacity can be beyond the working unit or workstring capacity. It can also compromise the stability of the facility (i.e. working unit or platform). Therefore, multiple trip are performed by cutting the casing into short lengths. During casing pulling operations when the casing is moving, debris may fall down around the casing causing it to get stuck and even be irretrievable. Multiple trips expose personnel to several cut-and-pull operations and increases the risk of HSE issues. In addition, the retrieved pipe needs to be handled safely and disposed off properly.

The casing cut can be done using explosives, chemicals, mechanical cutters or using abrasive cutters. Regardless of which type of cutting technique is used, usually the cut is performed when the casing is under tension. Some of the challenges associated with explosive cutters are transportation, handling and storage, uncertainties related to eccentricity or stand-off of the device and damage of the outer casing, dispersion of force from the device, and shape of the resulting cut. Radial cutting torches, which use thermite derivatives to melt casing radially, can cut the casing partially or cut the pipe behind casing. Chemical cutters utilize chemicals which react with steel. Bromine triflouride is an example of such a chemical which is extremely hazardous for surrounding and personnel with irreversible health effects. The efficiency of chemical cutters can be affected by the presence of scale, poor spray pattern, or eccentricity of casing. Mechanical cutters are either electrical pipe cutters (see Fig. 8.1) or hydraulic pipe cutters (see Fig. 8.2). One of the advantages of mechanical cutters is the centralizer which holds the cutter in the pipe center and the risk of damaging the outer casing due to eccentricity is reduced.

For abrasive cutting techniques, abrasive cutting particles are injected into a water jet and wear away the production tubing, casing, drill pipe or drill collar. As this technique has advantages for cut and removal, especially wellhead cut and removal, the subject is covered in more detail, later in this chapter.

Fig. 8.1 Mechanical pipe cutter which is powered electrically. (Courtesy of Baker Hughes)

Fig. 8.2 Mechanical pipe cutter which is powered hydraulically. (Courtesy of Schlumberger)

8.1.2 Casing Milling

In this operation, casing is milled when a length of casing needs to be removed. Such circumstances may include slot recovery or sidetracking. The process of opening a window is typically done by a mill, however, milling with the use of an abrasive fluid jet has been studied [2]. In a P&A operation the required length is usually longer than required for sidetracking. Therefore, typically a section casing milling operation is carried out.

8.1.3 Casing Section Milling

One of the main reasons limiting the use of rigless P&A operations is poor casing cement or uncemented casing. For conventional practice, a window is section milled and the operation is called *section milling*. The aim of section milling is to grind away a portion of casing and cement. While section milling the casing, the hole needs to be kept clean by removal of produced swarf and other debris. The term *swarf* is used for metal fillings or shavings created by the milling tool during the casing removal process. The opened window needs to be under-reamed to expose new formation. Then, a cement plug is placed. Section milling is a time consuming operation and difficult to execute safely and efficiently. The current rate of milling

Table 8.1 Section milling data gathered during P&A operation of a well on the NCS

Casing size	Milling fluid	Length of window	Type of cutters	Number of runs	Milling rate	Weight of removed metal
13 3/8-in.	KCl based polymer fluids or MMH based fluids	164 ft	Tungsten carbide	1–2	8.5 ft/h	72 lbm/ft

operations for 7-in. casing is typically around 7–9 (ft/h), with additional time taken for tripping, hole cleaning and cleaning of BOP cleaning. The operation increases risk and introduces different challenges. The fluids designed for section milling must have sufficient weight and viscosity to suspend and transport swarf to surface while keeping the opened hole stable. Sometimes, the required viscous profile of the designed fluids increases the ECD to exceed the fracture gradient, resulting in breaking the formation. This phenomenon may lead to fluid loss and subsequently swabbing and loss of well control. Presence of fluid loss also causes poor hole cleaning and risk of packing off the Bottom Hole Assembly (BHA) which can lead to sticking of the milling or under-reaming BHAs. Section milling is also affected by the location of casing couplings and casing accessories such as centralizers and scratchers. With current milling tools, there is a risk of splitting and buckling the casing, which effects performance and the ability to successfully mill the required interval. Swarf and skimmed casing, and debris can also damage the BOP and effect its functionality. At surface, the transported swarf must be separated and captured by use of handling equipment. Swarf needs to be handled and disposed off properly and as it has sharp angular surfaces, it introduces HSE challenges. Swarf therefore requires a special handling system with trained personnel who need to be equipped with protective equipment. The milling operation requires the use of a drilling rig which is costly. As the cutting tool is worn out after only some feet of milling, frequent trip out and in is often required, which is time consuming. An additional limitation of section milling is generated vibrations. Table 8.1 shows the milling data gathered from a well on the NCS. Due to challenges associated with conventional milling operations, some techniques or methods have been suggested as alternative solutions. These techniques include upward milling, PWC, melting the downhole completion and plasma-based milling.

8.1.4 Upward Milling

Section milling is a proven technique which gives full access to the original formation for creating a rock-to-rock barrier. However, swarf transportation to surface and swarf handling on surface are time consuming and costly operations which are associated with HSE risks. If swarf could be left behind in the wellbore, section milling without swarf to surface, the section milling technique could be more efficient. Upward milling is a new form of section milling technique where the milling operation is performed while moving upward, cutting the swarf into small bits, where the swarf falls down into the wellbore. This system consists of a taper mill, auger section, section mill, emergency release disconnect, jet sub, left-hand rotating mud motor, drill collars, torque isolation assembly, spring loaded pads, spiral stabilizer, and intensifier [3, 4]. Figure 8.3 shows the main tools of upward milling bottom-hole assembly, from top (component 1) to bottom (component 11).

The key components of the upward section mill assembly are shown in Fig. 8.3. At the bottom of the planned milled window, the assembly opens its knives and creates the cut through the casing. Then, it mills upward to the desired depth, and finally retracts the knives at the top of the window. In conventional milling, the knives are retracted when pulling the workstring upward. However, the retraction mechanism for knives in the upward milling method is challenging as the knives cannot retract through upward movement.

Emergency release disconnect—This is a designed weak link in the system to release the assembly if the knives do not retract or the BHA becomes stuck. By over-pulling the workstring, the designed weak link is activated and the BHA is

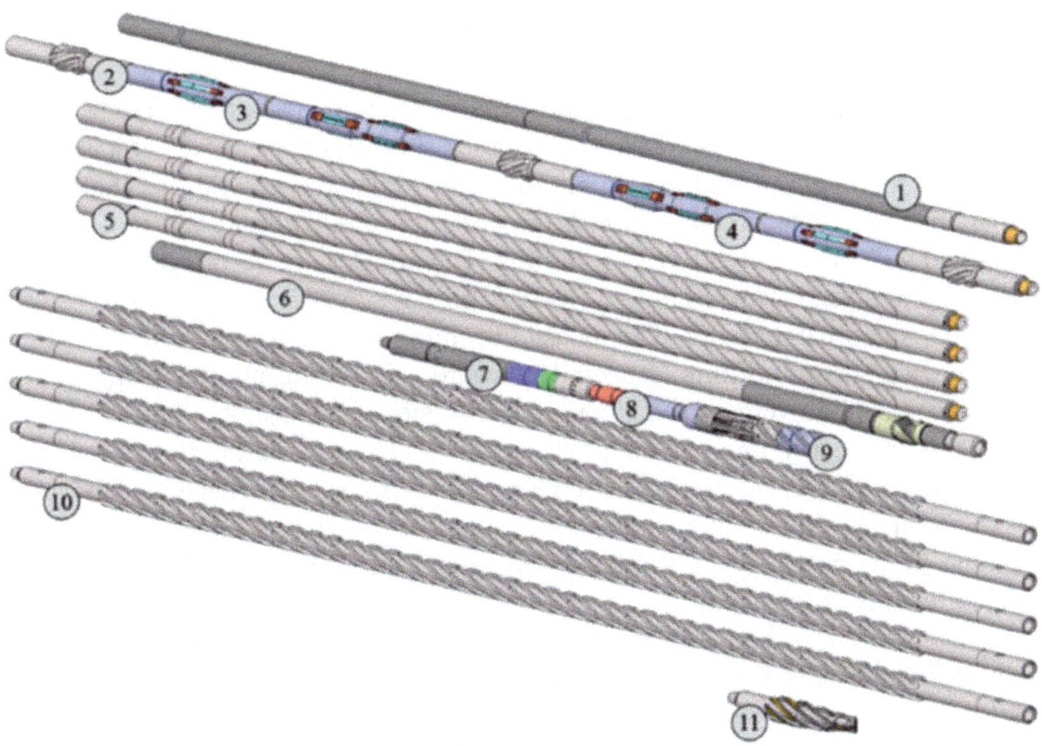

Fig. 8.3 Main components of an upward milling BHA without swarf to surface; (1) intensifier, (2) spiral stabilizer, (3) spring loaded pads, (4) torque isolator, (5) drill collars, (6) left-hand mud motor (7) jet sub, (8) disconnect, (9) section mill, (10) auger section, and (11) taper mill [3]

Table 8.2 Advantages and possible limitations of upward milling technique

Advantages	Possible limitations
• No HSE issues related to swarf handling • Time and cost efficient • No steel as part of permanent barrier	• High inclination can affect the swarf movement to the well leg

released. However, this is a last option to release the BHA. Other scenarios such as reciprocating the workstring to retract knives and pushing the knives to the bottom of window are tested prior to activation of the release disconnect feature.

Drill collars—As reaming is a part of the operation, it is important to ensure that the torque isolation assembly remains inside casing. Therefore, drill collars are installed above the left-hand mud motor.

Intensifier—This is a hydraulic spring to increase the impact force while enabling smooth load transition from the applied over-pull, at surface, to the knives, downhole, when milling upwards.

Left-hand mud motor—Right-hand rotation upward milling increases the risk and chance of unscrewing casing collars, especially at intervals where uncemented casing exists. Therefore a left-hand motor provides the downhole left-hand rotation and the required torque for the section mill and auger. A design feature for such a motor is high-torque and low-speed. This type of motor may be used in combination with coiled tubing to carry out rigless P&A operations.

Jet sub—To divert the flow of mud to the annulus while allowing swarf and cuttings to fall down the well, a jet sub is used below the left-hand mud motor. The main function of the jet sub is to avoid circulating fluid along the knives, if it is out of control and the swarf and cuttings might move upward and cause serious challenges and risks. The nozzle design and nozzle configuration are important to open the knives and generate enough force for milling. The nozzles are designed for different flowrates and fluid densities.

Torque isolator assembly—This is used to minimize the heavy vibration which occurs during section milling, especially upward section milling. By using such a component, axial movement and a continuous torsional constraint are provided.

Auger—In order to improve the process of swarf movement into the rat hole and prevent bridging, auger sections are used (see Fig. 8.3). Casing inside diameter, auger outside diameter, fluid flowrate, density of fluid and system operational procedures are some of the parameters considered during the design process of the auger and which affect the efficiency of it.

Taper mill—Swarf and cuttings can bridge across the cut-out and block the path for other swarf to fall down the wellbore. A taper mill is installed below the section mill to clean out such swarf bridges.

When milling upwards, to prevent backing-off of casing collars, a left-hand rotation is necessary. It can be facilitated either by a left hand mud motor or by left-hand workstring to surface. An alternative flow path is also a requirement as swarf or cuttings need to be deposited below the milled section. Table 8.2 presents advantages and possible limitations of the upward milling technique.

8.2 Perforate, Wash and Cement Technique

8.2.1 Concept Behind the Technique

Generally speaking, this technique was first taken into use in the 70's to establish annular barrier whereby casing is perforated, washed and cemented [5]. Briefly, a perforation gun is run to the barrier depth where there no cement or poor cement behind casing. The casing is perforated, Fig. 8.4, and the gun is either left in hole or retrieved. In the next step, a washing tool is Run in Hole (RIH) and washes the annular space behind the perforated casing to remove the debris, settled mud and mud film [6, 7]. The washing process is carried out downward and a several times to obtain fresh formation. At surface, removed metal and debris can be seen and monitored on shakers which gives better control of the washing process. When washing is completed, an integrity test is performed to check the quality of washed and removed zone. If the integrity test is successful, the washing tool may either be left below the bottom perforations to function as a mechanical foundation for the

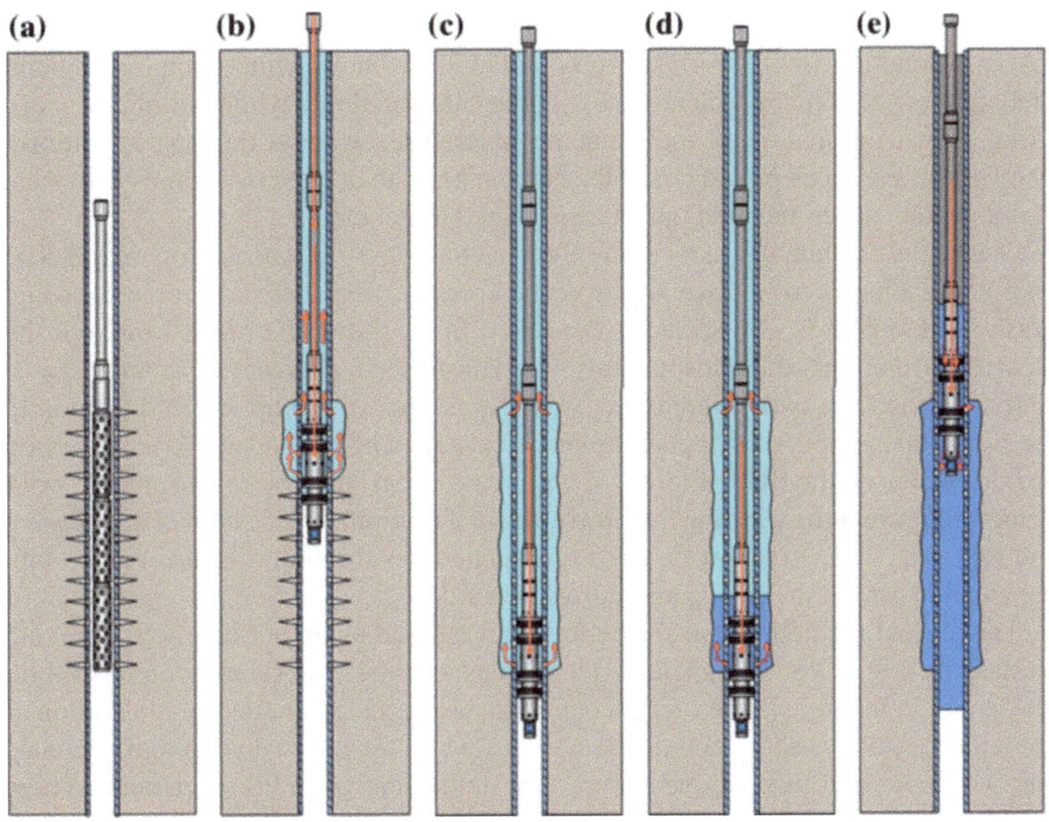

Fig. 8.4 Perforate and wash part of PWC technique; **a** casing is perforated, **b** washing tool is RIH and washes the annular space behind the perforated interval, downward, **c** BHA is placed below the bottom perforations, **d** spacer is pumped and work string is pulled, upward, **e** spacer is extended above the top perforations. (Courtesy of Archer Oiltools)

Fig. 8.5 Perforation gun and washing tool are RIH in single trip. (Courtesy of Archer Oiltools)

cement plug, which is placed in the next step, or it is used as BHA for the cementing stage. For the washing tool to serve as a foundation, a packer is incorporated in the tool and once activated, stays in place. For the next step, spacer is pumped through the perforations. To do so, a new BHA may be used if the wash tool has already been released after the washing process. While pumping spacer, the work string is pulled out of hole. The process is known as pump-and-pull. The spacer is placed below the bottom perforations and extended to above the top perforations.

Nowadays, the perforation gun and washing tool are run in a single trip and when perforating is completed, an activation mechanism is engaged and drops the gun into the well rat hole. The single trip method saves time (Fig. 8.5).

For the next step of the operation, the BHA is placed below the bottom perforations and cement slurry is pumped, Fig. 8.6. After pumping some volumes to remove the spacer around the BHA, work string is pulled out of hole while pumping cement. The pumping of cement is continued with the calculated rate while pulling the work string with an optimal speed until cement and BHA reach above the top perforations. The BHA needs to be pulled out of the cement plug, to at least two stands above the top of cement. Then the well must be circulated to get clean.

One of the challenging parts of the PWC technique is the washing operation. The goal of washing is to remove any materials present behind the perforated casing. Wash fluid, which is a modified water-based fluid, should be pushed through the created perforations, and transport any materials presents out from the annular space. Currently, there are two different methods of washing: swab cup tool and jet tool. In the swab cup method, rubber plastic cups are installed below and above the injection nozzle present on the BHA, Fig. 8.7. Cups create a seal between casing and the work string and prevent the washing fluid traveling in the annular space between the casing and tool, Fig. 8.8a. In this way, wash fluid penetrates through the perforations into the annulus behind the casing and moves upward.

The jet tool method uses a jetting tool to wash and clean out debris by spraying wash fluid, Fig. 8.8b. The angle of jet nozzles and the exit velocity of wash fluid play an important role in the success rate of the washing technique. Centralization of the jet tool while washing could be a concern while for the swab cup tool, the cups partly act as a centralizer. Table 8.3 presents field data from a P&A operation where the PWC technique was used.

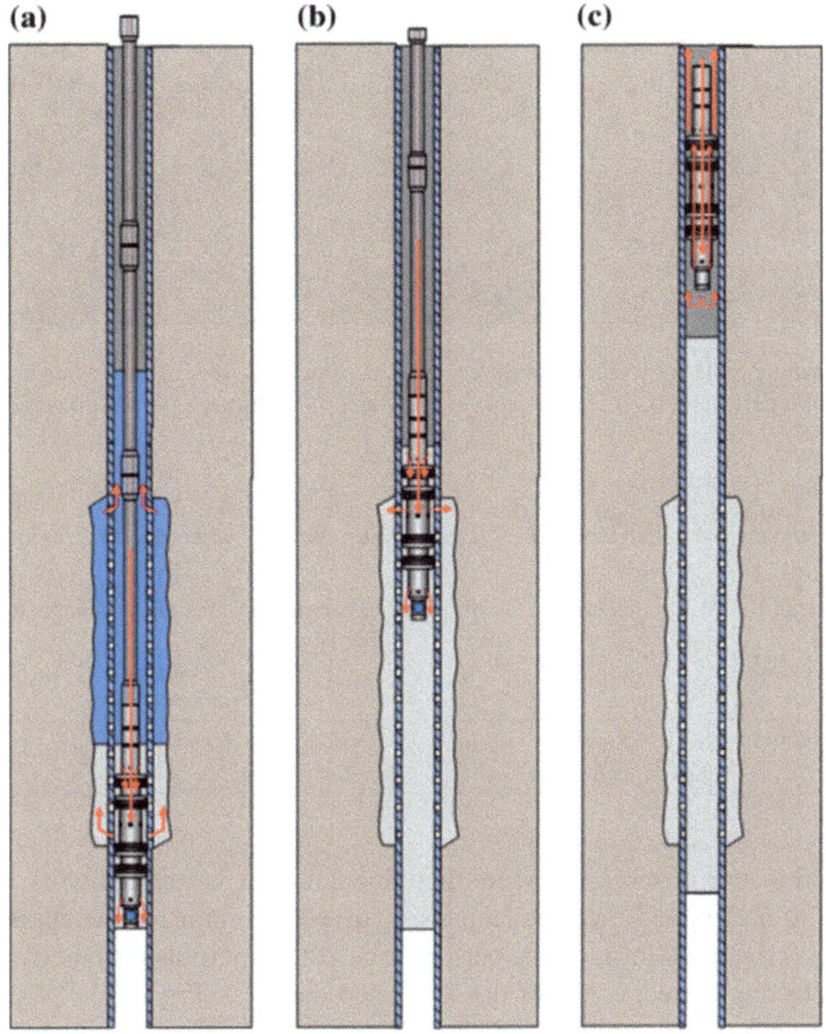

Fig. 8.6 Cementing part of PWC technique; **a** BHA is placed below the bottom perforations, pumping few volumes of cement, **b** pump-and-pull while cementing, **c** pump cement and circulate out the cement in BHA, pull the BHA out of cement, at least 2 stands above top of cement. (Courtesy of Archer Oiltools)

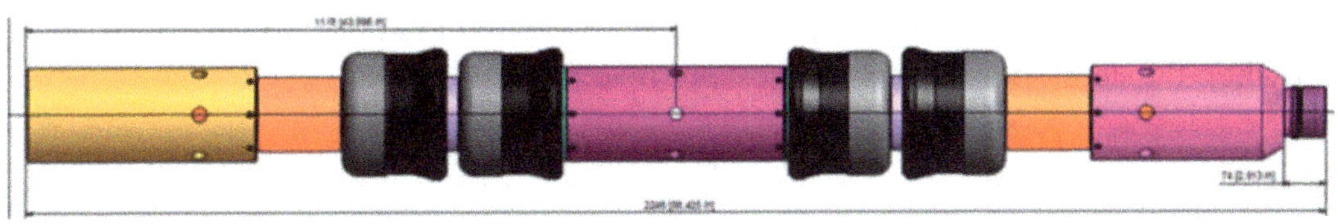

Fig. 8.7 Swab cup tool used in PWC technique. (Courtesy of Archer Oiltools)

When pumping cement slurry through the perforations, the cement should fill the annular space behind the perforations. Displacement efficiency of spacer and placement of cement is a strong function of exit velocity and inclination of the casing, Fig. 8.9. The displacement efficiency, during washing and cementing, is a

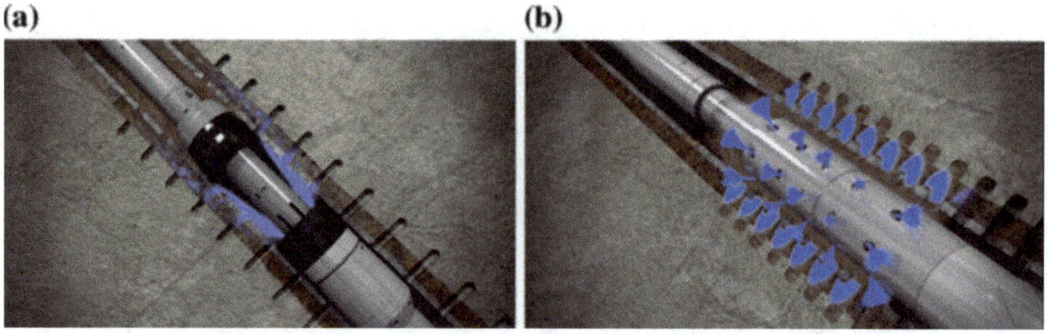

Fig. 8.8 Washing tools for PWC technique; **a** swab cups create a seal inside casing and force the wash fluid into perforations, **b** jet tool sprays the wash fluid through perforations. (Courtesy of Hydrawell AS)

Table 8.3 Field data obtained from a P&A operation on the NCS where PWC technique has been used

Casing size	Length of window	Wash tool	Number of trip in	Perforation size	Perforation phasing	Weight of removed metal	Inclination	Used time
9 5/8-in.	164 ft	Swab cups	Single trip	>1-in.	NA	2%	63°	36 h

concern and matter of research. More theoretical and experimental work should be performed to understand the mechanisms involved. To improve the cement placement and force cement through the perforations, different tools have been designed. Creating a cyclone effect is one of the suggested methods, Fig. 8.10.

There are advantages and possible limitations for the PWC technique, Table 8.4. Lack of qualification methods are the most challenging limitations. With current technologies, to qualify a PWC job, the cement inside the casing is drilled out and casing cement placed during the PWC job is logged by employing sonic logs. However, holes created during perforating challenge the reliability of logging data, in addition to uncertainties associated with sonic logs and the interpretation of logging data in general. If the annular barrier is qualified, cement is placed inside the casing and when the cement has set, it is pressure tested and tagged.

8.3 Explosives to Establish Annular Barrier

In P&A, establishing the annular barrier is one of the main challenges. In order to overcome the challenge, it has been suggested to use explosives for expanding the casing to create a seal or foundation for the annular barrier to be placed on. The amount of explosive to be used, is selected in such a way that the casing will be ballooned but not ruptured. The challenge is to select the correct amount of explosive

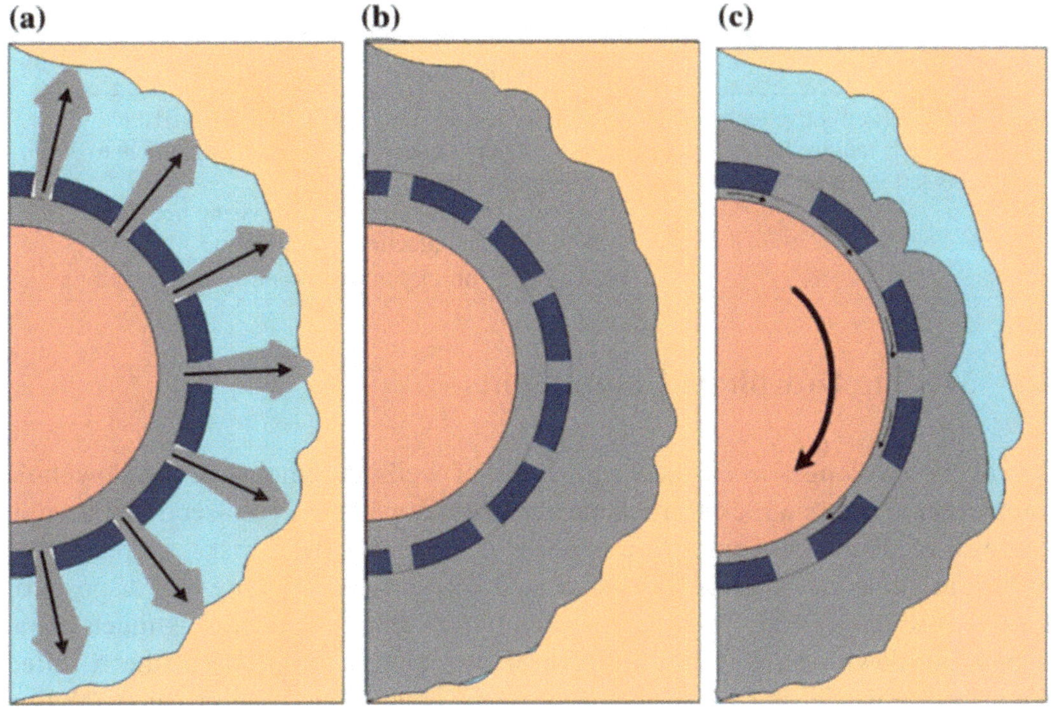

Fig. 8.9 Cementing of perforated casing in PWC technique; **a** cement is pumped through perforations, **b** the ideal cement job to be expected, **c** due to inefficient displacement and inclination cement slurry may not be able to fully displace spacer

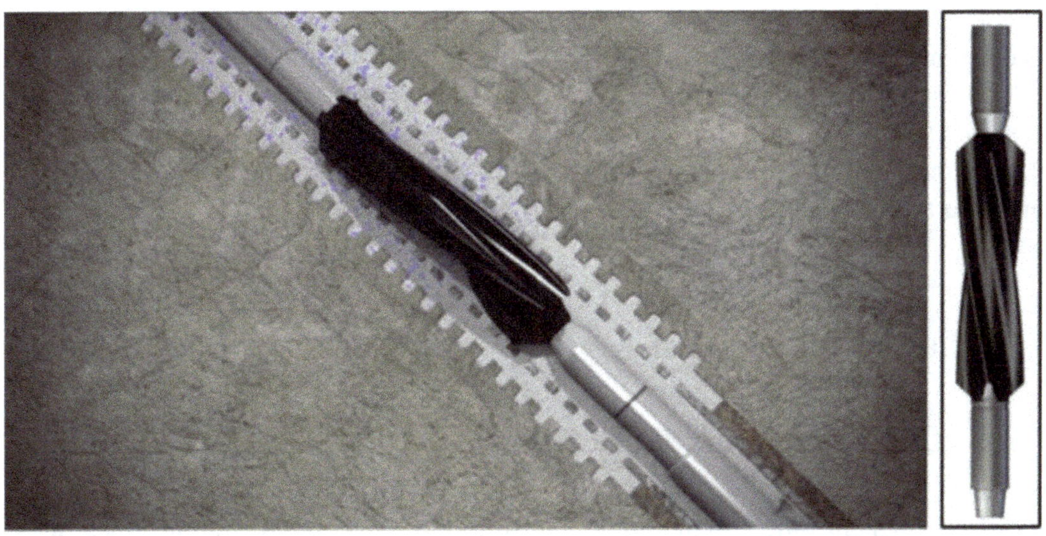

Fig. 8.10 Creating a cyclone effect for a better cement placement for PWC technique. (Courtesy of Hydrawell AS)

required as the casing string may not have its original thickness due to corrosion. This technique has been lab and yard tested but today has not been applied in the field.

Table 8.4 Advantages and possible limitations of PWC technique

Advantages	Possible limitations
• Time and cost effective technique • No milling is required • Metal is left in place	• Effectiveness of washing must be verified • No convenient qualification tool or technique to verify established annular barrier • Effective perforation size and phasing need more theoretical and practical investigation • Casing eccentricity during washing and cementing

8.4 Melting Downhole Completion

One of the challenges associated with P&A of wells is removal of the downhole completion to create a rock-to-rock barrier, also known as a cross-sectional barrier. Retrieval of downhole completion exposes personnel to HSE risks, increases the operational time, and carries cost associated with proper handling and disposal of the retrieved equipment. Therefore, a possible solution is to leave as much metal as possible downhole. But the presence of downhole completion at the required depth for the barrier is another challenge to be considered. One possible solution that may solve the issue and create a permanent barrier could be to melt all of the downhole completion and create a rock-to-rock barrier. For this method, the downhole completion and surrounding formation are melted in a controlled manner by use of thermite. In a thermite reaction, aluminum alloys and iron oxide (rusted steel) react and extreme amount of heat is generated. The oxygen required for the reaction is provided by the iron oxide [8]. Consider the reaction of thermite and the reaction mechanisms in Chap. 4.

The use of thermite for cutting tubing, drillpipe and bottomhole assemblies has already been employed in the field [9]. When considering melting the downhole completion and creating a barrier by modifying in situ materials, the barrier verification might be a challenge as discussed in Chap. 9.

8.5 Plasma-Based Milling

8.5.1 Concept Behind the Technology

During permanent plug and abandonment of Oil and Gas wells, the presence of the production tubing introduces challenges associated with logging cement behind the production casing, and cutting and pulling or section milling part of the production casing. Therefore, in conventional P&A methods, the production tubing needs to be retrieved, which is time consuming, costly and associated with risk. The limitations with cutting and pulling casing revolve around two main issues, the ability to effectively cut and retrieve casing and the manual handling of pipe at the surface. Current technology generally requires at least two BHA runs, one with a cutting assembly to cut the pipe at the required depth, then an additional run to retrieve (fish) the pipe above the cut. There are tools available that allow cutting and pulling in one run but a significant time reduction is not yet achieved. Many situations exist which make the pipe unrecoverable, even if the cut is fully successful. In such cases, section milling may be necessary. Challenges introduced by section milling have already been discussed in this chapter. The challenges related to section milling are proliferated by the type of production facility and working unit used for P&A. For offshore P&A activities, rigless P&A utilizing LWIV is a goal. The reason is a significant reduction of daily rental cost. Plasma-based technology may address some of these challenges. The development of plasma-based milling technology for through tubing well abandonment might be a potential solution. Generally speaking, plasma-based milling technology aims to disintegrate steel into small particles and transport the particles to surface [10].

8.5.2 Scientific Background of the Technology

In 1920's, Irving Langmuir described a fundamental state of matter which, unlike the other three fundamental states of matter, does not freely exist, where an ionized gaseous substance becomes highly electrically conductive. In this state, the behavior of matter is dominated by long-range electric and magnetic fields. In 1928, Irving coined the term "plasma" for the new matter state. Lightning and fire are examples of plasma. Plasma can be produced artificially by subjecting some gases to a strong magnetic field or by heating them [11, 12]. The most common gases used for generation of plasma include: air, argon, nitrogen, hydrogen and carbon dioxide. A Plasma jet can be used for different processes such as plasma cutting, plasma arc welding, plasma spraying, etc. Plasma cutting is a process of cutting an electrically conductive material utilizing an accelerated jet of superheated electrically ionized gas, plasma, having a large kinetic energy [12]. Figure 8.11 shows a schematic of a thermal plasma DC torch based on a cathode ionizing a gas stream.

Downhole conditions and materials imply that, the plasma-based milling technology cannot utilize state-of-the-art conventional plasma torch technology. The most important difference compared to conventional plasma torch technology is that the electrical arc with temperatures of tens of thousands of degrees Kelvin heats the surface of target material directly. In addition, its radiation component is also more efficient, with minimalized heating of intermediate gas. The intermediate gas flow in conventional plasma torches reduces the efficiency of heat transfer into the rock. Moreover, the arc creates area-wide, relatively homogeneous heat flow from a spiral arc on the whole surface for a high-intensity disintegration process.

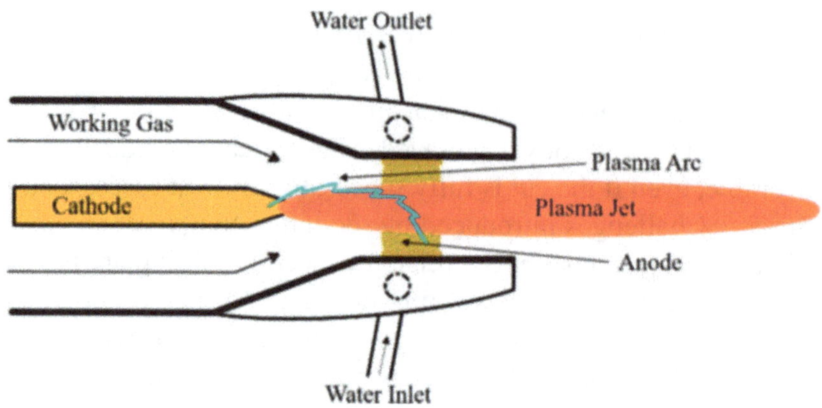

Fig. 8.11 A non-transferred plasma cutter based on hot cathode

Figure 8.12 shows a process of tubing and casing section milling using plasma-based tools. The tool is deployed through the tubing to the target zone where the plug is to be set (Fig. 8.12a). The electric arc is ignited, plasma is created and the tool moves upwards while milling the tubing (Fig. 8.12b). After tubing milling, the tool is moved back to its starting position and then removes casing and cement layers (Fig. 8.12c). After the removal of both tubing and casing, the tool is pulled out of hole (Fig. 8.12d). The section is then ready for cement plug placement (Fig. 8.12e).

The combination of a high temperature large cross-section plasma torch and rotating electric arc is another generation of plasma generators, which might be an effective tool for casing milling. The process using plasma technology is based on a mixture of hybridized plasma, chemical and thermochemical processes resulting in fast metal degradation and removal. The main process responsible for the rate and effectivity of steel degradation and removal is high temperature oxidation supported by melting and evaporation. Nowadays, several studies and techniques deal with the effect of water steam and temperature on the steel removal rate for a wide range of input parameters. One can conclude that temperature and heat transfer were found to be the key factors in increasing the constant rate needed for the required thermo-chemical and thermo-physical processes. The proportional contribution of the processes results in a steel removal effect, which varies with changing temperature and brings the following basic features [10]:

- The oxidative part of the targeted steels' structural degradation is an exothermic process - i.e. it supplies additional energy for all steel removal sub-processes.
- Oxidation and evaporation rate of steel raises with increasing plasma temperature, power density through the unit area at the plasma-steel interface and plasma enthalpy.
- Oxidation and evaporation rate of steel is most efficient in water steam and air-steam mixtures from an energetic point of view (in comparison with other industrial gases).
- There is a narrow temperature window in the range of 3055–3390 °C where enthalpy liberated from oxidative processes raises by a factor of 3. It means that

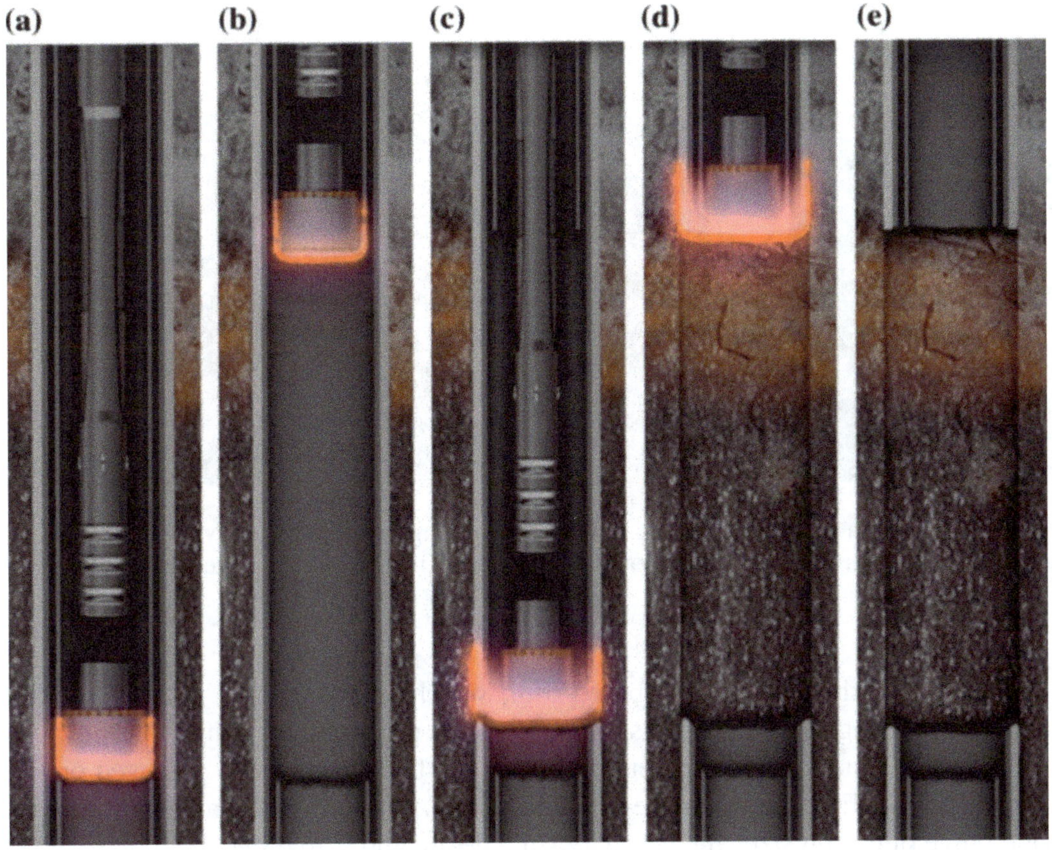

Fig. 8.12 Casing section milling of tubing and casing with plasma-based tool. (Courtesy of GA Drilling)

three times more energy is supplied to the steel removal processes without increasing external power of the plasma generator. This window should be valid for all types of steel alloys since at such high temperatures all the compounds are in gaseous phase.

- Above a steel surface temperature of 6100 °F, a total dissociation and evaporation occurs. Plasma particles impact on the steel surface in the form of active ionic atoms resulting in metal etching effect. It is important not to forget that oxidation is still active during melting and evaporation processes.

Because of steel oxidative processes, a large amount of energy is released during oxidation reactions and recycled to the steel removal processes. In closed vessel conditions, the total energy consumed for steel removal is at least by 30–40% lower than the theoretical value needed for steel melting. Penetration rate is a strong function of total power put into the steel degradation processes and the environment [10, 13]. Theoretically, by increasing input power the steel removal rate should be increased slightly linearly up to its saturation point, which can be obtained only by experimentation.

Fig. 8.13 **a** Plasma-based tool entering a multistring casing sample; **b** upper view on the sample after the experiment; **c** sample after diagonal section in order to reveal obtained steel-cement removal [14]

Figure 8.13 shows a plasma-based tool, which is acting on a mono-structure multistring casing sample whereas casing cements support the casings. As shown in Fig. 8.13, the inner casing and cement layer have been completely removed on the chosen section. Experiments have proved that a 3.5-in. tool is capable of milling a wide range of casing sizes including 4½-in., 5½-in. and 7-in. [14].

Scaled testing in pressures up to 1450 psi has been reported. Based on the challenges associated with section milling challenges, several parameters like ROP, steel types and cuttings types from plasma milling processes have been analyzed. Experiments carried out at different boundary conditions show that the efficiency of cutting steel can be characterized, empirically, by one special parameter. This parameter, ε, describes the energy needed for total removal of a mass of steel under the physical conditions. The parameter, ε, *h*as a statistical character as it summarizes the liberated energy coming from exothermic iron oxidation processes and the real electric energy supplied to the plasma generator. Therefore, it is evident that ε is always lower than the consumed electric energy. It was also found that ε is dependent on the degree or type of steel oxidation and hydrodynamic circumstances [14]. In order to determine the ROP, testing with a plasma generator has been carried out on two types of steels: carbon steel S355 and alloy steel with 20% Cr and 12% Ni. The value of ε is calculated from [14]:

$$\varepsilon = \frac{U \times I \times t}{m} \tag{8.1}$$

where $U \times I$ is the electrical power to plasma generator, m is the mass of the removed steel from the sample plate, and t is the time of the process.

A functional correlation has been reported between the steel removal rate (*SRR* [kg/h]) and plasma voltage U [V], current intensity I [A], plasma torch efficiency h [0–1] and net energy requirement per unit mass of removed steel ε [MJ/kg]:

$$SRR = \frac{U \times I \times 3.6 \times 10^{-3}}{\varepsilon} \times h \qquad (8.2)$$

In real casing conditions, in a water environment at low pressures, the value of ε is found to be in the range of 3–4 MJ/kg. When considering power output 250 kW, plasma torch efficiency 70% and net energy requirement per unit mass of removed steel 3 MJ/kg, the value of SRR is 210 kg/h [14]. This value means ROP 2.0–4.5 m/h for 9 5/8-in. casing section milling (depending on the wall thickness). This ROP is comparable to present-day section milling techniques, however the real difference is the fact that the plasma-based tool is able to mill various casing dimensions (as well as multiple strings) using one tool. This means a reduction in tripping and a significant increase in overall productivity.

For S355 steel, scanning electron microscopy (SEM) analysis clearly indicates the dominant presence of iron (II) oxide in the cuttings, Fig. 8.14. Structural analysis proved a heterogeneity between the formed oxidized and diffusive metallic layers in the cuttings. This resulted in differences in the thermal expansion coefficients of metal-oxide systems at the border of metallic and oxide layers. Therefore, hydrodynamic removal of such weakened multilayers could be realized relatively easily. In the

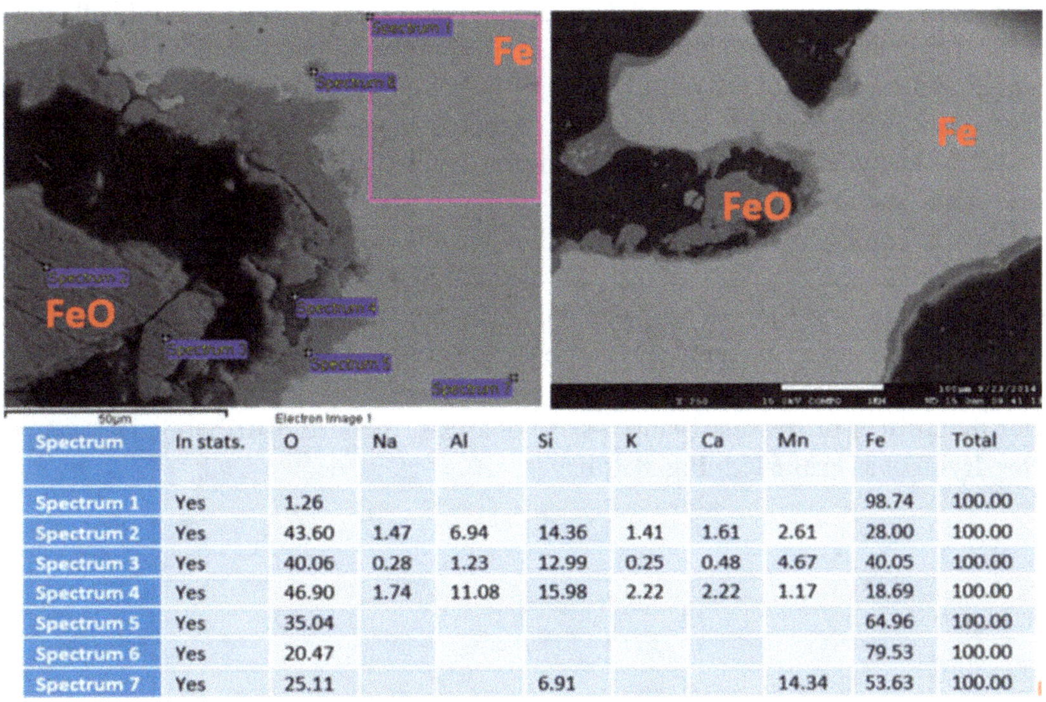

Spectrum	In stats.	O	Na	Al	Si	K	Ca	Mn	Fe	Total
Spectrum 1	Yes	1.26							98.74	100.00
Spectrum 2	Yes	43.60	1.47	6.94	14.36	1.41	1.61	2.61	28.00	100.00
Spectrum 3	Yes	40.06	0.28	1.23	12.99	0.25	0.48	4.67	40.05	100.00
Spectrum 4	Yes	46.90	1.74	11.08	15.98	2.22	2.22	1.17	18.69	100.00
Spectrum 5	Yes	35.04							64.96	100.00
Spectrum 6	Yes	20.47							79.53	100.00
Spectrum 7	Yes	25.11			6.91			14.34	53.63	100.00

Fig. 8.14 Samples of SEM image and EDX analysis of cuttings' material formed during plasma-based steel removal process of S355. (Courtesy of GA Drilling)

case of alloy steel, the aforementioned differences in thermal expansion properties of metal-oxide multilayers are significantly higher due to a higher grade of chemical heterogeneity in the microstructure. Figure 8.14 shows samples of SEM image and Energy-dispersive X-ray spectroscopy (EDX) analysis of cuttings´ material formed during plasma-based steel removal process of S355 steel.

Apparently, the plasma-based technology is capable of removing carbon steel as well as steel alloys without significant obstacles. Recently, plasma milling in a high-pressure environment has been presented. Subsequently, the following topics associated with the plasma-based milling process of production tubing and/or casing were researched [15]:

- Radial reach of plasma to cement in a high pressure (HP) environment up to 6000 psi
- Effect of the water-based fluids on the milling process in HP of 3600 psi
- Effect of the Oil-Based Mud (OBM) on the milling process in HP of 3600 psi
- Tests of possible damage to casing when milling eccentric tubing in HP of 3600 psi.

Cement removal at pressures of up to 42 MPa using electrical plasma has been tested at laboratory scale. In the case of implementation for either water-based or oil-based fluids, no interference effects on the milling process are reported but due to the presence of drilling fluid contaminating the cement, the removal process seems to be enhanced. The degradation is increased due to different thermal conductivity of present materials. Likely, chemical reactions with a plasma-forming medium are more significant and drilling fluid degradation is stronger or drilling fluid is flushed by the dynamics of implementation of the plasma forming medium into the process. In addition, it is possible to retrieve data of increased electrolysis when the process takes place in a "muddy" environment. The electrolysis level increase is different for WBM and OBM. This gives an important input to the knowledge regarding the structure of the milled casing.

Experimentally it has been shown that plasma-based milling technology can remove production tubing with control line and clamps. Since control line removal is a challenge using conventional technologies, this ability is another advantage.

A well documented advantage is related to the production of small particles instead of swarf. Figure 8.15a shows a typical example of cuttings collected from the casings after the milling processes. Using a sieve analysis, the size distribution of cuttings after drying was evaluated, Fig. 8.15b.

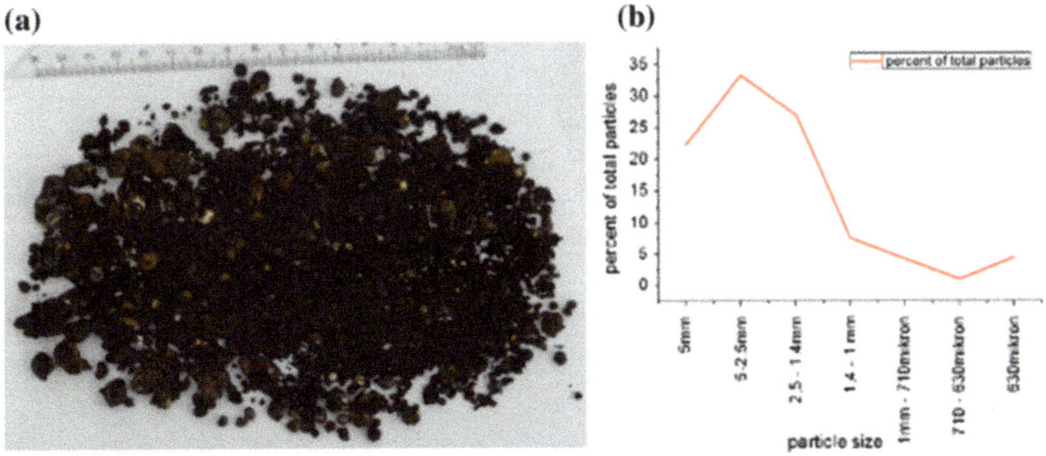

Fig. 8.15 **a** Cuttings generated during plasma milling processes in water environment (scale in mm); **b** cuttings size distribution [16]

The smaller particles are formed from small fragments of oxidized particles with an irregular shape. A fraction of bigger particles contains a larger number of globular particles having smooth surfaces. The ratio of cement particles is approximately the same for each size group. SEM-EDX analysis has been carried out for each size groups and it has been concluded that oxidation processes penetrate the steel volume. Figure 8.16a shows spherical particles identified as a ferrite material with small amounts of oxygen in the structure. Higher content of oxide is shown in the dark parts on the particles. Figure 8.16b, c show a visible inner structure of the oxide fragment. Advantages and possible limitations of plasma-based milling technology are listed in Table 8.5.

Fig. 8.16 **a** Spherical cutting particle having feritic structure; **b** and **c** SEM photo of oxidized cuttings surface. (Courtesy of GA Drilling)

Table 8.5 Advantages and possible limitations of plasma-based milling technology

Advantages	Possible limitations
• Rigless operation as the system is designed as a coiled tubing deployed solution • High milling ROP and subsequently cost effective • No swarf generation • Non-contact approach which improves reliability by minimizing the wear and tear of the tool or challenges associated with sticking • Fully automated coiled tubing milling process goes hand in hand with the enhanced safety of operational staff • No need to remove Christmas tree	• Not field proven yet and therefore, not commercially available • The Plasma Bit requires a purpose-built CT-reel conveyed umbilical • Ability to deliver sufficient electric power with transfer lines

8.6 Wellhead Cut and Removal

For a P&A operation, in Phase 3, the wellhead needs to be handled safely and efficiently. Depending on the well location and the corresponding authority regulations, the wellhead can be cut and removed or left in place with a cover protection. Considering deep or ultra-deep subsea wells, wellhead cut and removal may not be necessary as there might be no other activities (e.g. the fishing industry) in the area. However, it is a common practice to cut, below the baseline, and remove wellhead of land and platform wells.

Wellhead cut and removal can become a complex and costly operation, especially for subsea wells as a mobile offshore drilling unit, not necessarily a drilling rig, needs to be employed. Experience shows that the total time spent on mechanical wellhead removal of a subsea well can take between 6 and up to 40 h though a typical operation may take approximately 19 h. Therefore, it is necessary to consider wellhead cut and removal and its impact on the AFE. Different types of wellhead cutting are available including explosive cutting, hot cutting, mechanical methods, abrasive methods, and laser cutting. Some of these techniques are already in use whereas others are a relatively young state of the art technology. These technologies are explained in this section.

8.6.1 Explosive Cutting

Explosive technology has been used for control of blowing wells, removal of conductors for well abandonment, removal of platform piling for salvage, and the removal of debris which may present a hazard to navigation and the fishing industry [17]. In this technique, shaped charge cutters are used to produce slot type cuts rather than producing holes in a classical manner. In the classical manner, conical lined shaped

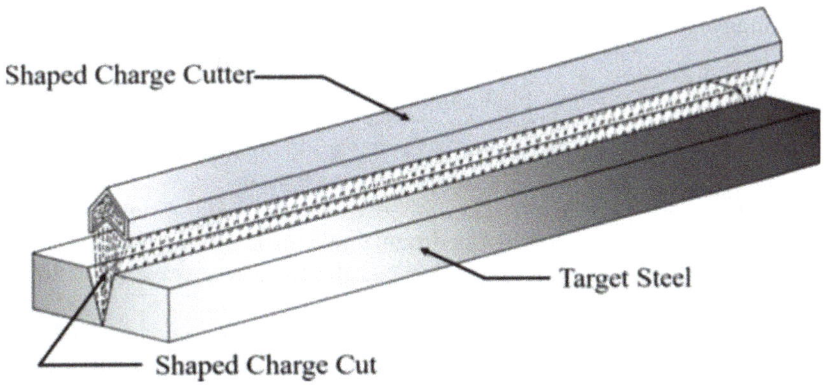

Fig. 8.17 Drawing of a shaped charge cutter and the provided cut in the steel target [17]

charges are used to produce perforations when completing oil and gas wells. The principal of charge cutters and conical shaped charges are the same but the charge cutters provide a linear cutting action (see Fig. 8.17). To create a cut in circular geometries (such as pipes and wellheads) circular cutters, which consist of two 180° hermetically sealed charges, are used (see Fig. 8.18). The circular charges can be used inside or outside circular geometries.

Generally, an explosive cutter system consists of three main parts: command unit, detonator, and charge. The command unit sends a signal via a shielded electrical cable to a detonator, and the detonator initiates the charge directly or via a cortex link. There are some advantages and possible limitations associated with use of explosive cutting for wellhead cut and removal, Table 8.6 (Fig. 8.19).

Fig. 8.18 Inside circular cutter [17]

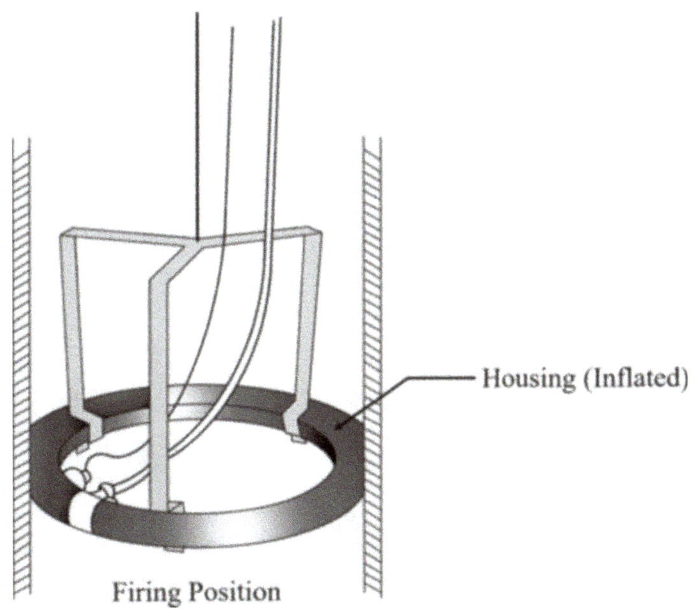

Table 8.6 Advantages and possible limitations associated with use explosive cutting for wellhead cut and removal

Advantages	Possible limitations
• Easy to handle and install • No limitation in size of cut • Fast cutting performance	• No guarantee of the completion of the cut • No control on cutting stages • Restrictions imposed by some authorities for wellhead cutting (environmental concerns) • Due to unclean cut, the removal process of wellhead could be difficult • Associated safety issues

Fig. 8.19 Schematic presentations of unclean cut created by explosives. (Courtesy of Blast Design)

8.6.2 Hot Cutting

The petroleum industry is familiar with different hot cutting methods including oxygen-gas cutting, oxygen-arc cutting, thermic lance, plasma arc cutting, pyrotechnic cutting, and flame jet cutting. The hot cutting technique for land-based and underwater (wet) cutting is almost the same. However, due to presence of water, a gas pocket needs to be created between the torch and target. One main reason to create the gas pocket is that water dissipates the heat more than air and the cut efficiency is dramatically reduced. General advantages and possible limitations of hot cutting are listed in Table 8.7.

In the flame cut process, an oxygen-fuel flame burns in the gas pocket and heats a spot on metal. A jet of pure oxygen, which is located in the center of the heating flame, blows against the spot on the metal to oxidize it with pure oxygen. As the torch is moved, the cut is formed [18]. Hydrogen is the prime fuel gas used for underwater cutting. The oxygen-acetylene flame is another type of flame which generates more heat compared to the oxygen-hydrogen flame. The flame equipment is bulky and requires added skills. In addition, it will only cut through steel and cannot cut through stainless steel nor nonferrous metals such as aluminum, and bronze. This lack of cutting ability is due to the low degree of oxidation of such materials. The flames cut efficiency is a function of water depth. Therefore, the technique is not used as it was used in the old days. The advancement of arc cutting technology has resulted in reduced use of flame cutting.

The arc cutting technique is almost similar to the flame cut but instead of a flame, a plasma arc is the source of heat. The arc heats the metal and oxygen is blown through the electrode to oxidize the metal. Compared to the flame cutting technique, arc cutting is faster and easier to handle and use. However, it can only cut through carbon or alloy steel. A variation of arc cutting is plasma-arc cutting.

The plasma-arc cutter generates a large amount of heat which acts on a spot on the steel surface. A gas flow blows away the molten metal, Fig. 8.20. The plasma-arc is able to cut through thick metal devices with high speed. It can cut through steel, aluminum, copper, and stainless steel alloys, cement and multiple casings.

Table 8.7 Advantages and possible limitations of hot cutting

Advantages	Possible limitations
• Easy to handle and install • Full control at all cutting stages • No limitation in size of cut • Guarantee of the complete cut	• Requires diver or ROV • Restrictions imposed by some authorities for wellhead cutting (environmental concerns) • Poor cutting performance • Associated safety issues with regards to explosion of fuels and gases

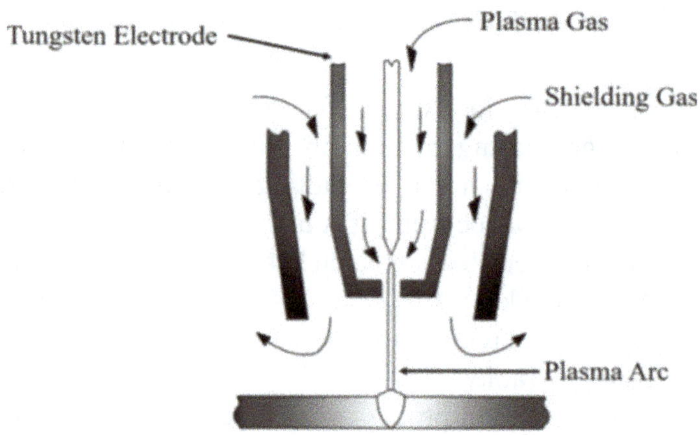

Fig. 8.20 Plasma-arc cutting of steel [18]

8.6.3 *Mechanical Methods*

Generally speaking, mechanical cutting methods have limitations specially when there is no cement in the annular space between conductor and casing string. The lateral movement of one string creates a challenge for cutting the next string. Mechanical cutting is divided into different categories including diamond wire cutting system, milling cutter, sawing (guillotine saw), and grinding.

8.6.3.1 Diamond Wire Cutting System

The system utilizes a series of machines, which are operated remotely, to create external cuts. The system uses a diamond embedded wire (e.g. a chain saw-like mechanism) to cut. The cutting operation can be done on steel, concrete or composite materials. A diamond wire cutting system consists of a clamping frame, cutting frame with wire driving pulleys and motor, wire feeding system, wire tensioning system, umbilical assembly, and diamond wire cable. As the cutting operation is mechanical, there is no operational limit concerning water depth. In addition, environmental-friendly, full control of the cutting operation, no limitation in size of cut, and fast cutting performance are other advantages of the system. One of the main limitations of this system is that only external cuts can be performed (see Fig. 8.21) [19]. In addition, the wire can get stuck when unstable structures are cut. These types of cutter make the cut above the baseline, seabed or ground, which is less of interest.

8.6.3.2 Milling Cutter

In milling cutting, a hydraulically actuated cutter is activated to create the cut while rotating (see Fig. 8.2). The mechanical cutter is equipped with carbide-tipped tungsten blades. When attempting multiple cemented casing strings, the blades may be worn out and trips in and out are required. Eccentricity of the inner string can result

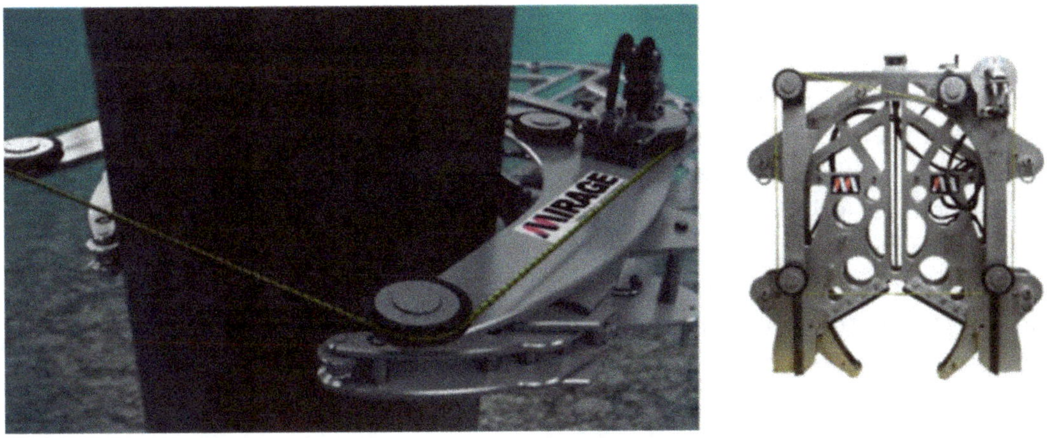

Fig. 8.21 Diamond wire saw. (Courtesy of Mirage Machines)

in an incomplete cut. This method is easy to handle, with fast cutting performance. However, a large amount of swarf is generated which needs to be handled. Replacing the blades can be time consuming, and the risk of over-torque may result in a tool stuck in the well.

8.6.3.3 Sawing (Guillotine Saw)

Guillotine pipe saws are designed for cold cutting and the most common type is reciprocating hydraulic driven saws with automatic feeding (see Fig. 8.22). This type of cutters can perform both on dry and wet environments and the operation can be controlled remotely [20]. Guillotine saws perform external cuts and their blade can get stuck when unstable constructions are subjected to cutting. These type of cutters are fast in cutting but they cut the pipe above the baseline, seabed or ground.

Fig. 8.22 Guillotine saw performing surface sectioning. (Courtesy of Oceaneering)

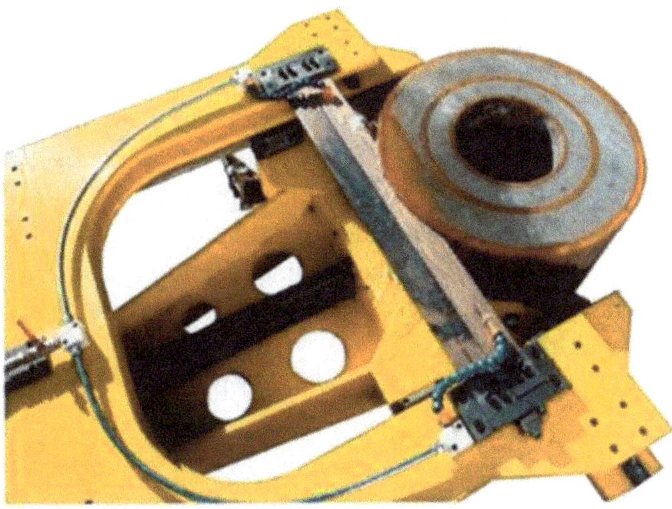

8.6.3.4 Grinding

Grinding is a type of mechanical machining where a cutting tool removes layers of the target material. The cutting tool is significantly harder than the target material. The electrochemical grinding cutting system is one type of grinding system. An electrochemical grinding cutting system consists of pumps, Direct Current (DC) generators, drive unit and manipulator, umbilical, and the cutting tool. Grinding cutters are environment friendly, safe and reliable with no limitation in size of cut. In addition, the cutting stages are under full control. However, it is a hot work method, slow process and vulnerable to casing compression.

8.6.4 Abrasive Methods

Abrasive methods have long been used in industrial and manufacturing processes to create cuts through rock, steel, and reinforced concrete [21]. Abrasive methods used in the petroleum industry to create cuts are categorized as sand cutting and abrasive water jet cutting. This categorization is based on the pressure used to create the cut. In a sand cutting technique, a high volume of particles are pumped at low pressure; however in abrasive water jet cutting, a low volume of solid particles are pumped at high pressure [22].

8.6.4.1 Sand Cutting

The process of tubing erosion caused by high-velocity sand has been a known well integrity issue. Development of mobile, high-pressure, high-horsepower pumping equipment, and controlling the rheological behavior of sand slurry resulted in sand cutting techniques in the 1960's [23]. In this technique, a fluid which contains abrasive solids is pumped through a set of nozzles with high differential pressure. The differential pressure is typically between 14 and 28 (MPa) with a flowrate between 350 and 450 (l/min). When the abrasive solids pass the nozzles, pressure is converted to kinetic energy and consequently high velocity is imparted to the solids. The solids with high velocities impact on casing, cement or formation and erode the target material in an organized pattern. Figure 8.23 shows the principle of sand cutting equipment. The equipment includes a high-pressure pump, blender unit with sand catch tank, hydroblast tool, and cutter heads with nozzles. The cut performance depends on nozzle differential pressure, sand concentration, nozzle stand-off distance and back-pressure.

The theoretical power available in the jet stream at the exist of nozzle may be expressed as [24]:

$$Power = QWh \tag{8.3}$$

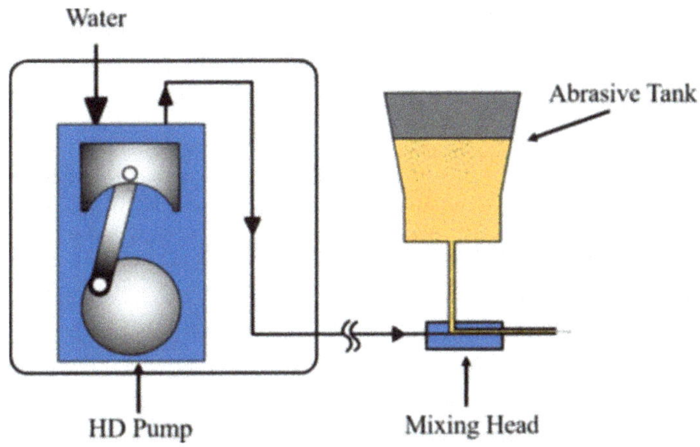

Fig. 8.23 Principle of a sand cutting unit

where Q is the flowrate of the sand-fluid mixture in ft^3/s, W is the specific weight of the sand-fluid mixture in lb/ft^3, and h is the drop in pressure head across the jet nozzle in ft.

Setting the weight of sand-fluid mixture consists of weight of sand and fluid:

$$W = W_s + W_f \tag{8.4}$$

where W_s is the weight of sand per ft^3 of sand-fluid mixture and W_f is the weight of carrier fluid per ft^3 of sand-fluid mixture.

By substituting Eq. (8.4) in Eq. (8.3) gives:

$$Power = Q\left(W_s + W_f\right)h \tag{8.5}$$

It can be assumed that during sand cutting, the energy imparted to casing and cement by jet stream is due to presence of sand and the energy of carrier fluid is negligible. So, W_f can be set at zero. Therefore, energy per unit of time or power imparted by sand in the jet stream is given by:

$$Power = QW_sh \tag{8.6}$$

The flowrate, Q, of the nozzle can be expressed as:

$$Q = AV \tag{8.7}$$

where V is the velocity of jet stream in ft/s and A is the area of nozzle orifice in ft^2. By substituting $V = \sqrt{2gh}$, then Eq. (8.7) can be written as:

$$Q = A\sqrt{2gh} \tag{8.8}$$

Substituting Eq. (8.8) in Eq. (8.6), the power can be given as:

$$Power = A\sqrt{2ghW_s h} \quad (8.9)$$

Or

$$Power = AW_s\sqrt{2g}(h^{3/2}) \quad (8.10)$$

The pressure head can be expressed in term of pressure drop and weight of the sand-fluid mixture as:

$$h = \frac{P}{W} \quad (8.11)$$

where P is the pressure drop in lb/ft^2. Therefore, substituting Eq. (8.11) in Eq. (8.10) gives:

$$Power = W_s A\sqrt{2g}\left(\frac{P}{W}\right)^{3/2} \quad (8.12)$$

Example 8.1 Assume that a sand cutter, with single nozzle, is used to cut a casing. The pressure drop across the nozzle increased from 1,000 to 2,000 psi. Calculated the theoretical cutting power of the sand-fluid stream.

Solution The theoretical cutting power varies with the 3/2 power of the pressure drop across the jet nozzle. Therefore, for constant values of A, Ws and W, increasing the pressure drop across the jet nozzle from 1,000 to 2,000 psi increases the cutting power of the sand-fluid stream by $2^{3/2} = 2.83$ times.

Sand cutting is an environmentally friendly technique, which is economical, fast and powerful. But it is difficult to monitor the progress and requires large volumes of sand or slag. Cutting multistring casing is also challenging. Therefore, abrasive water jet cutting has been developed.

8.6.4.2 Abrasive Water Jet Cutting

Abrasive Water-Jet Cutting (AWJC) technique uses high pressure at the nozzle but low volume of sand-fluid. The pressure at the nozzle ranges from 48 to 250 (MPa) and the flowrate ranges from 40 to 100 (l/min). The principle of AWJC technique is the same as sand-cutting, which means utilizing the kinetic energy of abrasive particles carried by a carrier fluid in a high velocity jet to erode the target material. Velocity of particles and distribution of abrasive particles within the carrier fluid are important parameters for the efficiency of the cutting process. One of the challenges associated with abrasive cutting is blockage of the nozzle by oversized grit particles. To minimize the risk, a certain flow is kept at all times to prevent blockage of the nozzle. In addition, Polymeric additives are optionally used to suspend the particles

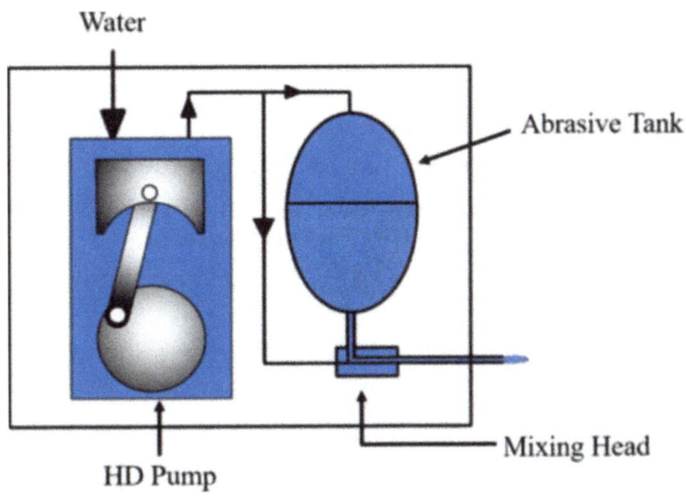

Fig. 8.24 Principle of an AWJC

in the carrier fluid and minimize the grit segregation rate if surface equipment fails and the pumping operation is halted.

A conventional AWJC unit consists of a cutting tool, manipulator, abrasive mixing or dispensing unit, high pressure water pumps, air compressors, hydraulic power unit, control panels, and cut monitoring systems (Fig. 8.24). The manipulator controls the positioning and movement of the nozzle. Presence of water in the interval of nozzle and target material reduces the efficiency of cutting by taking the kinetic energy of particles. Therefore, air compressors are used to blow air and create an atmosphere around the jet. Creating the atmosphere around the nozzle is more challenging where the cut depth increases.

During wellhead retrieval operation, the cutting tool is lowered into the well, centralized and anchored at the required depth. The AWJC unit can be placed on a vessel or MODU for offshore activities. The abrasive fluid is pumped to the nozzle by a water pump which is usually diesel engine driven. The cutting progresses as the manipulator rotates the nozzle. The AWJC technique offers a cold cutting solution, shock free cutting action, no torque between tool and target material, and proven remote operation. However, the size of topside support equipment, limited control over the reach,[1] volume of abrasive require on board, and the required number of crew to operate are some of the limitations of AWJC technique.

When considering the rate of penetration of abrasive cutters, power and velocity of the jet stream are the contributing parameters. Therefore, power equations and velocity equations are reviewed as follows.

Power Equations—In AWJC technique, the rate of penetration of hydraulic jet is proportional to power or energy of the jet at the interface of abrasive fluid and the target. The energy of jet stream is decreased with distance from the nozzle exit. As the distance between the nozzle exit and point in question increases, the energy diminishes to a value equal to the threshold cutting power. So the phenomenon can be expressed as:

[1]Reach is the cut length.

$$\frac{dL}{dt} = K_p(P_L - P_{th} - P_{losses}) \quad \left[\frac{\text{ft}}{\text{s}}\right] \tag{8.13}$$

where dL is the distance between the nozzle exit to point in question in [ft/s], P_L is the power contained in the jet stream at the point L in [(ft-lb$_f$)/s] or [hp], P_{th} is the threshold cutting power in [(ft-lb$_f$)/s] or [hp], P_{losses} is the hydraulic losses caused by casing, cutting restriction, and back-pressure in [(ft-lb$_f$)/s] or [hp], and K_p is the constant of proportionality for power equation in [1/lb$_f$] which is obtained from experimental data.

The power contained in the jet stream at point L distance from the nozzle exit is expressed as:

$$P_L = \frac{1}{2}\bar{m}_L\overline{V}_L^2 \quad \left[\frac{\text{ft} - \text{lb}_\text{f}}{\text{s}}\right] \tag{8.14}$$

where $\bar{m}_L$ is the mass rate of jet stream in [lb$_m$/s] and $\overline{V}_L$ is the jet velocity at the distance L in [ft/s]. Due to diffusion of the jet stream with distance, the mass rate is proportional to the initial mass rate at the nozzle exit. The mass rate is also proportional to the ratio of nozzle diameter to distance of point in question from the nozzle exit. Therefore, the mass rate is expressed as:

$$\bar{m}_L = C_m\bar{m}_0\frac{D}{L} \quad \left[\frac{\text{lb}_m}{\text{s}}\right] \tag{8.15}$$

where $\bar{m}_0$ is the initial mass rate at zero distance in [lb$_m$/s], D is the nozzle opening diameter in ft, or in., L is the distance from nozzle exit to the point of question in ft., or in., and C_m is an empirical dimensionless constant ($C_m = 5.2$). The jet stream velocity at distance L is proportional to initial velocity of the stream at the nozzle exit and to the ratio of nozzle diameter to distance of point in question from the nozzle exit. Therefore, the velocity equation is expressed as:

$$\overline{V}_L = \frac{C_v\overline{V}_0 D}{L} \quad \left[\frac{\text{ft}}{\text{s}}\right] \tag{8.16}$$

where $\overline{V}_0$ is the initial velocity of the jet at the nozzle exit in [ft/s], and C_v is an empirical dimensionless constant ($C_v = 6.4$). Substituting Eqs. (8.16) and (8.15) in Eq. (8.14) gives:

$$P_L = \frac{C_m C_v^2\bar{m}_0\overline{V}_0^2 D^3}{2gL^3} \tag{8.17}$$

where g is the conversion constant in $\left[\frac{\text{lb}_m - \text{ft}}{\text{lb}_f - \text{s}^2}\right]$. From continuity equation, it can be written:

$$\bar{m}_0 = \rho A V_0 \tag{8.18}$$

where A is the area of nozzle and can be written as:

$$A = \frac{\pi D^2}{4}$$

and

$$\overline{V}_0 = \sqrt{2g\frac{\Delta P}{\rho}144} \tag{8.19}$$

By substituting Eqs. (8.18) and (8.19) into Eq. (8.17), it gives:

$$P_L = \frac{BD^5(\Delta P_0)^{\frac{3}{2}}}{L^3\rho^{1/2}} \quad B = 3\pi C_m C_v^2(2g)^{\frac{1}{2}} \tag{8.20}$$

where ΔP_0 is the pressure differential across the nozzle in [psi], and ρ is the sand-fluid density in [lb_m/ft^3]. Combining Eqs. (8.20) and (8.13) will result:

$$\frac{dL}{dt} = K_p\left(\frac{BD^5(\Delta P_0)^{\frac{3}{2}}}{L^3\rho^{1/2}} - P_{th} - P_{losses}\right)\left[\frac{\text{ft}}{\text{s}}\right] \tag{8.21}$$

Velocity Equations—The rate of penetration *dL*/*dt*, of the hydraulic jet is proportional to the velocity of abrasive fluid at the interface of the fluid and the target material. So, the rate of penetration in terms of velocity can be expressed as:

$$\frac{dL}{dt} = k_v'(V_L - V_{th} - \Delta V_{bp})\left[\frac{\text{ft}}{\text{s}}\right] \tag{8.22}$$

where V_L is the velocity of abrasive fluid at the interface of the fluid and target material in (ft/s), V_{th} is the threshold velocity or the minimum velocity required to create the cut in (ft/s), ΔV_{bp} is the velocity of loss of the jet resulting from the return flow of the abrasive in (ft/s), and k_v' is the constant of proportionality for the velocity equation and is obtained experimentally

By substituting Eq. (8.16) into (8.22):

$$\frac{dL}{dt} = k_v'\left(\frac{C_v\overline{V}_0 D}{L} - V_{th} - \Delta V_{bp}\right) \tag{8.23}$$

By rearranging Eq. (8.23) and solving for dt:

$$dt = \frac{k_v L dL}{C_v\overline{V}_0 D - LV_{th} - L\Delta V_{bp}} \text{ [s]} \tag{8.24}$$

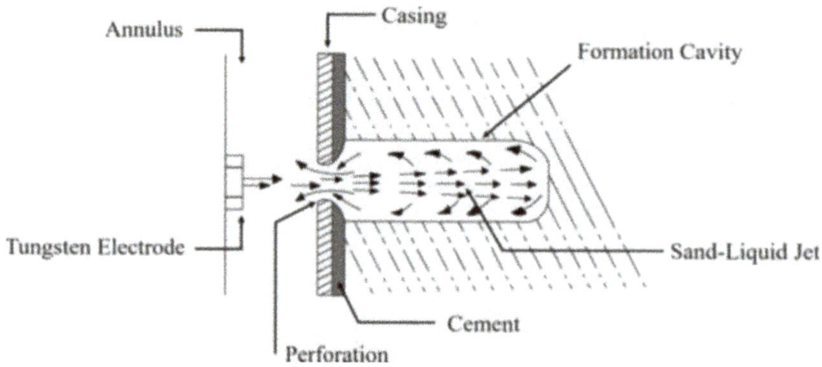

Fig. 8.25 Abrasive cut process for a cemented casing

where k_v is reciprocal of $k_v^{'}$ So, the integration of Eq. (8.24) yields [23] (Fig. 8.25):

$$t = k_v\left[\frac{C_v\overline{V}_0 D}{(V_{th}+\Delta V_{bp})^2} ln_e\left(\frac{C_v\overline{V}_0 D}{C_v\overline{V}_0 D-(V_{th}+\Delta V_{bp})L}\right)-\frac{L}{(V_{th}+\Delta V_{bp})}\right] \tag{8.25}$$

The research work conducted by researchers shows that the threshold cutting velocity is directly proportional to the hardness of target material:

$$V_{th} = cH \tag{8.26}$$

and

$$H \propto \frac{1}{L_{max}} \tag{8.27}$$

whereas

$$L_{max} = \frac{C_v D\overline{V}_0}{V_{th}+\Delta V_{bp}} = \frac{C_v D\overline{V}_0}{cH+\Delta V_{bp}} \tag{8.28}$$

where c is the proportionality constant, V_0 average fluid velocity of the jet at the nozzle exit in (ft/s), L_{max} is the maximum penetration in (ft), H is the relative abrasion hardness of material and is proportional to the reciprocal of the maximum penetration.

When considering AWJC, although casing back-pressure and size of opening created by the jet cutter have a significant effect on the cut efficiency, the effect of hydraulic jet stand-off, effect of sand concentration, and communication effect of materials by induced fractures or formation permeability are important parameters.

Advantages of AWJC includes but are not limited to fast cutting performance compared to the other cutting methods, environmentally friendly and no special permission is required to conduct it, and no torque between the tool and target. However, the drawbacks are limited control on the reach (cut length), cutting efficiency decreases with water depth, large volume of abrasive fluid is required on board, large topside spread compared to the other cutting methods, and number of crew to carry out the operation.

8.6.5 Laser Cutting

Light Amplification by Stimulated Emission of Radiation, broadly known as Laser, was coined by Gordon Gould in 1957. Generally speaking, lasers are devices which convert different kinds of energy to electromagnetic beams of monochromatic and coherent waves. Monochromatic means the output electromagnetic waves have a single output wavelength or in other words it means one color output. Coherent means that all the waves are in phase with one another. The generated waves span through the different regions including gamma, X-ray, ultraviolet, visible light, infra-red, microwave, and radio waves.

If the generated stream of electromagnetic beams have high enough energy, then they can create a cut on steel and rock samples. However, high-powered laser technology is required for such operations. The intensity of a laser beam depends on the wavelength of the beam. Common components of a laser are active medium, energy input (known as pump source), and feedback (laser cavity). An electron is pumped into a highly excited state and transit to a metastable region. As the electron loses its energy to return to its initial conditions, it generates photons in different directions. This process is known as spontaneous emission.

The efficiency of a laser cutter depends on several laser properties including discharge type, peak power, wavelength, average power, intensity, repetition rate, and pulse with the discharge type [25–27]. The laser discharge can be pulsed or continuous. In pulsed discharge type, the optical power appears in pulses for a certain period of time at some repetition rate. However, in continuous type of discharge, the optical power appears continuously.

The main challenge associated with the utilization of laser cutters at downhole conditions is the presence of wellbore fluids. Downhole fluids are opaque, near-opaque, or even dark which are not conducive to laser cutting.

References

1. Hartman, C.J., J.L. Cullum, and J.E. Melder. 2017. Efficient and safe casing removal with downhole hydraulic pulling assembly. In *Offshore technology conference*, OTC-27618-MS. Houston, Texas, USA: Offshore Technology Conference. https://doi.org/10.4043/27618-MS.
2. Vestavik, O.M., T.H. Fidtje, and A.M. Faure. 1995. Casing window milling with abrasive fluid jet. In *SPE annual technical conference and exhibition*, SPE-30453-MS. Dallas, Texas: Society of Petroleum Engineers. https://doi.org/10.2118/30453-MS.
3. Joppe, L.C., A. Ponder, and D. Hart, et al. 2017. Create a rock-to-rock well abandonment barrier without swarf at surface! In *Abu Dhabi international petroleum exhibition & conference*,

SPE-188332-MS. Abu Dhabi, UAE: Society of Petroleum Engineers. https://doi.org/10.2118/188332-MS.

4. Nelson, J.F., J.-T. Jørpeland, and C. Schwartze. 2018. Case history: a new approach to section milling: leaving the swarf behind! In *Offshore technology conference*, OTC-28757-MS. Houston, Texas, USA: Offshore Technology Conference. https://doi.org/10.4043/28757-MS.
5. Garcia, J.A., and C.R. Clark. 1976. An investigation of annular gas flow following cementing operations. In *SPE symposium on formation damage control*, SPE-5701-MS. Houston, Texas: Society of Petroleum Engineers. https://doi.org/10.2118/5701-MS.
6. Ansari, A.A., D. Ringrose, and Z. Libdi, et al. 2016. A novel strategy for restoring the well integrity by curing high annulus-B pressure and zonal communication. In *SPE Russian petroleum technology conference and exhibition*, SPE-181905-MS. Moscow, Russia: Society of Petroleum Engineers. https://doi.org/10.2118/181905-MS.
7. Ferg, T.E., H.-J. Lund, and D.T. Mueller, et al. 2011. Novel approach to more effective plug and abandonment cementing techniques. In *SPE Arctic and extreme environments conference and exhibition*, SPE-148640-MS. Moscow, Russia: Society of Petroleum Engineers. https://doi.org/10.2118/148640-MS.
8. Hay, E., and R. Adermann. 1987. Thermite sparking in the offshore environment. In *Offshore Europe*. Aberdeen, United Kingdom: Society of Petroleum Engineers. https://doi.org/10.2118/16548-MS.
9. Cole, J.F. 1999. Pyro technology for cutting drill pipe and bottomhole assemblies. In *SPE/IADC drilling conference*, SPE-52824-MS. Amsterdam, Netherlands: Society of Petroleum Engineers. https://doi.org/10.2118/52824-MS.
10. Kocis, I., T. Kristofic, and M. Gajdos, et al. 2015. Utilization of electrical plasma for hard rock drilling and casing milling. In *SPE/IADC drilling conference and exhibition*, SPE-173016-MS. London, England, UK: Society of Petroleum Engineers. https://doi.org/10.2118/173016-MS.
11. Ebersohn, F.H., J.P. Sheehan, A.D. Gallimore, et al. 2017. Kinetic simulation technique for plasma flow in strong external magnetic field. *Journal of Computational Physics* 351: 358–375. https://doi.org/10.1016/j.jcp.2017.09.021.
12. Krajcarz, D. 2014. Comparison metal water jet cutting with laser and plasma cutting. *Procedia Engineering* 69: 838–843. https://doi.org/10.1016/j.proeng.2014.03.061.
13. Gajdos, M., I. Mostenicky, and I. Kocis, et al. 2016. Plasma-based milling tool for light well intervention. In *SPE/IADC middle east drilling technology conference and exhibition*, SPE-178179-MS. Abu Dhabi, UAE: Society of Petroleum Engineers. https://doi.org/10.2118/178179-MS.
14. Gajdos, M., T. Kristofic, and S. Jankovic, et al. 2015. Use of plasma-based tool for plug and abandonment. In *SPE offshore Europe conference and exhibition*, SPE-175431-MS. Aberdeen, Scotland, UK: Society of Petroleum Engineers. https://doi.org/10.2118/175431-MS.
15. Kristofic, T., I. Kocis, and T. Balog, et al. 2016. Well intervention using plasma technologies. In *SPE Russian petroleum technology conference and exhibition*, SPE-182120-MS. Moscow, Russia: Society of Petroleum Engineers. https://doi.org/10.2118/182120-MS.
16. Gajdos, M., I. Kocis, and I. Mostenicky, et al. 2015. Non-contact approach in milling operations for well intervention operations. In *Abu Dhabi international petroleum exhibition and conference*, SPE-177484-MS. Abu Dhabi, UAE: Society of Petroleum Engineers. https://doi.org/10.2118/177484-MS.
17. De Frank, P., R.L. Robinson, and B.E. White, Jr. 1966. Explosive technology a new tool in offshore operations. In *Fall meeting of the society of petroleum engineers of AIME*, SPE-1602-MS. Dallas. Texas, USA: Society of Petroleum Engineers. https://doi.org/10.2118/1602-MS.
18. Mishler, H.W., and M.D. Randall. 1970. Underwater joining and cutting - present and future. In *Offshore technology conference*, OTC-1251-MS. Houston, Texas: Offshore Technology Conference. https://doi.org/10.4043/1251-MS.
19. Brandon, J.W., B. Ramsey, and J.W. Macfarlane, et al. 2000. Abrasive water-jet and diamond wire-cutting technologies used in the removal of marine structures. In *Offshore technology conference*, OTC-12022-MS. Houston, Texas: Offshore Technology Conference. https://doi.org/10.4043/12022-MS.

20. Olstad, D., and P. O'Connor. 2010. Innovative use of plug-and-abandonment equipment for enhanced safety and efficiency. In *IADC/SPE drilling conference and exhibition*, SPE-128302-MS. New Orleans, Louisiana, USA: Society of Petroleum Engineers. https://doi.org/10.2118/128302-MS.
21. Huang, Z., G. Li, and M. Sheng, et al. 2017. *Abrasive water jet perforation and multi-stage fracturing*, 2nd edn. Gulf Professional Publishing. 9780128128077.
22. Sorheim, O.-I., B.T. Ribesen, and T.E. Sivertsen, et al. 2011. Abandonment of offshore exploration wells using a vessel deployed system for cutting and retrieval of wellheads. In *SPE Arctic and extreme environments conference and exhibition*, SPE-148859-MS. Moscow, Russia: Society of Petroleum Engineers. https://doi.org/10.2118/148859-MS.
23. Brown, R.W., and J.L. Loper. 1961. Theory of formation cutting using the sand erosion process. *Journal of Petroleum Technology* 13 (05): 483–488. https://doi.org/10.2118/1572-G-PA.
24. Pittman, F.C., D.W. Harriman, and J.C. St. John. 1961. Investigation of abrasive-laden-fluid method for perforation and fracture initiation. *Journal of Petroleum Technology* 13 (05): 489–495. https://doi.org/10.2118/1607-G-PA.
25. Adeniji, A.W. 2014. The applications of laser technology in downhole operations - a review. In *International petroleum technology conference*, IPTC-17357-MS. Doha, Qatar: International Petroleum Technology Conference. https://doi.org/10.2523/IPTC-17357-MS.
26. Batarseh, S.I., B.C. Gahan, and B. Sharma. 2005. Innovation in wellbore perforation using high-power laser. In *International petroleum technology conference*, IPTC-10981-MS. Doha, Qatar: International Petroleum Technology Conference. https://doi.org/10.2523/IPTC-10981-MS.
27. Pooniwala, S.A. 2006. Lasers: the next bit. In *SPE Eastern regional meeting*, SPE-104223-MS. Canton, Ohio, USA: Society of Petroleum Engineers. https://doi.org/10.2118/104223-MS.

Chapter 9
Barrier Verification

The main goal of a P&A operation is to restore the cap rock functionality by establishing competent barriers. When a barrier is established, its functionality needs to be verified. There are different test procedures to verify integrity of permanent barriers. Some include verification of annular barrier (barrier between casing and formation), some include verification of permanent plug inside casing, and some others include verification of barriers in open holes. The main challenge for barrier verification is the lack of direct relationship between laboratory verification and field performance testing. In the laboratory testing of cement, the following parameters are evaluated: mechanical properties (e.g. compressive strength, tensile strength, Young's modulus, etc.), shear bond strength, hydraulic bond strength and tensile bond strength, fluid migration analysis, static gel strength analysis, etc. In addition, most of laboratory experiments replicate best case scenario with respect to contamination of cement slurry. However, there is no simple way of accurately testing these parameters at expected downhole conditions. In fact, the only tests available to verify cement plugs in the field are hydraulic pressure testing, weight testing and tagging with workstring. The annular barrier is tested indirectly by logging. This chapter will review these field test methods.

9.1 Annular Barrier Verification

The concept of cross-sectional barrier has already been defined in previous chapters. In order to verify the cross-sectional barrier, barrier behind casing needs to be verified where casing exist. There are different methods to qualify the integrity of annular barrier. Acoustic logging of annular barrier, passive noise logging, temperature logging, and hydraulic pressure testing are the most commonly used verification method which will be reviewed in this chapter.

M. Khalifeh and A. Saasen, *Introduction to Permanent Plug and Abandonment of Wells*, Ocean Engineering & Oceanography 12,
https://doi.org/10.1007/978-3-030-39970-2_9

9.1.1 Acoustic Logging of Annular Barrier

Acoustic logging of annular barrier technique is the prime method used in petroleum industry for verification of casing cement but also may be used for other types of barrier materials. Therefore, this technique will be discussed more in detail.

In acoustic logging method, an acoustic signal is emitted by a transducer, the emitted signal goes to a journey through casing fluid, casing steel, barrier behind casing, adjacent formation and all the way back to two receivers, Fig. 9.1. The receivers pick up the reflected acoustic signal and engineers process the collected data to check the quality of annular barrier.

The history of sonic logging goes back to 1950s when during formation evaluation by sonic logs the occurrence of skipped cycles were noticed where openhole and cased hole were evaluated [1]. Since then, the technology has advanced whereas ultrasonic tools have been developed and are in use. Nowadays, cement bond logs are run to determine cement to casing bonding, cement to formation bonding, and

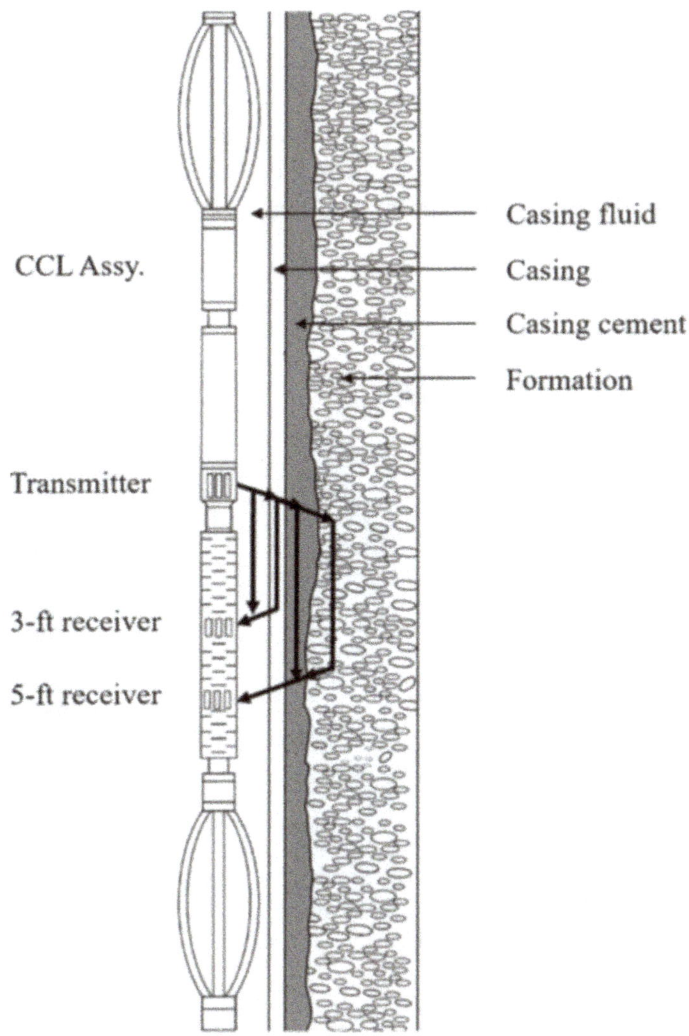

Fig. 9.1 A cement bond log (CBL) and variable density log (VDL) tool

evaluate cement conditions. The evaluation of cement conditions include detection of channeling, compromised cement (by gas cut, dehydration, etc.), top of cement, and micro-annuli.

Acoustic logging is a process of recording of some acoustic property of formation and wellbore. The result of process is displayed on an acoustic log which presents traveling time of acoustic waves versus depth in a wellbore. The acoustic logging of annular barrier can be through sonic, and ultrasonic (pulse-echo) measurements.

9.1.1.1 Sonic Measurements

Sonic tools are working in a frequency range of 10–30 kHz. An electrical signal is sent to a piezoelectric transducer, and the transducer generates an omnidirectional acoustic signal. Piezoelectric transducers are capable to receive electricity and convert it to acoustic signal and vice versa. Acoustic signals are sound waves which are categorized into compressional wave (P-wave), shear wave (S-wave), and plate wave, Fig. 9.2. Compressional wave (sometimes called longitudinal wave) is a wave that the motion occurs in the same direction or opposite direction to wave propagation. Compressional waves can travel through solid, liquid and gas. Shear wave (sometimes called an elastic S-wave) is a wave that the motion is perpendicular to the wave propagation. Large amplitude shear waves can travel only through solid phases. Plate wave (sometimes called Lamb wave) propagates in solid plates but slightly slower than compressional wave in steel plates.

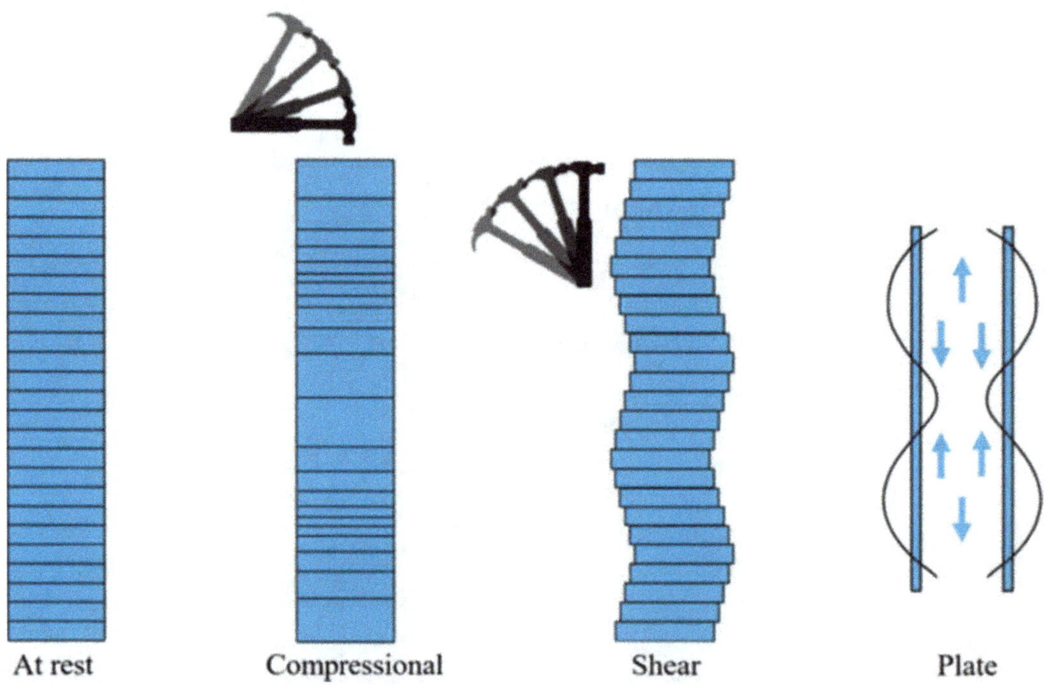

Fig. 9.2 Wave propagation modes [2]

Piezoelectric transmitter fires acoustic signal, compressional wave, and the wave propagates through casing fluid, casing steel, barrier material behind casing, and formation. The wave is reflected back through all of these medium toward the receivers (two piezoelectric receivers). Two receivers, placed 3 and 5 ft away from transmitter pick up the reflected signal, Fig. 9.1. The 3 ft receiver picks up the signal from casing arrival and casing fluid and the 5 ft receiver picks up the formation arrival. The recorded data are presented on a log which consists of a log heading, main log and repeat section, determination of logging pressure, before and after survey calibrations and parameter box, and log tail. A log heading is composed of four parts: API heading, remark box, well schematic, and tool schematic with sensor distances, Fig. 9.3.

The main log presentation consists of three tracks: Track 1 for quality control, Track 2 show the quality of bonding between cement and casing, and Track 3 which shows formation arrival (see Fig. 9.4). Depending on the logging tool used for logging annular barrier, more tracks may be distinguished, Fig. 9.4.

A repeat section is to examine the logging operation and the recorded length is usually 100 m which covers the zone where good sealing is expected. Sonic logs are always run under pressure in order to differentiate microannulus from channeling. As the pressure affects the amplitude of sound, the logging pressure is determined and presented on the log. The parameter box and sonic summary, before and after calibration, are presented on the log. The last section on a log presentation is a log tail which repeats the top part of the heading.

When transmitter emits an acoustic signal, an elapsed time is passed until the receiver can detect the first part of the wave arrival, which is exceeding a preset amplitude threshold. The elapsed time is known as *transit time*. Amplitude is the strength of the first arrival wave. When time passes, the amplitude recorded by receiver increases up to a level and then decreases. Time taken by an emitted signal to travel from transmitter to receiver is called *travel time*. Figure 9.5 shows the above-mentioned terminologies. When the emitted signal travels through different medium (e.g. casing fluid, casing, material behind casing), it loses the energy. The loss of sound energy, strength, is called *attenuation*. So, the higher the attenuation the lower the amplitude.

An acoustic logging tool may be configured with some complementary logging tools. The most common complementary logs include casing collar locator (CCL) and Gamma ray logs. Figure 9.6 shows two sonic logs equipped with CCL, GR detector and centralizers; however, tool centralization is valid in wells with inclination less than 50°. CCL is used to locate casing collar for depth calibration. GR detector is also used for depth and formation calibration. In fact, Track 1 does not provide any information regarding the cement quality but it is used for quality control.

A transmitter fires a compressional wave and it propagates spherically. A part of the compressional wave travels downwards along casing fluid along casing steel and a part reaches the casing and travels downward along the casing. Another portion of the wave which entered to casing steel travels further toward material behind casing and formation. If there is solid material behind casing, at the interface of casing and the solid material shear waves is generated. So, the generated shear wave and

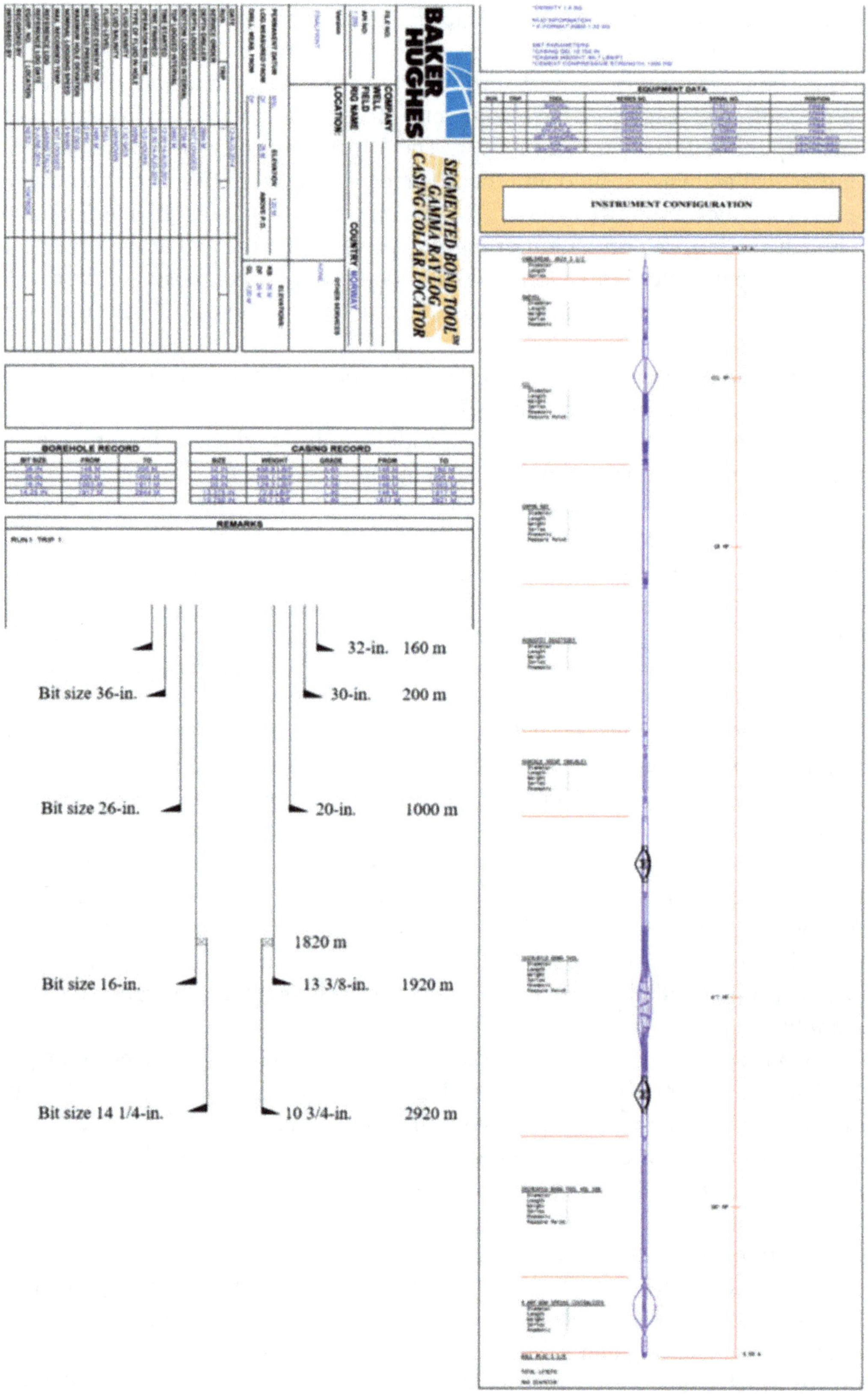

Fig. 9.3 A log heading

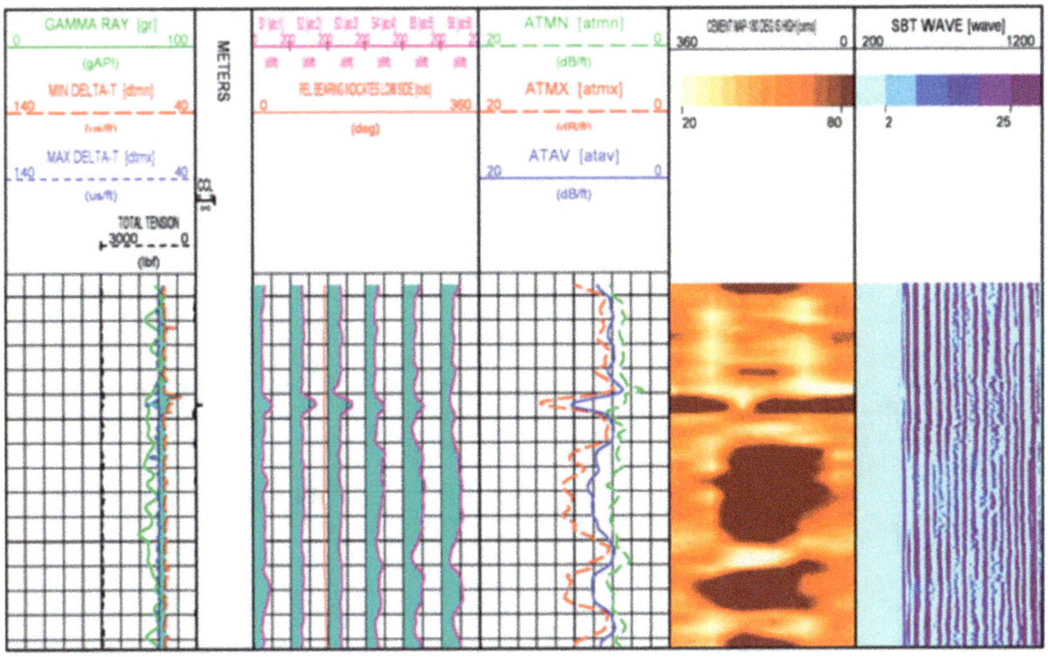

Fig. 9.4 A CBL-VDL log with radial mapping which includes five tracks

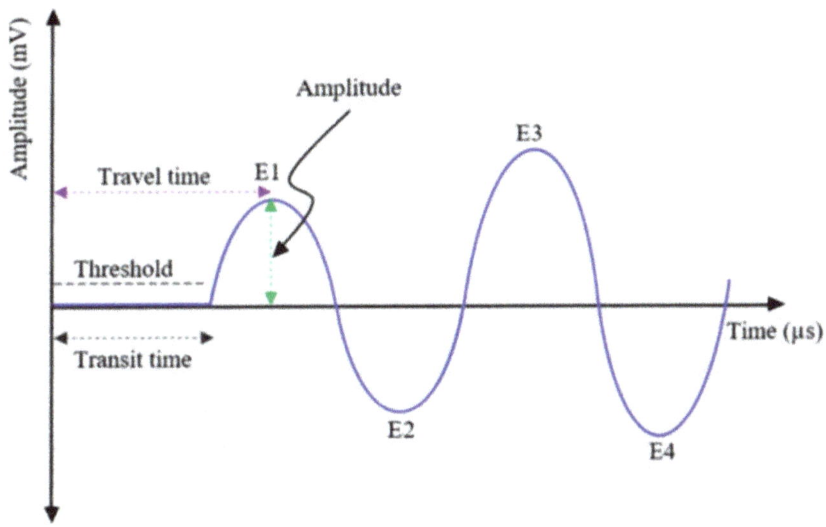

Fig. 9.5 Recorded amplitude of an acoustic signal

compressional wave continue to travel through the solid material behind casing and formation, and also downwards. But if there is no solid material behind casing, shear wave is not produced and only the compressional wave will travel through annulus behind casing to formation and downwards. In order to get the wave propagation inside the casing plate, the waves are emitted in a predetermined critical angel.

Example 9.1 In order to understand the concept behind acoustic logging of casing, you are asked to run an experiment. Pick up a pozzolan or steel coffee cup with a

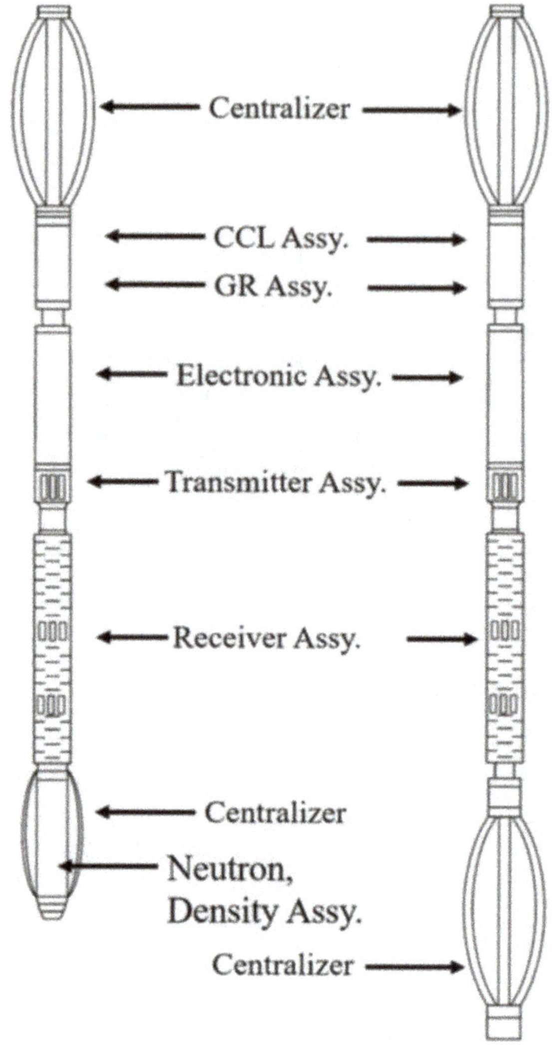

Fig. 9.6 A cement logging tool assembly [3]

teaspoon. Ask an assistant to hold the cup from its handle part with no hands around it, and hit the cup with teaspoon from inside and listen to the noise, Fig. 9.7a. For the second time, ask the assistant to hold the cup between his/her hand tightly. Again, try to hit the cup with the teaspoon from inside and listen to the noise, Fig. 9.7b.

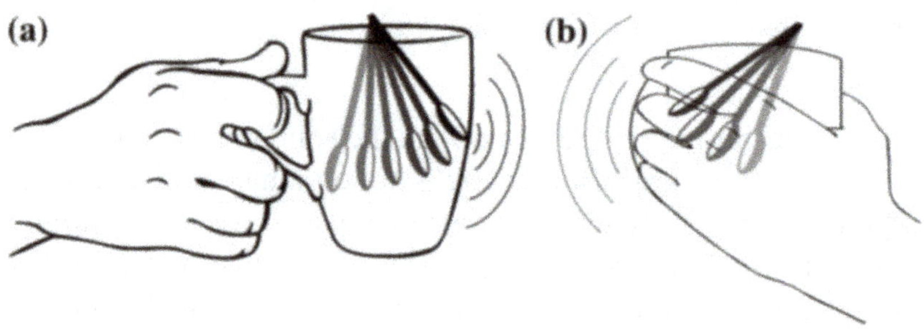

Fig. 9.7 Simulating reflected noise from an uncemented and cemented casing: **a** cup has no barrier behind it, **b** hand acts as barrier behind cup and absorbs the noise energy

Now assume that the cup is casing, teaspoon is transmitter, and hands around the cup is cement behind casing. Based on the experiments explain and interpret your observations.

Solution When holding the cup form its handle and hitting the cup wall with teaspoon, the sound energy is high. In other words, the reflected sound has high energy which is known as ringing. The same phenomenon happens when there is free casing, the casing rings.

When holding the cup tightly between hands, the emitted sound is absorbed by the hand and lower sound energy is reflected. When there is solid barrier behind casing, the emitted sound travels further and less energy is reflected back.

As described earlier, the transmitted acoustic signal is picked up by two receivers: 3 and 5 ft. 3 ft receiver picks up the reflected sound from casing and presented on Track 2, which is known as CBL log. The CBL log shows the quality of bonding between casing and the material behind it. When CBL log shows high amplitude (high energy), then the casing reflects back most of the transmitted sound because of poor bonding. In other words, the casing is ringing because of poor bonding between solid material adjacent to casing, whereas sound is not absorbed. But if there is good bonding between solid material in annulus behind casing and casing, the sound travels further to formation and it is picked up by the 5 ft receiver. The recorded data are presented on Track 3, which is known as VDL log. The VDL log records the amplitude of transmitted sound though casing fluid, casing, barrier behind casing, and formation (Fig. 9.8). When VDL log shows formation arrival, then there is a solid material in the annulus behind casing up to formation.

Information of the compressional wave velocity allows engineers to measure compressional acoustic impedance (Z) of materials. Every material has its natural acoustic impedance property and by estimating the acoustic impedance of an unknown material, the material might be distinguished. This is the main driver of sonic logs for verification of annular barrier. The acoustic impedance of a homogenous, non-dissipative

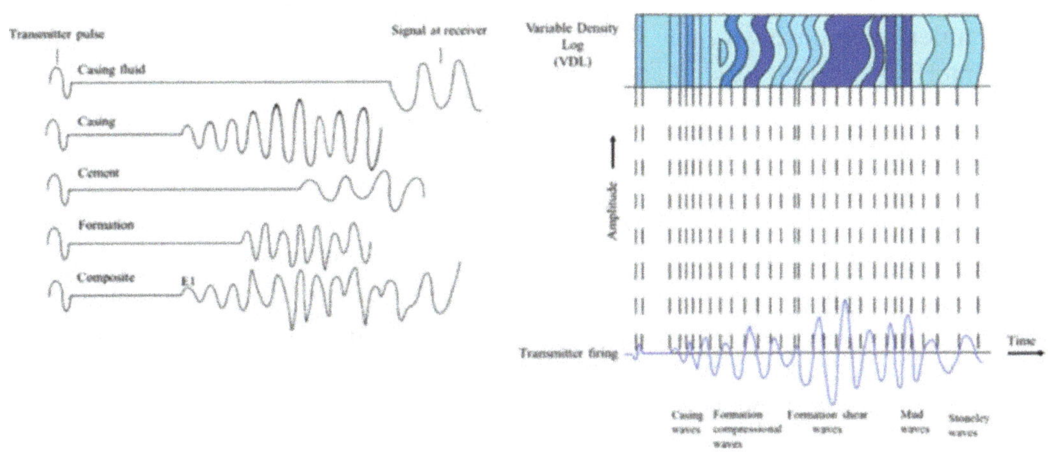

Fig. 9.8 Signal arrivals from casing fluids is the latest as sound travels slower in liquid phases, casing arrival is the first to arrive. Composite amplitude is presented on VDL log

medium, is given by:

$$Z = \rho v_P \tag{9.1}$$

where ρ is the material density (kg/m^3) and v_P is compressional wave velocity (m/s). The acoustic impedance is expressed in $10^6 \left(\frac{\text{kg s}}{\text{m}^2}\right)$ which known also as mega-Rayleigh (MRayl).

When designing, executing and interpreting sonic logs, there are well parameters, organizational and operational factors, and human factors, which can affect the final result. The well parameters include, but are not limited to: temperature and pressure, wellbore-fluid properties, casing size and thickness, cement thickness, and surrounding formation. Organizational and operational factors may include selection of service provider, pre-job meeting and discussions, surface pressure (equipment and procedure used), detection setting, and log quality control procedures. All of these factors affect the reliability of the final logs and their interpretation.

Considering utilization of sonic logging tools in P&A, for verification of annular barrier, their reliability may be questioned if the logging tool is calibrated according to well conditions for primary cementing. This is due to change in properties of annular barrier over time, casing thickness changes, and formation subsidence. Therefore, calibration and re-evaluation of logs based on current well condition are expected. Advances of sonic logging tools has been in progress since their development and utilization. Re-evaluation of annular barrier, in place from primary cementing, with recently modified sonic logging tools could give a better understanding of annular barrier condition.

Example 9.2 In recent years, PWC technique has been developed and employed to avoid section milling. After barrier establishment with PWC, the internal cement is drilled out and annular barrier is logged with sonic tools. How can the perforated casing create difficulties and uncertainty in sonic logging?

Solution When casing is perforated, plate waves cannot effectively travel through casing unless they have very low shot density. In addition, generation of shear waves at the interface between casing and cement is disrupted. Therefore, increased attenuation of sound waves is expected which is caused by the holes created during perforation. One possible solution is to remove the effect of holes during Fast Fourier Transform processing.

Sonic logs have their advantages including: wireline operation, non-destructive technique, and safe operation. However, there are some limitations associated with utilization of sonic tools. Sonic logging provides a qualitative evaluation of cement quality and it does not show the direction of anomaly. In other words, the CBL-VDL graphs show an average of circumferential measurements. These tools are also sensitive to liquid-filled micro-annulus. Therefore, ultrasonic tools have been developed and are employed more often than sonic tools.

9.1.1.2 Ultrasonic Measurements

Ultrasonic pulse-echo (PE) techniques were introduced to the industry for cement evaluation in the early 1980s. These cement-mapping tools operate at much higher frequencies than acoustic tools, typically between 200 and 700 kHz [2, 4, 5]. The principle of the ultrasonic technique is to cause a small area of the casing to resonate across its thickness. In pulse-echo techniques, a transducer, acting as both a transmitter and a receiver, mounted in a rotating head, sends out a short pulse of ultrasound and picks up the echo containing resonance, Fig. 9.9 [6]. As ultrasonic tools provide peripheral rotation, a radial map (Track 4 in Fig. 9.4) is generated as result and defect location can be distinguished. In order to apply real time corrections for impedance calculations, ultrasonic tools provide real time measurements of fluid impedance in a built-in mud cell. The rate of decay of the resonance will be lower if there is fluid behind the casing whereas cement will damp the resonance faster [7].

A limitation with PE measurements is that they are only able to investigate the presence of cement behind a single casing string. Another known weakness of this ultrasonic sonic measurement technique is its sensitivity to the presence of small gas bubbles [8].

Viggen et al. [9] attempted to model ultrasonic pitch-catch measurements in a through-tubing logging configuration. In pitch-catch techniques, there is one transmitting transducer and one or more receiving transducers. They used a finite element model of a double-casing geometry with a two-receiver pitch-catch setup. In their study, they found that a cascade of leaky flexural Lamb wave packets appears on both casings caused by leaked wave fronts. Their study shows that the received pulse

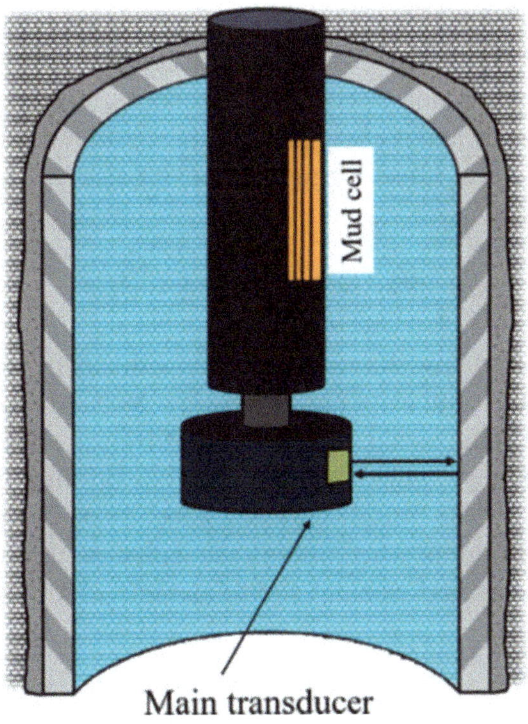

Fig. 9.9 Ultrasonic measurement, pulse-eco technology

from the second wave packet contains information about the bonded material in the outer annulus as well as the interface between cement and formation.

Viggen et al. [10] attempted to analyze outer casing echoes through simulations of ultrasonic pulse-echo measurements through-tubing. Their work examined the hypothesis that anomalies behind a second casing string can cause significant variations of pulse-echo. Their finding shows that variations of the outer casing interface echo with the outer casing thickness and the B-annulus material may be too subtle to be reliably applied to through-tubing logging. In addition, they found that eccentricity of the casing and transducer angle influence the travel time of the interface echo.

A recent development targets the utilization of electro-magnetic acoustic transducers (EMAT) for the generation of guided acoustic waves in the casing [11]. For this technology, a Lorentz force[1] is used to generate and measure acoustic waves directly. EMAT operate with a coil, a magnet and a conductive casing, and it can function both as transmitter and receiver. EMAT generate two fundamental wave modes; shear horizontal and lamb flexural (plate). In the shear horizontal wave mode, the particle motion is perpendicular to the direction of wave propagation; however in the lamb flexural mode, the particle motion is normal to the casing surface. The study of these wave modes is a direct measurement of the shear modulus of the solid material behind casing with a higher resolution compared to conventional acoustic techniques whilst eliminating sensitivity to the wellbore fluid or the need for physical contact of the transducers with the casing [12].

9.1.2 Noise Logging Measurements

When a leak occurs through cement defect, noise may be generated which depends on the defect size and geometry, leakage rate, and surrounding materials. If the generated noise is above a threshold level, it can be detected and analyzed; either by passive noise logging or active noise logging.

9.1.2.1 Passive Noise Logging

Fluid flow through a leakage pathway generates noise with two measureable parameters; intensity and frequency. Noise intensity, also known as acoustic intensity, is defined as the energy carried by the sound wave per unit area. In the context of leakage, noise intensity depends on fluid flowrate and differential pressure driving it, while noise frequency depends on the geometry of leakage pathway. As a rule of thumb, when fluid flows with ease through a large area, a low frequency noise is generated whereas fluid flowing with difficulty through a narrow space generates a

[1] It is an electromagnetic force that acts on a charged particle which is moving with a velocity through and electric magnetic field.

high frequency noise. There are a variety of tools able to record noise generated by leaks over a wide frequency range with a high resolution and high sensitivity.

9.1.2.2 Active Noise Listening

Active listening is an acoustic technique whereby a very short acoustic pulse is fired at the region being examined and the reflected signal is thereafter recorded. After a short waiting time, the same region is examined again by an identical signal and the reflected signals are thereafter subtracted from each other. If there is no difference in the reflected signals, it means that there is no motion or other changes in the material behind the casing. In other words, motion in the material during emitting the first signal and the second signal will result in an incoherence in the signals at the same depth. The strengths of this technique compared to conventional noise logging is that it is sensitive to a wider range of flowrates, provides a quantitative estimate of flow velocity, the distance to channels can be estimated and gas migration through a column of liquid in the channel can be detected.

9.1.3 Temperature Logging

Temperature logs are used to detect the temperature anomalies behind casing caused by cement hydration or leakage of fluids. Of temperature logs, cement hydration detection, communication indicator, radial differential temperature, active temperature logging and distributed temperature sensing are the most known temperature logging techniques [8].

9.1.3.1 Cement Hydration Detection

Temperature logs are used to detect the temperature anomalies behind casing caused by cement hydration or leakage of fluids. Cement hydration occurs over a period of six to twelve hours after initial mixing of cement and is an exothermic chemical reaction which generates considerable heat. It is the temperature rise inside the well due to the heat conducted by the casing from the cement that is readily detected by temperature logs. Temperature logs recorded at a suitable point in time can be used to detect the TOC, however, complete verification of the seal quality of a primary cementing operation is challenging. Detecting cement hydration must be performed before the temperature increase due to hydration has dissipated and this technique is not therefore suitable for inspecting the casing cement later in the well life.

9.1.3.2 Communication Indicator

When fluids leak through a defect, the temperature of the surroundings is affected. In thermodynamics, the Joule–Thomson effect describes the temperature change of a real gas or liquid when it is forced through a restriction. A condition of the Joule–Thomson effect is that the enthalpy, H, remains constant:

$$H = U + PV \tag{9.2}$$

where U is internal energy, P is pressure and V is volume. According to the Joule–Thomson effect, the change in PV shows the work done by the fluid. When a fluid passes through a defect, the PV is increased and to keep the H constant, U is decreased. This means that cooling due to expansion is expected if gas flow occurs. A conventional temperature log measures the fluid temperature inside the well and the recorded data is plotted versus depth. By comparing the obtained data with the geothermal temperature gradient the depth of anomalies can be identified that might be related to fluid leakage through defects in the cement. Temperature gradient differences can also be created by injected fluid flowing in the channels.

9.1.3.3 Radial Differential Temperature

Radial Differential Temperature (RDT) logging is a modified version of conventional temperature log for detecting channels. This method utilizes two sensors (in addition to a sensor in the center of well) to measure the pipe wall surface temperature around its circumference. The difference in temperature between casing wall and sensor in the center of pipe is measured and plotted versus depth. Deviation of recorded temperatures from geothermal temperature can be used to locate channels near to the casing.

9.1.3.4 Active Temperature Logging

Active temperature logging uses short-term local inductive heating of the metal in the casing to give the reservoir fluid a thermal signature that can be detected during production. As a result of inductive heating of the casing, a thermal anomaly is induced both inside the well and in the fluid moving behind the casing which can be detected by sensors once the fluid is produced into the well. Figure 9.10 shows an active temperature logging tool equipped with inductor, distributed temperature sensors; T1, T2 and T3, collar locator, gamma ray detector and water resistivity recorder.

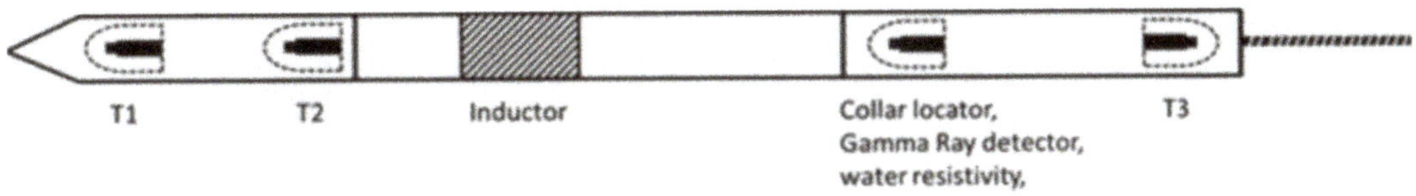

Fig. 9.10 Active temperature logging tool with its inductive heater

9.1.3.5 Distributed Temperature Sensing—Fiber-Optic Sensing

Pulses of light generated by a laser sent through an optical fiber are reflected repeatedly from the fiber walls. The fiber and its coating form a wave guide with total internal reflection such that light is not lost through the fiber walls. A sensor or combination of sensors can be placed along the fiber and record measurands such as; pressure, temperature, seismic, mechanical stresses, chemicals, and flow [13, 14]. Figure 9.11 shows three main types of fiber-optic sensor arrangements; single point sensor, multi-point sensors, and distributed sensors. The single point sensor measures the parameter of interest at a single point in space typically at the end of the fiber. The multi-point sensor measures the measurand at a number of fixed, discrete points along a single fiber-optic cable. The distributed sensor measures the measurand with a certain spatial resolution at any point along the fiber-optic cable. In the latter case, the fiber cable itself is the sensor and backscattered light carries information. Distributed sensing has the potential to identify leakage pathways and thereby cement defects. Two different distributed sensing systems are Distributed Temperature Sensing (DTS) and Distributed Acoustic Sensing (DAS).

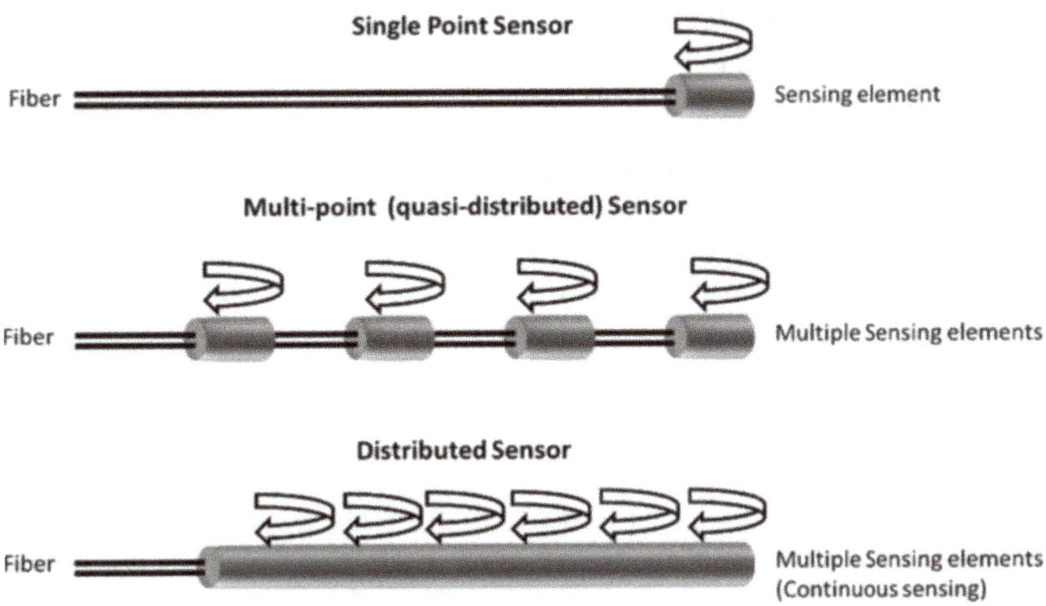

Fig. 9.11 Different modes fiber-optic sensing

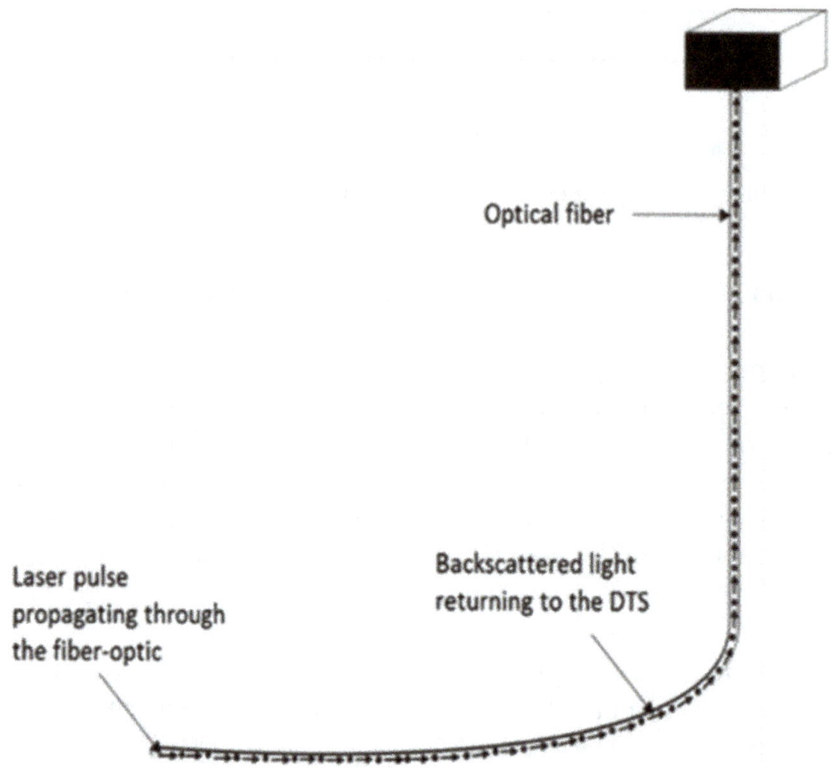

Fig. 9.12 Graphical schematic of DTS system

In the DTS system, a short pulse of light is launched into the fiber. The forward propagating light generates Raman backscattered light at two distinct wavelengths, from all points along the fiber, Fig. 9.12. These two wavelengths are named "Stokes light" and "anti-Stokes light" and are generated due to inelastic scattering of a photon. Anti-Stokes light is temperature-dependent, while the Stokes light is weakly temperature-dependent. The local temperature of the optical fiber is calculated from the ratio between the amplitude of the Stokes and the anti-Stokes detected light.

One of the challenges for temperature logging tools is high-temperature wells where the generated temperature during cement hydration or temperature anomaly caused by leakage is difficult to identify from the high geothermal temperature.

9.1.4 Hydraulic Pressure Testing

Inability to log annulus behind the second steel string, Fig. 9.13a, is one of the main challenges associated with the current acoustic logging technologies. So, utilization of rig might be inevitable to examine the annular barriers, red boxes shown in Fig. 9.13a. Therefore, rig is required to retrieve the production tubing to log the annular barrier behind the production casing, the solid red boxes in Fig. 9.13b. But

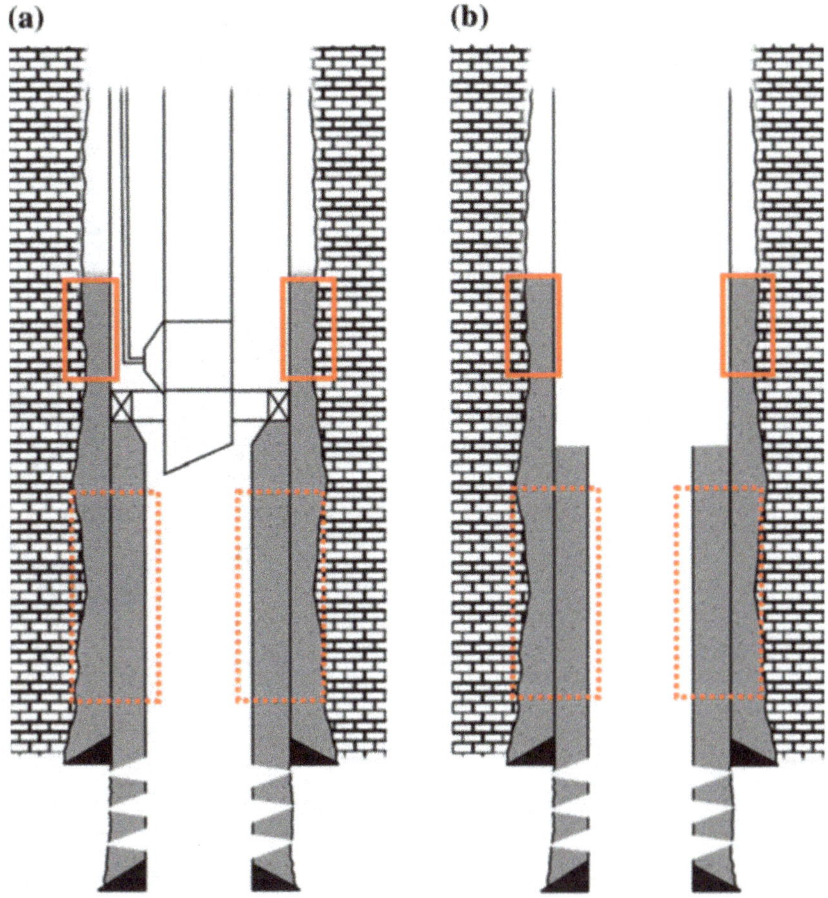

Fig. 9.13 Acoustic logging tools are not capable to log through two strings of steel; **a** rigless operations is not possible due to technology deficit of acoustic logging tools, **b** rig is required to retrieve the production tubing and only logging the red box marked with solid line. The dotted-line-box area cannot be verified even after removal of production tubing

logging the annular barrier behind the production casing, across the liner (the dotted red box in Fig. 9.13b), still remains unsolved due to technology deficit.

When utilization of acoustic logging tools for verification of annular barrier is impossible or may create extra work, hydraulic pressure testing (known as communication testing) might be an option [15]. Such circumstances may include, verification of casing cement when production tubing is in place, verification of annular barriers behind the second casing string, or when PWC technique is used to establish both internal and external barriers [16], Fig. 9.13a.

In hydraulic pressure testing method, a bridge plug is installed at the base, where base of annular barrier is supposed to be. The installed bridge plug is pressure tested and when it passed the pressure testing, above the bridge plug a small window is perforated. Another bridge plug which is equipped with a wireless pressure gauge is installed away above the created perforations. This bridge plug also needs to pass pressure testing. A new window needs to be perforated above the second bridge plug, Fig. 9.14. The distance between the windows of perforations depends on the required

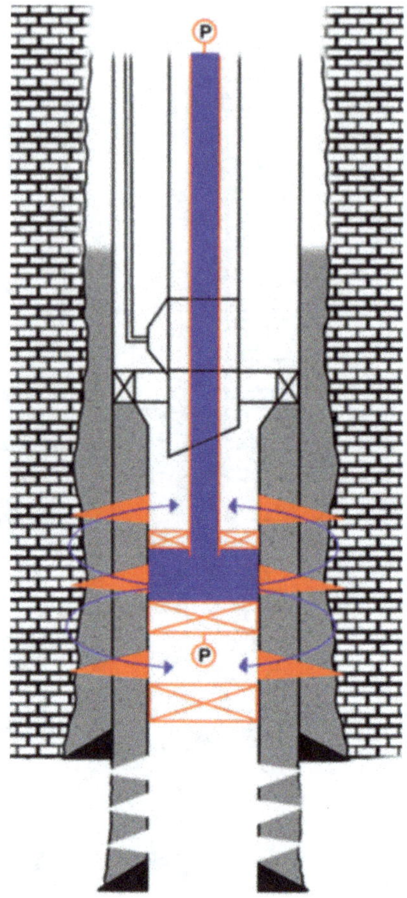

Fig. 9.14 Hydraulic testing, communication testing, of annular barrier behind second casing string

length of annular barrier to be verified. A workstring, equipped with a packer is run and the packer is engaged above the second perforations. A fluid is pumped through the workstring and pressure changes are monitored. If the downhole pressure gauge and surface gauge do not record any changes, the annular barrier is qualified.

The pressure testing is a cycle of extended leak-off and drawdown tests, Fig. 9.15. The pressure test data can be used for both investigating the hydraulic communication between the perforations, and corresponding the pressure data to expected formation strength.

Hydraulic communication test—When fluid is pumped through middle perforations, any pressure changes above the packer attached on workstring or between the bridge plugs means failure of annular barriers. However, no pressure changes means intact and subsequently verified annular barriers.

Extended leak-off test—The extended leak-off tests are conducted to make sure that perforations reached to the adjacent formation. This can be examined by increasing the injection pressure until leak-off occurs. If the leak-off and breakdown pressures are corresponding to expected formation strength, it means perforations penetrated to formation. In addition, extended leak-off test is conducted to make sure that annular barriers can hold the maximum anticipated pressure.

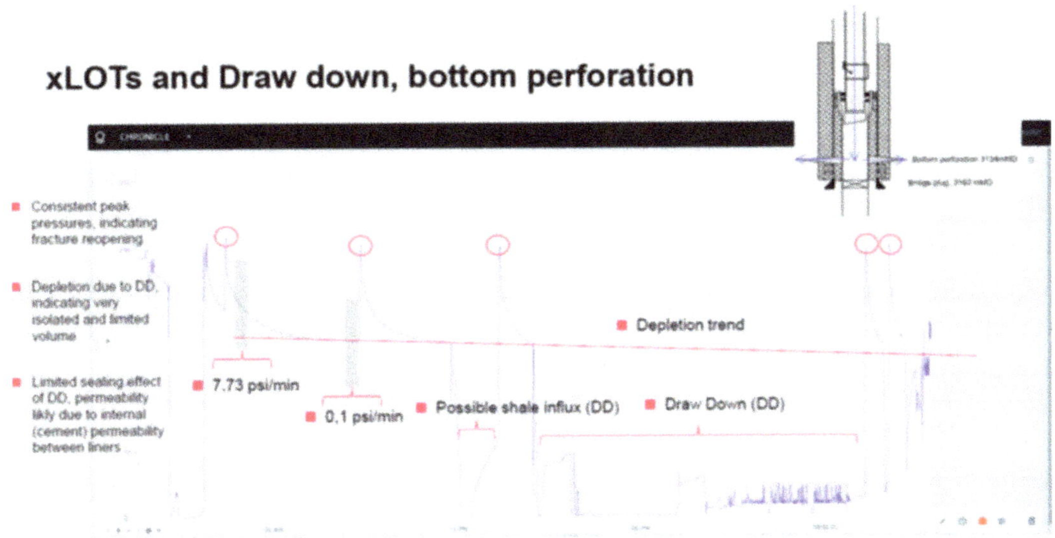

Fig. 9.15 Extended leak-off test and drawdown test to verify annular barriers, shown in Fig. 9.14, by hydraulic testing

9.2 Internal Barrier Verification/Plug Inside Openhole or Casing

9.2.1 Hydraulic Pressure Testing

Pressure testing is applied to plugs installed inside casing, openhole plugs which are extended to casing or plugs installed entirely in openhole, Fig. 9.16. Pressure

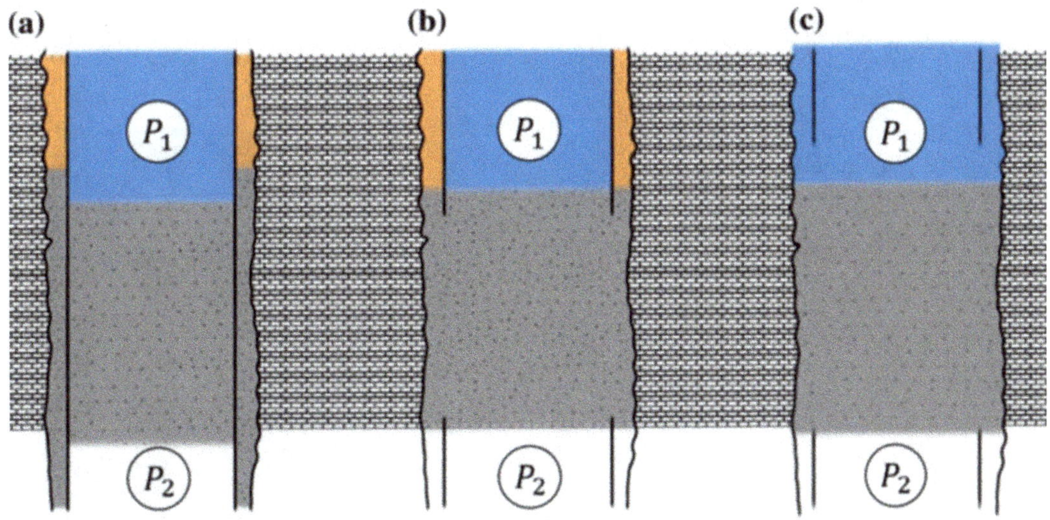

Fig. 9.16 Cement plug installed in wellbore; **a** cement plug is installed inside casing across a qualified annular barrier, **b** cement plug installed in an openhole but extended to casing, **c** cement plug entirely installed in an openhole

testing has other names such as pump pressure testing and hydraulic testing. Where mechanical plug is used as foundation for cement plug and the foundation passed the pressure testing, pressure testing of plug is meaningless.

There is a misunderstanding about information obtained by carrying out pressure testing. Pressure testing gives an insight about sealability of cement plug and sealing capability at the interface of cement plug and adjacent element [17]. But it does not necessary provide information about the hydraulic bond strength (for more information regarding hydraulic bond strength refer to Chap. 3) of entire plug length.

Depending on the direction of applied hydraulic pressure, positive pressure testing or negative pressure testing can be distinguished.

9.2.1.1 Positive Pressure Testing

In positive pressure testing, fluid is injected by surface pump whereas pressure above the plug is higher than the pressure below, P_1 is higher than P_2 (see Fig. 9.1) [18]. When the ΔP across fulfills the requirement asked by local authority, the pressure is monitored for some minutes and if a stable pressure is reading, the plug is a qualified plug. The positive pressure testing is carried out on plugs installed inside casing and across a qualified annular barrier, and plugs installed inside openhole but extended to casing string (see Fig. 9.16a, b). Pressure testing of plugs installed entirely in openhole is meaningless (see Fig. 9.16c). The reason is that when subjecting the openhole plug to the injected fluid, the fluid can penetrate the surrounding formation, Fig. 9.17.

There are some concerns associate with the positive pressure testing technique including, but not limited to uncertainty associated with sealing capability of casing connections, casing corrosion, and ballooning effect of casing. When hydraulic pressure is applied, the injected fluid can leak thorough casing connections and stable pressure reading may not be reached (Fig. 9.18). In this case, it is difficult to

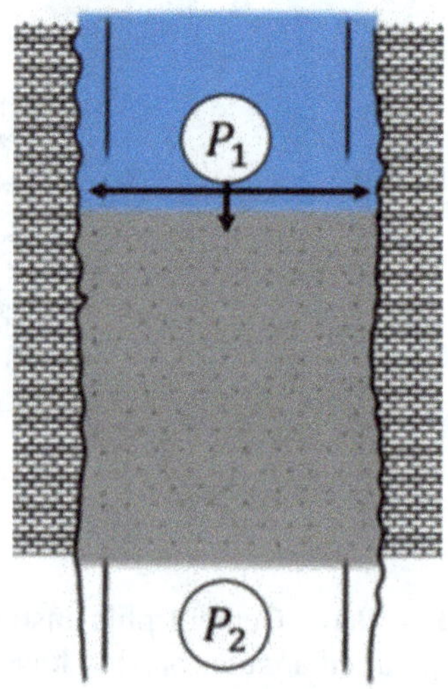

Fig. 9.17 Positive pressure testing of the plug installed entirely in openhole is meaningless

Fig. 9.18 Potential leak path through threads in positive pressure testing

identify the source of leak; casing connections or a failed plug. Where casing experiences small holes cause by corrosion or caused by mechanical wearing, the applied hydraulic pressure leaks through the casing and pressure monitoring does not show a stable reading. Ballooning effect is susceptible when there is liquid in the annular space behind casing and casing thickness has been affected over years. In this scenario, the casing can expand if the applied pressure exceeds the casing design criteria such as its elasticity (Fig. 9.19).

Positive pressure testing can also be used to estimate the shear bond strength between plug and the adjacent material by following equation:

$$Shear\ bond\ strength \geq \frac{P_p \times A_p}{\pi \times D_i \times L_p} \tag{9.3}$$

where P_p is the pump pressure, A_p is the surface area of plug, D_i is the inner diameter of the geometry plug placed inside, and L_p is the plug length. However, this is valid when the material sealability and the material mechanical strength is higher than the shear bond strength of the entire plug.

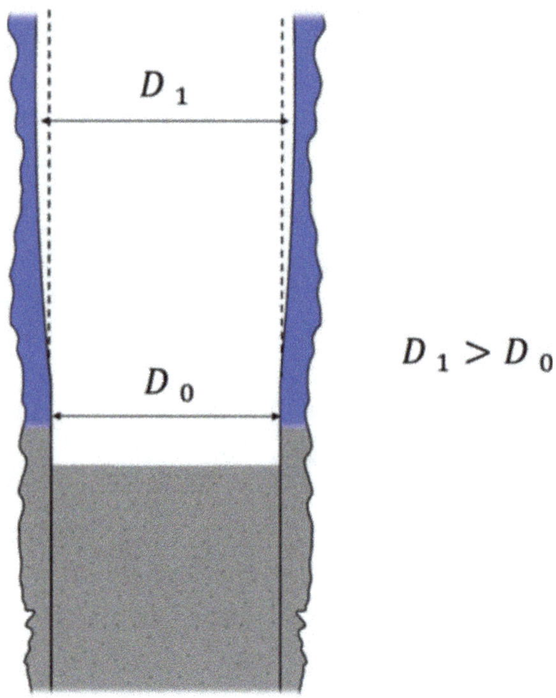

Fig. 9.19 Ballooning of casing while carrying out the positive pressure testing, D_0 is the origin casing diameter and D_1 is the diameter after ballooning

9.2.1.2 Negative Pressure Testing

In negative pressure testing (also known as inflow testing), the hydrostatic pressure above the plug is decreased so that pressure below the plug (P_2) will be higher than pressure above the plug (P_1), see Fig. 9.16. Then, the changes in pressure is recorded. A stable pressure means a sealed plug. If a transparent fluid is placed on top of plug, possible leak can be seen directly by use of downhole camera. Negative pressure testing is used where integrity of connections or casing string above the plug is questioned and positive pressure testing cannot be performed. In addition, when plug is entirely placed in openhole, which positive pressure testing is not feasible, negative pressure testing may be performed. The challenge associated with this method is that the current pressure, below the plug, might be lower than the expected final pressure. Therefore, plug is not qualified based on the estimated future pressure but the current pressure.

9.2.2 Weight Testing

When plug is installed, it is necessary that the plug keeps its position and does not move due to increase of pressure below it. Weight testing is a method to measure the positioning, bond strength to adjacent element, and also measures the plug location. Where cement plug is entirely placed inside openhole, it is not possible to positive pressure test it or even sometimes negative pressure test it. Thus, weight testing is carried out to check the positioning of the plug; however, weight testing does not provide any information about the hydraulic sealability of plug.

Weight testing measures the shear bond strength of plug to adjacent material. So, the required shear bond strength measured by drillpipe during weight testing is defined as drillpipe tag weight, W_{dp}, divided by circumferential area of cement plug and is given by:

$$Shear\ bond\ strength \geq \frac{W_{dp}}{\pi \times D_i \times L_p} \tag{9.4}$$

The main challenge to estimate the shear bond strength is the plug length, as the theoretical plug length is different from plug length retained from contamination.

Studies show that much larger amount of shear bond strength is achieved for short cement plugs when placed inside small diameter geometries. Studies also show that by increasing the plug length placed inside a constant diameter geometry, the required shear bond strength decreases.

Weight testing is operationally feasible in rig-based operations by use of drillpipe but it might be feasible to carry it out in rig-less operations by use of coiled tubing or wireline.

9.2.2.1 Drillpipe

It is a normal practice to weight test the plug when drillpipe is available on site. To avoid any challenge introduced by contaminated cement on top of plug, top of cement is drilled to reach hard cement. This operation is known as cement dress-off. The required weight is calculated carefully to avoid any damage to the cement plug. By using a part of drillpipe weight, the pre-determined weight, positioning of plug is tested (see Fig. 9.20a). In fact, weight testing provides the shear bond strength between plug and adjacent material. If the plug can hold the applied weight without being displaced, its positioning is qualified.

9.2.2.2 Coiled Tubing

Coiled tubing can be used for weight testing where drillpipe is not available. To dress-off cement plug, a downhole motor is used. One of the main limitations of coiled tubing to be used in weight testing is the maximum weight that can be created. In addition, coiled tubing may be susceptible to helical ramp or tortuosity (see Fig. 9.20b) and difficult to apply more weight.

(a) (b) (c)

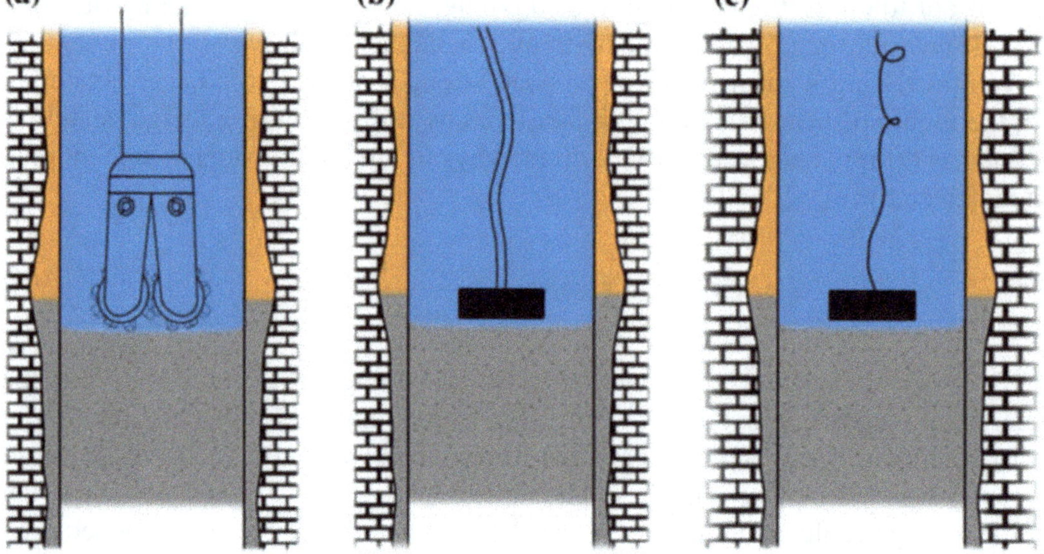

Fig. 9.20 Weight testing of cement plug placed inside casing; **a** drillpipe, **b** a heavy weight may be used with coiled tubing but coiled tubing may experience helical shape due to its design factors, **c** limited weight can be used for wireline

9.2.2.3 Wireline

Wireline may also be used for weight testing where there is no drillpipe nor coiled tubing unit. A downhole motor is used to dress-off the plug and then a limited weight is applied on the plug with no chance to apply additional weight (see Fig. 9.20c). Compared to drillpipe and coiled tubing, the use of wireline for weight testing is not accepted by many regulators due to the limitations of exerted weight. However, wireline can be used to confirm the depth of top of cement.

9.3 Hydraulic Pressure Equivalent to Drillpipe Tag Weight

As mentioned earlier, in some cases is difficult to perform positive pressure testing and weight testing needs to be carried out instead. In fact, weight testing and positive pressure testing impose the force on top of plug; weight testing is a mechanical way and positive pressure testing in a hydraulic way of doing it. So, it is possible to estimate the equivalent hydraulic pressure to drillpipe tag weight, Eqs. (9.3) and (9.4) can be equal:

$$\frac{P_p \times A_p}{\pi \times D_i \times L_p} = \frac{W_{dp}}{\pi \times D_i \times L_p} \quad (9.5)$$

and simplification of the equation gives:

$$P_p = \frac{W_{dp}}{\frac{\pi \times D_i^2}{4}} \tag{9.6}$$

Equation (9.6) shows that the required pump pressure to estimate the tag weight and the equivalent pump pressure is independent of plug length.

References

1. Tixier, M.P., R.P. Alger, and C.A. Doh. 1959. *Sonic logging*. Society of Petroleum Engineers.
2. Nelson, E.B., and D. Guillot. 2006. *Well cementing*, 2nd ed. Sugar Land, Texas: Schlumberger. ISBN-13: 978-097885300-6.
3. Rouillac, D. 1994. *Cement evaluation logging handbook*. Saint-Just-la-Pendue, France: Editions Technip. 2-7108-0677-0.
4. Gong, M., and S.L. Morriss. 1992. Ultrasonic cement evaluation in inhomogeneous cements. In *SPE annual technical conference and exhibition*. SPE-24572-MS. Washington, D.C.: Society of Petroleum Engineers. https://doi.org/10.2118/24572-MS.
5. Havira, R.M. 1982. Ultrasonic cement bond evaluation. In *SPWLA 23rd annual logging symposium*. SPWLA-1982-N. Corpus Christi, Texas: Society of Petrophysicists and Well-Log Analysts.
6. Acosta, J., M. Barroso, and B. Mandal, et al. 2017. New-generation, circumferential ultrasonic cement-evaluation tool for thick casings: case study in ultradeepwater well. In *OTC*. OTC-28062-MS. Rio de Janeiro, Brazil: Offshore Technology Conference. https://doi.org/10.4043/28062-MS.
7. Foianini, I., B. Mandal, and R. Epstein. 2013. Cement evaluation behind thick-walled casing with advanced ultrasonic pulse-echo technology: Pushing the limit. In *SPWLA 54th annual logging symposium*. SPWLA-2013-RRR, New Orleans, Louisiana: Society of Petrophysicists and Well-Log Analysts.
8. Khalifeh, M., D. Gardner, and M.Y. Haddad. 2017. Technology trends in cement job evaluation using logging tools. In *Abu Dhabi international petroleum exhibition & conference*. SPE-188274-MS, Abu Dhabi, UAE: Society of Petroleum Engineers. https://doi.org/10.2118/188274-MS.
9. Viggen, E.M., T.F. Johansen, and I.A. Merciu. 2016. Simulation and modeling of ultrasonic pitch-catch through-tubing logging. *Geophysics* 81 (04): 383–393. https://doi.org/10.1190/geo2015-0251.1.
10. Viggen, E.M., T.F. Johansen, and I.A. Merciu. 2016. Analysis of outer-casing echoes in simulations of ultrasonic pulse-echo through-tubing logging. *Geophysics* 81 (06): 679–685. https://doi.org/10.1190/geo2015-0376.1.
11. Kamgang, S., A. Hanif, and R. Das. 2017. Breakthrough technology overcomes long standing challenges and limitations in cement integrity evaluation at downhole conditions. In *Offshore technology conference*. OTC-27585-MS, Houston, Texas, USA: Offshore Technology Conference. https://doi.org/10.4043/27585-MS.
12. Patterson, D., A. Bolshakov, and P.J. Matuszyk. 2015. Utilization of electromagnetic acoustic transducers in downhole cement evaluation. In *SPWLA 56th annual logging symposium*. SPWLA-2015-VVVV, Long Beach, California, USA: Society of Petrophysicists and Well-Log Analysts.
13. Lumens, P.G.E. 2014. Fibre-optic sensing for application in oil and gas wells. Department of Applied Physics. Eindhoven, The Netherlands: Technische Universiteit Eindhoven. http://dx.doi.org/10.6100/IR769555.
14. Rambow, F.H.K., D.E. Dria, and B.A. Childers, et al. 2010. Real-time fiber-optic casing imager. SPE J 15 (04): 1,089–1,097. https://doi.org/10.2118/109941-PA.

15. Abdel-Mota'al, A. A. 1983. Detection and remedy of behind-casing communication during well completion. In *Middle east oil technical conference and exhibition*. SPE-11498-MS. Manama, Bahrain: Society of Petroleum Engineers. https://doi.org/10.2118/11498-MS.
16. Delabroy, L., D. Rodrigues, and E. Norum, et al. 2017. Perforate, wash and cement PWC verification process and an industry standard for barrier acceptance criteria. In *SPE Bergen one day seminar*. SPE-185938-MS. Bergen, Norway: Society of Petroleum Engineers. https://doi.org/10.2118/185938-MS.
17. CSI-Technologies. 2011. Cement plug testing: weight vs. pressure testing to assess viability of a wellbore seal between zones. https://www.bsee.gov/sites/bsee.gov/files/tap-technical-assessment-program//680aa.pdf.
18. Bois, A.-P., M.-H. Vu, and K. Noël, et al. 2018. Cement plug hydraulic integrity - the ultimate objective of cement plug integrity. In *SPE norway one day seminar*. SPE-191335-MS. Bergen, Norway: Society of Petroleum Engineers. https://doi.org/10.2118/191335-MS.

www.ingramcontent.com/pod-product-compliance
Lightning Source LLC
LaVergne TN
LVHW082119200926
848484LV00008B/1983
* 9 7 8 1 0 1 3 2 7 1 0 9 0 *